Hydropower in Authoritarian Brazil

During the later twentieth century, Brazil's right-wing military dictatorship built a vast network of hydropower dams that became one of the world's biggest low-carbon electricity grids. Weighed against these carbon savings, what were the costs? Johnson unpacks the social and environmental implications of this project, from the displacement of Indigenous and farming communities to the destruction of Amazonian biodiversity. Drawing on rich archival material from forty sites across Brazil, Paraguay, and the United States, including rarely accessed personal collections, Johnson explores the story of the military officers and engineers who created the dams and the protestors who fought them. Brazilian examples are analyzed within their global context, highlighting national issues with broad consequences for both social and environmental justice. In our race to halt global warming, it is vital that we learn from past experiences and draw clear distinctions between true environmentalism and greenwashed political expedience.

Matthew P. Johnson, PhD, is an environmental historian and a 2022–2024 Environmental Fellow at the Harvard University Center for the Environment.

Studies in Environment and History

Editors

J. R. McNeill, *Georgetown University*
Ling Zhang, *Boston College*

Editors Emeriti

Alfred W. Crosby, *University of Texas at Austin*
Edmund P. Russell, *Carnegie Mellon University*
Donald Worster, *University of Kansas*

Other Books in the Series

Ellen Arnold *Medieval Riverscapes: Environment and Memory in Northwest Europe, c. 300-1100*
Richard C. Hoffmann *The Catch: An Environmental History of Medieval European Fisheries*
Samuel Dolbee *Locusts of Power: Borders, Empire, and Environment in the Modern Middle East*
Andy Bruno *Tunguska: A Siberian Mystery and Its Environmental Legacy*
Lionel Frost et al. *Cities in a Sunburnt Country: Water and the Making of Urban Australia*
Adam Sundberg *Natural Disaster at the Closing of the Dutch Golden Age: Floods, Worms, and Cattle Plague*
Germán Vergara *Fueling Mexico: Energy and Environment, 1850–1950*
Peder Anker *The Power of the Periphery: How Norway Became an Environmental Pioneer for the World*
David Moon *The American Steppes: The Unexpected Russian Roots of Great Plains Agriculture, 1870s–1930s*
James L. A. Webb Jr. *The Guts of the Matter: A Global Environmental History of Human Waste and Infectious Intestinal Disease*
Maya K. Peterson *Pipe Dreams: Water and Empire in Central Asia's Aral Sea Basin*
Thomas M. Wickman *Snowshoe Country: An Environmental and Cultural History of Winter in the Early American Northeast*
Debjani Bhattacharyya *Empire and Ecology in the Bengal Delta: The Making of Calcutta*
Chris Courtney *The Nature of Disaster in China: The 1931 Yangzi River Flood*
Dagomar Degroot *The Frigid Golden Age: Climate Change, the Little Ice Age, and the Dutch Republic, 1560–1720*
Edmund Russell *Greyhound Nation: A Coevolutionary History of England, 1200–1900*
Timothy J. LeCain *The Matter of History: How Things Create the Past*
Ling Zhang *The River, the Plain, and the State: An Environmental Drama in Northern Song China, 1048–1128*

Hydropower in Authoritarian Brazil

An Environmental History of Low-Carbon Energy, 1960s–1990s

MATTHEW P. JOHNSON
Harvard University

CAMBRIDGE
UNIVERSITY PRESS

CAMBRIDGE
UNIVERSITY PRESS

Shaftesbury Road, Cambridge CB2 8EA, United Kingdom

One Liberty Plaza, 20th Floor, New York, NY 10006, USA

477 Williamstown Road, Port Melbourne, VIC 3207, Australia

314–321, 3rd Floor, Plot 3, Splendor Forum, Jasola District Centre, New Delhi – 110025, India

103 Penang Road, #05-06/07, Visioncrest Commercial, Singapore 238467

Cambridge University Press is part of Cambridge University Press & Assessment, a department of the University of Cambridge.

We share the University's mission to contribute to society through the pursuit of education, learning and research at the highest international levels of excellence.

www.cambridge.org
Information on this title: www.cambridge.org/9781009428699

DOI: 10.1017/9781009428743

First published 2024

A catalogue record for this publication is available from the British Library

A Cataloging-in-Publication data record for this book is available from the Library of Congress

ISBN 978-1-009-42869-9 Hardback

This book is dedicated to all my Brazilian and Paraguayan colleagues and friends who helped along the way, and to the memory of Lucas Mores, one of the brightest of them.

Contents

Figures

Acknowledgments

This book is based on a dissertation I completed in the Department of History at Georgetown University, the research for which was funded in part by a grant from the Fulbright U.S. Student Program. I revised the manuscript for publication as a postdoctoral fellow at the Harvard University Center for the Environment, under the mentorship of Gabriela Soto Laveaga and Victor Seow in the Department of the History of Science. Parts of Chapter 2 are based on ideas first presented in "Sacrificing Guaíra Falls: Geopolitics and the Environmental Impact of South America's Biggest Dam, 1962–1982," *Luso-Brazilian Review* 60, no. 1 (2023): 37–67, and parts of Chapters 4 and 7 are based on ideas first presented in "Uma geração sem terra: Injustiça ambiental em comunidades indígenas deslocadas por construções de hidrelétricas no Brasil, desde os anos 1980," *Historia Ambiental Latinoamericana y Caribeña (HALAC)* 11, no. 3 (2021): 209–233.

Many people helped with the research and writing that went into producing this book. For starters, I owe tremendous debts of gratitude to those who helped me build this book's bibliography. In Brazil and Paraguay, special thanks are due to the librarians and archivists at Eletronorte, the Companhia Hidrelétrica do Rio São Francisco, Itaipu Binacional, and the Centro da Memória da Eletricidade, as well as those at other government repositories such as the Arquivo Nacional in Rio de Janeiro, the libraries and archives of the Congresso Nacional in Brasília, and the library of the Instituto Nacional de Pesquisas da Amazônia in Manaus.

I am also thankful for the help of librarians and archivists at the repositories of the prominent nongovernmental institutions that

I consulted, such as the Fundação Getúlio Vargas in Rio de Janeiro, the Conselho Indigenista Missionário in Brasília, and the Comissão Pastoral da Terra in Goiânia.

I am also grateful for the time and help of those who assisted me in locating and accessing collections in smaller and sometimes informal repositories. In Manaus, Philip Fearnside set aside time to speak with me at length about Brazilian dams and opened up his personal collection of papers about the reservoirs he has studied. Malariologists Wanderli Tadei and Raul Amorim both spoke with me at length in a formal interview setting about the relationship between mosquitoes, malaria, and dams in the Amazon, providing invaluable information on the subject. In Presidente Figueiredo, a small town about an hour's drive north of Manaus, Egydio Schwade granted me access to the archive he maintains about the history of the Waimiri-Atroari and the industrial projects that have encroached on their land. His admirably organized collection had a wealth of information about the Balbina Dam.

In the Northeast, André Monteiro and Ruben Siqueira assisted in connecting me with helpful colleagues. In Juazeiro, Juracy Marques shared his research and put me in touch with two of his students, Adzamara Amaral and Luciano Silva de Menezes, who had also done research on the communities displaced by Sobradinho and Itaparica. Francisco de Assis Silva helped coordinate my visit to the Biblioteca de Dom José Rodrigues at the Universidade Estadual de Bahia, Juazeiro, which had a rich collection of useful documents. In Paulo Afonso, Valda Aroucha and Galilleu Torres Matias hosted me and helped me network during my time there, and I remain immensely grateful for their assistance and hospitality. They put me in touch with Ruy Tenório and Jamile Silveira, among others, at the Universidade Estadual de Bahia, Paulo Afonso, who shared helpful documents and information with me. Antônio Galdino kindly opened his personal collection of archived issues of *Folha Sertaneja*, a local newspaper with ample information about Paulo Afonso Falls. Galdino and tour guide Antônio Souza also spoke with me at length about the history of the falls and the long-extinct national park that once existed there. Nicolas Kalango took me to see the falls and dams and accompanied me on other adventures during my visit. Valda also put me in touch with Alzeni Tomaz and Uilton Tuxá, both of whom helped deepen my understanding of the region's Indigenous communities and their fights for social justice. Uilton shared important documents with me about the Tuxá's struggle for land and generously accompanied me on a visit to both Rodelas and D'zorobabé.

There I met with Tuxá leaders, who graciously welcomed me and recounted painful stories about broken promises and continued hardships.

In Curitiba, Fernanda Striker Fernandes, the librarian at the Centro de Formação Urbano Rural Irmã Araújo, helped me find important documents and put me in touch with Werner Fuchs, who graciously sat down with me for interviews. At the regional branch of the Fundação Nacional de Saúde, Elisangela and Enéas Cordeiro helped me get in touch with Allan Martins da Silva and Roderlei Araujo, who generously spoke with me in a formal interview setting about mosquitoes and malaria at Itaipu and other reservoirs in the state, opened their personal collections of documents to me, and put me in touch with helpful colleagues in Western Paraná. At the Núcleo de Documentação da Justiça Federal do Paraná, Valdir Chacon helped me locate and access an important set of documents about fisheries.

In Western Paraná, I was lucky to meet equally helpful and generous people. At the Centro de Pesquisa e Documentação sobre o Oeste de Paraná in Marechal Cândido Rondon, Rodrigo Ribeiro Paziani and Jussara Dornelles de Souza helped me locate an important collection of documents about the Sete Quedas National Park before it was flooded. In Guaíra, Mario and Diego Prado helped me network and put me in contact with helpful individuals, as did Bentinho, who also took me out on the reservoir during a time of drought to boat over what was once Guaíra Falls and check out the rocky outcroppings that protruded when water levels dropped. Ana Menél, Edson Galvão, Frei Pacífico, and Joel Onofre shared information and stories about the history of Guaíra and Sete Quedas, and José Cirineu Machado and other fishermen at the Colônia de Pescadores Z-13 also generously shared their stories. Suemy and Ana Helena Foletto kindly opened their personal collection of archived issues of *Jornal Ilha Grande*, a local newspaper that contains a wealth of information about Guaíra Falls.

In Foz de Iguaçu, Vilma Macedo generously shared the personal papers of her late husband, Juvêncio Mazzarollo, which included copies of local newspaper and magazine articles, and Alexandre Palmar also shared with me his digitized copies of local newspapers. At the Núcleo de Entomologia, Adão Celeste Ferreira, Adelino Fideles, and Alvir Swiderski shared helpful information and put me in touch with Maria do Carmen and Mário Pillecco, whose interviews were valuable for helping me understand the history of malaria outbreaks alongside Itaipu's reservoir. Flávio Kabroski, the president of the Colônia de Pescadores Z-12, sat down with me for an interview about the experience of fishermen in the reservoir and invited me to attend one of the fishing colony's meetings, where I explained my research project and met with other fishermen who shared their stories.

In Asunción, Paraguay, many people helped me with my research. Enrique Gaska, director of the Coordinación Nacional de Pastoral Indígena, opened the organization's library to me, shared his own articles about the Avá Guaraní communities displaced by Itaipu, and spoke with me at length about the subject. He also put me in touch with Mariblanca Barón, who graciously opened her library in Nueva Esperanza and shared stories about her decades-long work with the Avá Guaraní communities dispossessed by Itaipu. Although we never got a chance to meet in person, Ricardo Caballero Aquino, ambassador and director of the Archivo Histórico de la Cancillería, was very supportive and offered some helpful suggestions.

In both Asunción and Cuidad del Este, many malariologists at the Servicio Nacional de Erradicación del Paludismo helped me locate documents about malaria and Itaipu and shared important information in interviews. The abbreviated list includes Patricia Luraschi, Mónica Ozorio, Liz Paola Ruiz Diaz, Alexis Liz Palacios, Miguel Diarte, Cinthia, Cándido, and Amado Villalba, Christian Sisanovsky, Leonardo Taboada, and Mirtha Franco.

I was also fortunate to cross paths with two engineers, Luiz Fernando Rufato and João Paulo Aguiar, the chief engineers of two of this book's case studies (Tucuruí and Sobradinho, respectively) during the 1970s and 1980s. They both generously took the time to share their thoughts with me. Rufato spoke with me in an informal setting when I was beginning my research, and João Paulo sat down with me for a formal interview during the later stages of my research and shared helpful documents from his personal collection. I am grateful to them for sharing their stories and perspectives on these dams.

This book's bibliography was also built by generous colleagues in academia. Felipe Oliveira shared his copies of rare documents, some of which were destroyed in the tragic fire at Brazil's Museu Nacional. Bruno Biasetto also shared some of his documents, and Nathalia Capellini shared helpful information about electricity in Manaus from her own research on energy in Amazônia. I am also grateful to Dana Aronowitz and Amanda Rudd at the Georgetown University Library's Interlibrary Loan office for tracking down countless rare articles.

When the time came to write the manuscript, I was fortunate to be surrounded by many smart and generous colleagues who shared valuable input. Special thanks are due to the following people for their time and help with the manuscript.

For reading and commenting on earlier drafts of the full manuscript, J.R. McNeill, Meredith McKittrick, David Painter, Shawn Miller, Ling Zhang, Lucy Rhymer, and two anonymous reviewers with Cambridge University Press.

For reading and commenting on earlier drafts of multiple chapters, Lise Sedrez, Bryan McCann, Erick Langer, Timothy Newfield, Jackson Perry, Douglas McRae, and David Jones; and for reading and commenting on earlier drafts of individual chapters or subsections, Natascha Otoya, Emma Moesswilde, Bryna Cameron-Steinke, Rebecca Andrews, Dylan Proctor, Rob Christensen, Andrew Ross, Francisco Céntola, Ishmael Annang, Loren Galesi, André Felipe Cândido da Silva, Gabriel Lopes, Ingrid Casazza, Leonardo Lignani, Matthew Tarduno, and Hélène Benveniste.

For listening and offering helpful comments on parts of the research shared as oral presentations, Lise Sedrez, Dominichi Miranda de Sá, Stephen Bell, Euripides Antônio Funes, Rogério Ribeiro de Oliveira, José Augusto Pádua, Valéria Fernandes, Filipe Oliveira, Natasha Barbosa, Hana Mariana, Millena Farias, Bruno Capilé, Paula Fortini, Bruno Araujo, Letícia Andrade, Gabriel Paes, Gabriel Pereira, Alexia Shellard, Lucas Mores, Sandro Dutra e Silva, André Vasquez, Eric Carter, and Jamile Silveira; students and faculty in the Georgetown University Environmental History Research Group, the Laboratório de História e Natureza na Universidade Federal do Rio de Janeiro, the Laboratório de História Ambiental do Cerrado, and the Grupo de Pesquisa de Água e Saúde; and panel attendees at the Sociedad de Historia Ambiental Latinoamericano y Caribeña meeting in Liberia, Costa Rica, the World Congress of Environmental History in Florianópolis, Brazil, and the American Society for Environmental History meeting in Eugene, Oregon, United States.

Geoff Wallace designed the maps for this book with great skill and patience, and a handful of Georgetown and Harvard colleagues weighed in with helpful feedback on the visuals.

There are many more people still who helped indirectly and supported me while I was researching and writing this book, and I owe tremendous thanks to all of them.

Notes on the Text

ON ABBREVIATIONS

Many of the actors in this book are companies, government agencies, and nongovernmental organizations whose names are commonly referred to using acronyms.

For all such entities, I have spelled out the full name in both English and Portuguese at the first reference in each chapter (and at the first reference in the footnotes of each chapter) and then used the acronym thereafter.

For the company Itaipu Binacional, the standard abbreviation is Itaipu, and I refer to the company using this abbreviation except when there might be ambiguity about whether the dam or the company is the subject. In those circumstances, I use the full name.

For all archival and library collections, I use only the acronym in chapter footnotes. The bibliography lists all archives and libraries, with both acronyms and full spelling, as well as their geographic location.

ON NAMING AND SPELLING

Many of the subjects in this book include people and places from across borders, which invites multiple naming conventions. Four of the most common are addressed here.

Guaíra Falls, a binational waterfall on the Brazilian–Paraguayan border discussed in Chapters 2 and 5, is referred to as Sete Quedas in Brazil and Los Saltos del Guairá in Paraguay and is most commonly translated as Guaíra Falls in English. I use Guaíra Falls throughout, except when referencing the Sete Quedas Dam and the Sete Quedas National Park, and maintaining the original phrasing of primary sources.

Ernst Mann, a naturalist from Guaíra who makes brief appearances in this story and whose collection provided rich information about the Sete Quedas National Park, sometimes spelled his first name Ernest or Ernesto. I have used Ernst except when maintaining original phrasing in citations.

Avá Guaraní, an Indigenous group living along the Brazilian–Paraguayan border whose story is discussed in Chapters 4 and 7, is spelled Avá Guaraní (and sometimes Ava Guaraní) in Spanish, and they are also referred to as Avá Chiripa and Guaraní Paranaenses in Paraguay. In Brazil, their name is spelled Avá Guarani, and they are also referred to as Ñandeva. I use the Spanish spelling of Avá Guaraní when referring to the communities in Paraguay and the entire community broadly, and I use the Portuguese spelling of Avá Guarani when referring specifically to the communities on the Brazilian side of the reservoir.

Waimiri-Atroari, an Indigenous community living north of Manaus whose story is discussed in Chapter 6, is sometimes hyphenated and sometimes not. I have hyphenated the name throughout, except when maintaining original phrasing in titles.

Introduction

In 1968, the floodwaters of the recently completed Cocorobó Dam submerged the remnants of the site of the Canudos Massacre, one of the most traumatic and infamous episodes in Brazilian history. In the 1880s, a local preacher, Antônio Conselheiro, had amassed a large following and built a small settlement – which he named Belo Monte [Beautiful Hill] – near an area known as Canudos, an impoverished part of the semiarid interior of the northeastern state of Bahia. In 1889, military officers overthrew the monarchy that had ruled Brazil since independence (1822) and installed a republic. The new government soon came into conflict with the preacher and his followers, seeing them as a potential source of opposition.

In October 1896, after a dispute broke out between Conselheiro and a local lumber merchant, the state government sent troops to the town to forestall bloodshed. The troops, instead, moved against Canudos, but were repelled by its residents, prompting the state government to call on the federal government for help. Over the course of a year, the federal government sent four military expeditions to quash the previously peaceful dissidents. After three defeats, the heavily armed fourth expedition overran the town and slaughtered nearly all its inhabitants. Historians estimate the death toll was somewhere between 15,000 and 30,000, including both town residents and federal troops. The surviving residents fled the town, and the military retreated, leaving behind the ruined foundation of the town and the remnants of battle. The historic site sat mostly untouched for seventy years, when the rising floodwaters drowned it.[1]

[1] On Canudos, see the firsthand account of journalist Euclides da Cunha, who accompanied one of the military expeditions, *Os sertões: campanha de Canudos* (Rio de Janeiro:

The Cocorobó Dam was a long time coming. The reservoir was the brainchild of President Getúlio Vargas, who was famous for strengthening the federal government and promoting state-led economic growth. The National Department for Works to Combat Drought [Departamento Nacional de Obras Contra as Seca, DNOCS] designed and built the dam to provide irrigation water for farmers. Construction began in 1951, but soon stalled. The project was not even half complete in March 1964, when military officers orchestrated a coup that toppled the government and installed a dictatorship that lasted more than two decades. The generals accelerated construction. In December 1967, engineers and their teams finished the dam, and during the subsequent months, the reservoir began to fill.

The dam was controversial, though it was finished with little fanfare. The federal government maintained that its motivations for choosing the site were purely empirical. That is, the location was simply the best choice in technical terms, and the fact that it flooded a historic site was an unfortunate secondary effect. In contrast, critics argued that the government built the dam for the explicit purpose of inundating the site of Canudos, a tragedy in which the military and the republican government were the villains. Regardless of intent, the result was the same – the remnants of one of the country's most important historic sites were drowned.[2]

But the reservoir had an ironic twist. Although it erased the material remnants of the massacre, it breathed new life into the mythology surrounding it. During his time as a lay preacher, Conselheiro predicted that someday the desert would become the sea, and the sea would in turn become the desert. Of course, he expected that divine providence, not engineers working for the federal government, would be the architect of this profound transformation. But for Northeasterners familiar with the tales of Conselheiro and his prophecies, the effect was the same. His prediction seemed to be coming true; the desert was becoming the sea.

Laemmert Editores, 1902), translated in English by Samuel Putnam as *Rebellion in the Backlands* (Chicago: University of Chicago Press, 1944); Mario Vargas Llosa's fictionalized account, *La guerra del fin de mundo* (Barcelona: Seix Barral, 1981), translated in English by Helen R. Lane as *The War of the End of the World* (New York: Farrar, Straus and Giroux, 1984); Robert M. Levine, *Vale of Tears: Revisiting the Canudos Massacre in Northeastern Brazil, 1893–1897* (Berkeley: University of California Press, 1992); Marco Antônio Villa, *Canudos: O povo da terra* (São Paulo: Editora Ática, 1995); and Adriana Michele Campos Johnson, *Sentencing Canudos: Subalternity in the Backlands of Brazil* (Pittsburgh: Pittsburgh University Press, 2010).

[2] On Cocorobó Dam, see Comitê Brasileiro de Barragens (CBDB), *A História das Barragens no Brasil, Séculos XIX, XX, e XXI: Cinquenta Anos do Comitê Brasileiro de Barragens* (Rio de Janeiro: CBDB, 2011), 85–86.

The story has a second irony. The reservoir was part of a broader dam-building campaign carried out by the military government that recalled at least one notable characteristic of the infamous Canudos Massacre: its geographic dimensions. The origins of the historic conflict were tensions between the coast and the interior. In the 1890s, upwards of 80 percent of Brazilians lived on the coast, both in metropolitan areas and in the countryside immediately surrounding them. But coastal cities represented just a small fraction of the country's immense territory, and since independence, the government had been aspiring to integrate the country's distant hinterlands into the national fabric. For the republican government, its initial investment in the Canudos campaign was a means of demonstrating its authority in remote areas, though the repression thereafter transcended such logic. The government succeeded in showcasing its might, but at a huge cost: It revealed itself as a perpetrator of violence, which hardly qualified it for the respect and legitimacy it thought it deserved. Cocorobó's reservoir promised to expunge the material traces of this tragedy, but the military regime that built it was then planning a spate of big reservoirs – most to produce hydropower – that also engendered violence against marginalized communities in the interior.

The motivations for dam building were of a different nature than those that drove the Canudos Massacre, but other similarities make the comparison compelling. The percentage of Brazilians living along the coast had changed little between the 1890s and 1960s, and both offensives emanated from the federal government and were intended to integrate hinterlands into the national fabric. Although the Canudos Massacre was far more lethal than the dam-building campaign, reservoirs also entailed violence, displacing hundreds of thousands of people from their homes. Thus, there was tremendous irony in trying to erase the material remnants of one of the country's most tragic assaults on the interior with a tool that would also inflict violence against hinterland communities. While Cocorobó's reservoir was quietly covering the vestiges of Canudos, the administration that finished it was planning a series of even bigger interventions in the interior that would bring immense benefits for some and considerable suffering for others.

Between 1964 and 1989, the Brazilian military dictatorship and the civilian administration that succeeded it in 1985 built some of the world's biggest and most environmentally controversial hydropower dams, whose combined installed capacity totaled 45,000 megawatts. Most of the power and environmental controversies came from about twenty-five big dams with high output and big reservoirs (Figure I.1). Just five dams

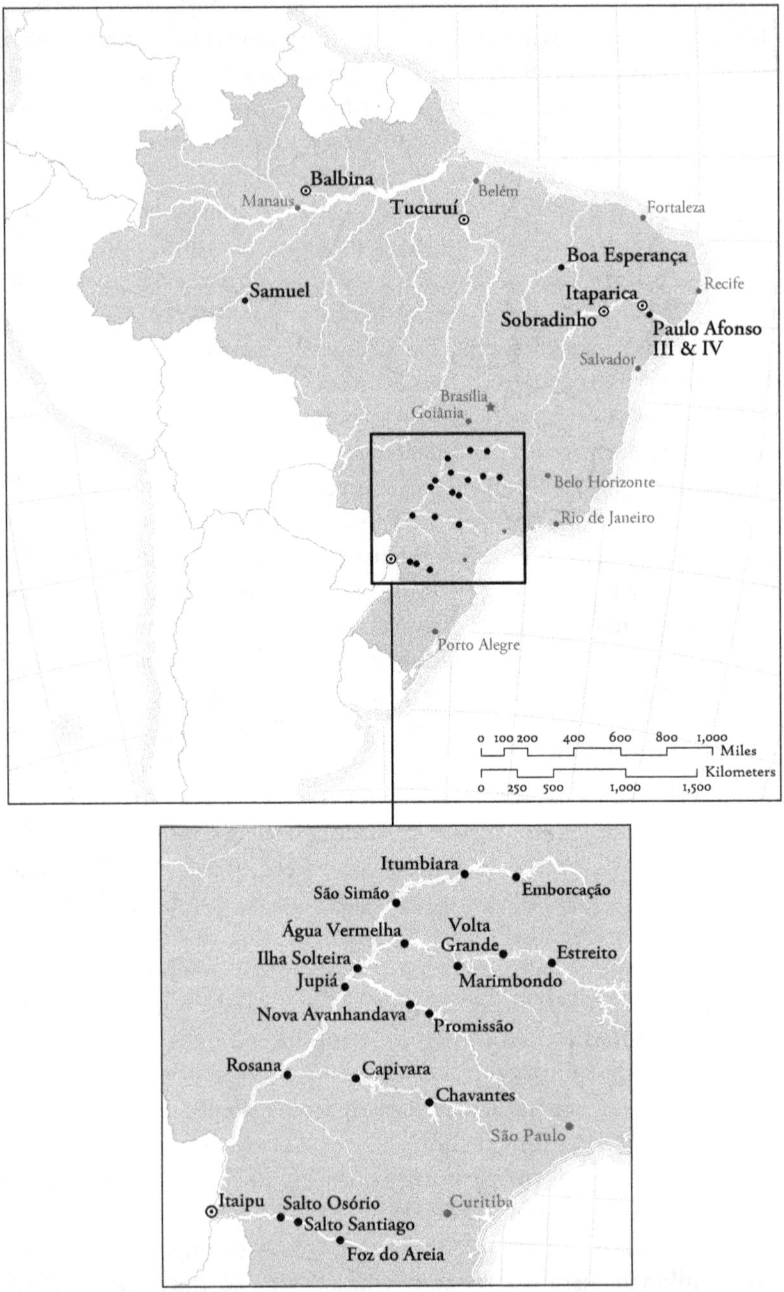

FIGURE I.1. Brazilian hydropower dams, 1964–1989. Showing all dams built during this period with an installed capacity greater than 500 megawatts and/or a reservoir greater than 200 sq. km. The dams featured in this book are highlighted with circled dots. Map by Geoffrey Wallace(G. Wallace Cartography & GIS).

alone accounted for more than half of this installed capacity, and their massive reservoirs bore the brunt of criticism related to social and environmental impacts. They are the focus of this book.[3]

The first two are Sobradinho and Itaparica, two big reservoirs built on the São Francisco River, a long and voluminous river referred to as the Brazilian Nile because it flows south to north through the semiarid northeast of Brazil. Sobradinho (1971–1978)[4] was the dictatorship's first big dam, and it created the largest reservoir in South America. The military regime started construction on Itaparica (1975–1988) shortly thereafter, and the subsequent civilian government finished it.

The third featured dam is Itaipu (1973–1982), a binational dam on the mighty Paraná River, which forms a stretch of the Brazilian–Paraguayan border in the southwestern corner of the country. The dam was built in the remnants of the Atlantic Rainforest and was famous for being the world's then-biggest power plant in terms of installed capacity. Only China's Three Gorges Dam (1994–2003) has since outstripped it, and Itaipu remains the world's most productive power plant in terms of the cumulative electricity it has produced over its lifetime.

The final two featured dams are in the Amazon Rainforest. The fourth is Tucuruí (1975–1985), a big dam erected on the lower stretches of the Tocantins River, on the eastern edge of the rainforest. It was the first giant reservoir built in the Brazilian Amazon, and it was Brazil's second-most productive power plant after Itaipu. The fifth is Balbina (1981–1989), a big dam on the Uatumã River, a tributary of the Amazon, whose headwaters are about 150 kilometers north of Manaus. It was the last and most controversial of the military government's big dams, and, after a series of delays, the civilian government finished it – both Itaparica and Balbina were almost done when the military regime stepped down in 1985, and the civilian politicians that replaced the generals shared the commitment to finishing both dams.

3 For further details and comparison with Brazilian dam building before and since, see Matthew P. Johnson, "Temples of Modern Pharaohs: An Environmental History of Dams and Dictatorship in Brazil, 1960s–1990s," (PhD Diss., Georgetown University, 2021), 17–31.

4 The timelines listed here refer to the year that the government officially greenlit the dams and the year in which their reservoirs finished filling. In some cases, dam sites had been studied for many years prior to official authorization, and in other cases, construction was delayed for years after authorization. In most cases, reservoirs filled months or years before turbines began generating power, and in all cases, the government continued to install additional turbines to generate power in the years after the reservoirs filled.

In contrast to the reservoir that flooded the remnants of the Canudos Massacre site, which stored water for irrigation, these big dams were designed to generate electricity to power cities and industrial centers.[5] When the construction dust settled in 1990, hydropower provided about 93 percent of the country's electricity, and Brazil had become one of the world's largest hydroelectric producers (it remains second only to China).[6] Hydropower remains Brazil's main source of electricity, and the military regime's dams continue to account for a considerable portion of that grid, about 30 percent of the country's total electricity generation, and about 45 percent of the country's hydropower generation.[7]

Brazil's dam-building boom was in turn part of a larger regional trend. Hydropower has been instrumental in Latin America's rapid urban and industrial growth, and during the latter decades of the twentieth century, dams generated upwards of 60 percent of the electricity consumed in the region. Mirroring Brazil's own trajectory (discussed in Chapter 7), this percentage has dropped to roughly 40 percent over the last two decades, but Latin America remains well above the world average of 16 percent, and in some countries – such as Costa Rica and Colombia – hydropower still accounts for more than 60 percent of electricity consumed.[8]

The output of these dams supported industrialization and urbanization and did so with comparatively few greenhouse gas emissions. Reservoirs emit methane and other greenhouse gases, as discussed in greater detail in the chapters ahead, but most do so at much lower rates compared to fossil-fuel-powered electrification. Globally, hydropower spares the

[5] For more on DNOC irrigation dams, see Anthony Hall, *Drought and Irrigation in North-East Brazil* (New York: Cambridge University Press, 1978), and Eve E. Buckley, *Technocrats and Politics of Drought and Development in Twentieth-Century Brazil* (Chapel Hill: University of North Carolina Press, 2017).

[6] This figure comes from the International Energy Agency (IEA)'s webpage on Brazil's energy profile, https://www.iea.org/countries/brazil (last accessed April 2023). The Ministério de Minas e Energia's Sistema de Informações Energéticos and Brazilian energy expert Antônio Dias Leite list the figure at around 87 percent in the early 1990s, though their calculations include only the Brazilian half of Itaipu's output, despite the fact that Paraguay sells most of its share to Brazil. See Dias Leite, *Energia do Brasil, 3ª edição* (Rio de Janeiro: Editora Nova Fronteira, 2014), 296.

[7] Author's calculations.

[8] The latest regional-level data is from 2015. Paraguay gets almost 100 percent of its electricity from hydropower, most of which comes from the Itaipu Dam. In Costa Rica, hydropower accounts for roughly 75 percent of the country's electricity. In Panama, Colombia, Venezuela, Suriname, and Uruguay, the figure is 60 percent or higher. All countries in the region except for Mexico, Nicaragua, Guyana, and French Guiana are well above the world average. See World Bank Online Database, https://data.worldbank.org/indicator/EG.ELC.HYRO.ZS (last accessed April 2023).

Earth's atmosphere more greenhouse gas emissions than any other source of low-carbon (i.e., renewable) electricity. The annual carbon savings of the collective output of dams worldwide are somewhere between three and eight gigatons (global annual carbon emissions are about 40 GT).[9]

Most of the Brazilian military regime's dams were low-emission power plants, but they also unleashed a host of other damaging social and ecological impacts. Big reservoirs flooded the lands of Indigenous communities and farmers, erased cherished landscapes, and set in motion a series of ecological transformations with deleterious consequences for those nearby.

This book is an environmental history of this dam-building boom that explores the military government's motivations for building big dams, its efforts to mitigate their environmental impacts during the planning and construction stages, and the social and environmental consequences after reservoirs filled. It builds on and contributes to scholarship in the fields of energy and environmental history that focuses on past experiences with low-carbon energy, which offer important lessons for countries ramping up investment in renewables in order to mitigate the impacts of climate change.[10] The case of Brazilian dams suggests that building up an arsenal of low-carbon energy will come with its own social and environmental costs and that citizens will have to be attentive to holding companies and governments accountable for social and environmental justice.

This book also builds on and contributes to the scholarship on authoritarian environmentalism. The high stakes of a looming climate

9 The lower figure comes from Luis Berga, "The Role of Hydropower in Climate Change Mitigation and Adaption: A Review," *Engineering* 2 (2016): 313–318, and the higher figure comes from author's calculations, extending the data presented in Chapter 3 to hydropower's worldwide installed capacity.

10 Hydropower and nuclear energy – the two biggest low-carbon energy technologies to date – have deep historiographies, though most scholars have not framed their work in the context of the technologies' low-carbon properties. For a sampling of recent literature on other forms of low-carbon energy, see Marianna Dudley, "When's a Gale a Gale? Understanding Wind as an Energetic Force in Mid-Twentieth Century Britain," *Environmental History* 26, no. 4 (October 2021): 671–695; Jennifer Eaglin, *Sweet Fuel: A Political and Environmental History of Brazilian Ethanol* (New York: Oxford University Press, 2022); Thomas D. Rogers, *Agriculture's Energy: The Trouble with Ethanol in Brazil's Green Revolution* (Chapel Hill: University of North Carolina Press, 2022); James Morton Turner, *Charged: A History of Batteries and Lessons for a Clean Energy Future* (Seattle: University of Washington Press, 2022); and Elizabeth Chatterjee, "The Poor Woman's Energy: Low-Modernist Solar Technologies and International Development, 1878–1966," *Journal of Global History* 18 (November 2023): 439–460.

catastrophe and the desultory efforts of governments throughout the world to divest from fossil fuels have compelled some scholars to look to China's recent environmental actions as a model for "authoritarian environmentalism" and to probe past dictatorships for evidence of successful environmental interventions.[11] The case of Brazilian dam building provides more evidence for the environmental failures of authoritarian governments than successes, with two important points of emphasis.

First, dictatorships, democracies, capitalist and communist governments, and imperial and newly independent governments alike have all supported big dams. Thus, in Brazil, authoritarianism was not the paramount factor motivating dam building, but the regime's search for legitimacy did shape decision-making, and repression did forestall activism that might have mitigated social and environmental impacts. Scholars are right to point out that authors critical of authoritarian environmental management often overlook similar states of affairs in

[11] For scholarship that specifically addresses the desirability of an eco-dictatorship, see David Shearman and Joseph Smith, *The Climate Change Challenge and the Failure of Democracy* (Westport: Praeger, 2007); Mark Beeson, "The Coming of Environmental Authoritarianism," *Environmental Politics* 19, no. 2 (March 2010): 276–294; Bruce Gilley, "Authoritarian Environmentalism and China's Response to Climate Change," *Environmental Politics* 21, no. 2 (March 2012): 287–307; Dan Coby Sahar, "Rejecting Eco-Authoritarianism, Again," *Environmental Values* 24 (2015): 345–366; and Yifei Li and Judith Shapiro, *China Goes Green: Coercive Environmentalism for a Troubled Planet* (Cambridge: Polity Press, 2020). For scholarship that explores the environmental dimensions of past and present authoritarian regimes, see Murray Feshbach and Alfred Friendly Jr., *Ecocide in the USSR* (New York: Basic Books, 1992); Judith Shapiro, *Mao's War against Nature: Politics and the Environment in Revolutionary China* (New York: Cambridge University Press, 2001); Paul Josephson, *Resources under Regimes: Technology, Environment, and the State* (Cambridge: Harvard University Press, 2004); Franz-Josef Brüggemeier et al., eds., *How Green Were the Nazis? Nature, Environment, and Nation in the Third Reich* (Athens: Ohio University Press, 2005); Frank Uekötter, *The Green and the Brown: A History of Conservation in Nazi Germany* (New York: Cambridge University Press, 2006); Timothy Doyle and Adam Simpson, "Traversing More than Speed Bumps: Green Politics under Authoritarian Regimes in Burma and Iran," *Environmental Politics* 15, no. 5 (2006): 750–767; Jeannie Sowers, "Nature Reserves and Authoritarian Rule in Egypt: Embedded Autonomy Revisited," *Journal of Environment and Development* 16, no. 4 (2007): 375–397; Marco Armiero and Wilko Graf von Hardenberg, "Green Rhetoric in Blackshirts: Italian Fascism and the Environment," *Environment and History* 19, no. 3 (2013): 283–311; Stephen Brain and Viktor Pál, eds., *Environmentalism under Authoritarian Regimes: Myth, Propaganda, Reality* (New York: Routledge, 2019); and Miguel Angel Del Arco Blanco and Santiago Gorostiza, "'Facing the Sun': Nature and Nation in Franco's 'New Spain' (1936–51)," *Journal of Historical Geography* 71 (2021): 73–82.

democracies,[12] but the evidence from Brazilian dam building suggests that dictatorship only worsened social and environmental outcomes.

Second, it is not clear that the world has yet seen a regime built on the principle of "authoritarian environmentalism," or the mobilization of authoritarian practices for the purposes of environmental protection. There are, of course, many cases of forceful environmental policies that might be considered "authoritarian environmentalism." However, in such cases environmental care has not been either the motive for authoritarianism or a guiding principle of governance, and such individual policies do not often reflect a strong overall record on environmental affairs.[13]

Similar to other authoritarian governments of the time, the Brazilian military dictatorship's relationship with environmentalism was reactionary and fragmented. The military regime's principal concern was economic development at all costs, and its environmental efforts were its grudging answer to a burgeoning environmental movement that surged at home and abroad in the 1970s and 1980s. Furthermore, this reaction came in response to concerns from groups least subject to repression (such as the Catholic Church and multinational banks) and gained momentum only after the military regime began gradually restoring democratic freedoms in the late 1970s. Further still, the greatest environmental merit of the dictatorship's dams, their carbon savings, was not factored into the decision to build them. Engineers broke ground at dam sites long before the rise of popular concern over climate change in the late 1980s.

Thus, affirming what Yifei Li and Judith Shapiro argue for China, the better phrase to capture the Brazilian dictatorship's environmental actions is "environmental authoritarianism," or the use of environmental policies to maintain or entrench authoritarian rule.[14] The past provides no examples of virtuous eco-dictatorships and ample evidence that pursuing authoritarianism for environmental ends would entail dangerous trade-offs. Governments can implement and enforce stringent environmental regulations within a democratic setting, and history – Brazil's included – provides scant evidence that embracing dictatorship offers a solution to the planet's environmental problems.

12 Brain and Pál, *Environmentalism under Authoritarian Regimes*, 3, 8.

13 For an example of forceful environmental policies, see Heejin Han, "Authoritarian Environmentalism under Democracy: Korea's River Restoration Project," *Environmental Politics* 24, no. 5 (2015): 1–20.

14 See Li and Shapiro, *China Goes Green*, 22–24, 188.

For the world's democracies, an informed citizenry is essential to addressing anthropogenic climate change and related issues of social and environmental justice. Most consumers in urban countries remain largely unaware of the people and environments in the distant hinterlands that sustain them. Since the nineteenth-century, much of the world has industrialized and urbanized, and most people now live in urban or suburban areas. Industrialization has also reached rural areas, and commodity chains increasingly connect these spaces to one another. The resource production chain is often rife with social and environmental injustices, and the first step toward responsible environmentalism is simply knowing what happens in distant hinterlands.

This disconnection is so entrenched that it has become the job of journalists and scholars of environmental history and related disciplines to illuminate the otherwise hidden pathways that connect city dwellers and suburbanites to the most basic resources that sustain them, such as energy, water, and food. Scholars have focused on water, foodstuffs, and building materials and have more recently turned their attention to energy.[15] This book builds on this tradition with the goal of familiarizing Brazilian consumers with the rivers that electrify their cities and factories and encouraging consumers elsewhere to think through parallels between hydropower and the energy sources that sustain their own livelihoods.

One of the most powerful transformations that Brazilian dams brought about for people living near them was the silence that followed the flooding of giant waterfalls. In one case, at Guaíra Falls, the roar was so loud that people in the adjacent town were accustomed to the sound at all times, and the subsequent silence after Itaipu flooded it was one of the reservoir's most disturbing features. This book's most fundamental goal is to bring back this noise, though not in dam-adjacent communities but rather in the minds of those in distant metropolitan areas consuming electric power. I want to replace the soft hum of electrified cities with

[15] For a classic articulation of this goal from an environmental historian, see William Cronon, *Nature's Metropolis: Chicago and the Great West* (New York: N.W. Norton & Co., 1991), which covers foodstuffs and lumber. On water, see Marc Reisner, *Cadillac Desert: The American West and Its Disappearing Water* (New York: Penguin, 1986); on food, see Sidney Mintz, *Sweetness and Power: The Place of Sugar in Modern History* (New York: Penguin Books, 1986); and Michael Pollan, *The Omnivore's Dilemma: A Natural History of Four Meals* (New York: Penguin Books, 2006); on energy, see Andrew Needham, *Power Lines: Phoenix and the Making of the Modern Southwest* (Princeton: Princeton University Press, 2014), and Christopher Jones, *Routes of Power: Energy and Modern America* (Cambridge: Harvard University Press, 2014).

the roar of powerful waterfalls; the silence of turning on a light switch and charging a cell phone with the shouts and protests of displaced communities. Those sounds, in turn, should serve as a constant reminder that metropolitan and electrified rural areas remain deeply connected to distant environments and the communities that live there despite our ostensible distance from them.

Our trip into the hinterland calls for a roadmap. This book's main structure is chronological, though some chapters have overlapping timelines and are organized thematically. Chapter 1, "Setting the Scene," provides requisite background information. It discusses Brazil's military dictatorship, environmental movement, and industrial growth, as well as the ways in which the military regime used dam building for political ends, as governments elsewhere have also done. Building dams has been remarkably nonpartisan. During the twentieth century, governments of all types in all parts of the world built big dams to gain political support because they generate local economic benefits and are potent symbols of both industrial growth and proficient governance. In Brazil, the military regime aligned its dam-building ambitions with the goals of its National Integration Program, which sought to build infrastructure in areas it deemed sensitive to subversion. The symbolic benefits of building dams were especially profitable in Brazil, and elsewhere in the so-called developing world, because dams represented the country's ability to overcome long-standing economic inequalities and racist stereotypes of inferiority that then characterized the global economic order. The military regime hoped to capitalize on such benefits to legitimize its tenure. Having introduced important background information, the remainder of the book is dedicated to supporting its principal arguments.

This book makes three interrelated arguments. The first is that political pressures encouraged the military regime to build big dams with huge social and environmental impacts and discouraged it from investing in efforts to mitigate them. Engineers had to consider basic technical and economic criteria in assessing potential dam sites, but it was political pressures that compelled the military regime to commit to building massive reservoirs regardless of their environmental footprints. That political considerations rivaled and sometimes overruled economic criteria in choosing dam sites is an important reminder that dam building's social and environmental impacts were not an inevitable consequence of economic growth, but rather the result of contingent political decisions.

Chapter 2, "Building 'the Big Dam'," presents three cases that support this argument. The most salient example was Itaipu, built in partnership with the Paraguayan government, then also under military dictatorship. Brazilian engineering legend Octávio Marcondes Ferraz championed an alternative with a much smaller social and environmental footprint, but both governments foreclosed public debate and refused to even consider his alternative in the interest of resolving a tense border dispute. In other cases, too, political concerns influenced site selection, and the pressures of the National Integration Program encouraged the military regime to build all its dams quickly, which meant that construction outpaced reliable environmental studies and sound mitigation measures.

This book's second argument is that the few environmental protection measures the military regime did implement were designed to protect power plant infrastructure and showcase environmental concern without altering project designs or slowing construction. Chapter 3, "Pharaonic Environmentalism," contains the evidence. During the 1970s, the rise of environmentalism globally and within the World Bank – which was then helping finance Brazil's dams – urged the military regime to develop an image as a good environmental steward so it could maintain access to loans. The government-owned electricity company Eletrobras and its subsidiaries contracted renowned ecologist Robert Goodland to study their dams, but only after already committing to dam sites with massive environmental footprints. The state-owned companies gave him little time to undertake his studies, pressured him to censor his conclusions, and applied his findings selectively. The safeguards the government did implement were those designed to prolong the life of the power plants and those that would dramatically showcase its environmental consciousness, such as animal rescue missions. The military government hoped these measures would bolster its image as a regime that could build durable mega dams and also protect the environment. The generals succeeded at the former but faltered at the latter.

This book's final argument is that the dictatorship's pharaonic environmental protection programs failed to forestall a series of social and ecological disasters at dam sites. Chapter 4, "Negotiating with Floodwaters," is about the disastrous impacts of reservoirs on displaced communities. The generals mostly ignored the social costs of their big dams because they felt pressure to build them quickly and cheaply, and because they believed their pharaonic environmentalism would satisfy their critics. The military regime provided scant resources to help displaced communities transition to new homes and subsistence practices, and many were left to start anew with almost no financial compensation. For the generals, it helped that most of these people were poor and from racially marginalized groups that had little political clout. Nevertheless, the Catholic Church helped organize these communities, some of whom succeeded in earning more just compensation.

The military regime's limited interventions also did little to ameliorate damaging ecological changes that reservoirs unleashed. Chapter 5, "Environmental Transformations," covers these dramatic episodes. It details the loss of cherished land and waterscapes that had been protected as national parks, the booms and busts of reservoir fisheries, and mosquito-related torments in communities along the margins of reservoirs. To be sure, not all changes spelled disaster. Fisheries boomed in the decades following the formation of reservoirs, and malaria outbreaks on the edges of reservoirs were mild by historic standards. But in most places, disregard for the environment led to a series of local disasters that drew attention to the high environmental costs of building big dams.

Chapters 4 and 5 focus on the most well-known and visible impacts and are therefore not complete accounts of the social and environmental impacts of these dams. In some cases, the government mandated those displaced from reservoirs to resettle on Indigenous land, which generated conflicts. Building reservoirs also set in motion dramatic ecological changes downstream and in estuaries, and dam-building companies also cleared large swaths of forest, sometimes on Indigenous land, to build the transmission lines to connect power plants to cities. None of these stories are elaborated on in these chapters, which focus on the most immediate and traceable impacts in order to be concise and digestible and because stories of additional impacts neither add to the book's arguments nor discredit them.

The dictatorship's last and most controversial dam offers a compelling example of all three arguments outlined above. Chapter 6, "The Notorious Balbina Dam," tells the story of this contentious reservoir, which was built between 1981 and 1989. Political pressures were instrumental in the decision to build the dam, whose floodwaters inundated both Indigenous land and large stretches of the Amazon Rainforest. Instead of investing in meaningful environmental safeguards, the government planned an ostentatious greenwashing campaign. The result was social and ecological calamities on par with those at earlier dams.

But there was one principal difference that made Balbina unique: timing. Balbina came on the heels of a freshet of other controversial dams that had turned many Brazilians against them. Moreover, the military regime stepped down in 1985, during construction, and the civilian government that replaced it finished the dam. The restoration of democratic freedoms enabled dam critics to pressure the government for more effective safeguards. Though the civilian government did not suspend the project, it did implement better belated remediation programs than the military regime had done for its reservoirs. Balbina was thus the last of its kind and became a watershed moment in the history of Brazilian dams.

Thereafter, the government changed its approach to building dams. Chapter 7 tells the story of Brazilian dam building from the 1990s to the 2010s, focusing on both the national movement against dams, which coalesced in the early 1990s, and the legislation that the government passed in response to the demands of these activists and other environmentalists. These laws made sound environmental impact studies and an environmental operating license requisites for building big infrastructure projects such as dams. The result has been that many of the country's most controversial dam sites remain dam-free, and many of the dams that were built were erected only after design modifications to reduce their impact. To be sure, some of the reservoirs that democratic administrations have built since the dictatorship have had deleterious impacts despite such safeguards, and big dams remain controversial in Brazil. Nevertheless, Indigenous groups and environmentalists succeeded in using legal channels to curtail the most damaging impacts. In this respect, the story of the military regime's big dams has had an encouraging ending.

The fate of displaced communities over the subsequent decades is a grimmer tale. The dictatorship's big reservoirs displaced hundreds of thousands of people, some of whom undoubtedly recovered and found ways to prosper in their new settings. However, at least two Indigenous communities – the Avá Guaraní and the Tuxá – never got that chance, because the government has denied them land, the most essential resource for their well-being. More than thirty years later, these Indigenous communities are still fighting to regain their land.

1

Setting the Scene

Dictatorship and Industrial Growth

INTRODUCTION

According to the generals, their 1964 coup was "carried out in order to build,"[1] and they wasted no time in making good on the promise. During its tenure, the military dictatorship orchestrated one of the biggest construction booms in Brazil's history. The generals hired teams of eager engineers to build roads, bridges, power plants, and other big infrastructure projects. More than anything else, massive hydropower dams came to symbolize the dictatorship's directive to build, and none was bigger than Itaipu, the giant binational dam on the Paraná River. In November 1982, shortly after its reservoir filled, one of the country's leading newspapers – which was no longer censored but still submissive to the government – ran a full-page spread about Itaipu's director José Costa Cavalcanti, "the great dam builder," as the headline called him. The article featured text from an interview in which the director reflected on the prestige that building Itaipu afforded the military regime. Costa Cavalcanti remarked, "Itaipu is, without a doubt, a testament to the capacity of the government … it certainly bolsters the country's image abroad, raising confidence in our entrepreneurial capacity and stimulating national pride."[2] The military regime's great dam builders were fulfilling the generals' mandate to build, and their work showcased the capabilities of the government and helped legitimize dictatorship.

[1] Original Portuguese is "A Revolução foi feita para construir." See Emílio Garrastazu Médici, *I Plano Nacional de Desenvolvimento (PND), 1972/74* (Brasília: Congresso Nacional, 1971), 13, General Collection, BCN.

[2] "Costa Cavalcanti: O Grande Barrageiro," *O Estado de Paraná*, November 14, 1982.

This chapter sets the scene for this book, providing the requisite background for the chapters that follow. It begins with a short overview of the military regime, focusing on repression and the gradual restoration of democratic freedoms, highlighting the role the latter played in facilitating the country's burgeoning environmental movement. It then turns to the dictatorship's plans for industrial growth and energy production. The chapter closes with an overview of the symbolism that surrounds big dams and an introduction to the influential generals and engineers responsible for orchestrating the dictatorship's dam-building campaign.

This chapter also lays the groundwork for the book's first argument, that political pressures encouraged the military regime to build big dams quickly and with little regard for their social and environmental impacts. The Brazilian military had a long record of helping overthrow governments, and over time it constructed the narrative that its role in these coups was legitimate because in each instance the period of formal military rule had been short-lived. After the 1964 coup, in contrast, the generals postponed returning the country to civilian leadership, which meant they had to look elsewhere for validity.

This chapter argues that building giant dams in distant hinterlands was appealing to the military government because they produced an essential resource for economic growth, aligned with the goals of the National Integration Program, and generated huge symbolic benefits, all of which helped legitimize military rule. The argument is not that these political considerations alone motivated the dictatorship to build dams. Dams are an eminently practical means of producing electricity, and the Brazilian government had been investing in hydropower since the 1930s. The claim here is that politics were a powerful supplementary motivation that accounts for the accelerated and enlarged nature of the military's dam-building campaign, which produced some of the biggest and most controversial dams in the world with minimal environmental safeguards.

DICTATORSHIP

On March 31, 1964, a military coup overthrew the government of João Goulart, and the military seized control of the country.[3] The officers

[3] There is a debate in the literature as to whether the most appropriate term for the period of military rule is "military dictatorship" or "civilian-military dictatorship." The latter term highlights the role of civilian actors who supported the coup and military rule thereafter. The consensus is that either term is fine so long as writers acknowledge the support of

marketed their takeover as a righteous effort to forestall a communist dictatorship, but as was the case with other military coups of the time, the threat was greatly exaggerated. The context of the Cold War was influential; the Brazilian military was staunchly anti-communist, and the 1959 Cuban Revolution had made officers fearful that a similar uprising could occur in Brazil.

Goulart was a left-leaning nationalist among a group of reform-minded statesmen who had worked with former president Getúlio Vargas and were following in his footsteps.[4] During his term, Goulart attempted to implement modest reforms in landholding, taxes, education, and housing aimed at simultaneously resolving Brazil's long-standing social inequalities and a short-term economic downturn. These moderate reforms – Goulart was after all a large landowner himself – upset his wealthy and powerful adversaries, which included the military. His opponents did not have the congressional majority needed for an impeachment hearing, so the military conspired to remove him illegally.[5]

The United States indirectly supported the coup, offering various forms of aid. During the years leading up to the coup, the US Central Intelligence Agency channeled millions of dollars to right-wing political candidates with ties to the generals plotting the revolt. US military personnel also established close relationships with the generals themselves, who kept US intelligence informed of their plans. During the coup, the United States kept a fleet of ships ready to help the military regime and recognized its rule immediately.[6] After the coup, the United States

civilian actors. This book uses the term "military dictatorship" throughout and acknowledges civilian cooperation. For an overview of this debate, see Marcelo Ridenti, "The Debate over Military (or Civilian-Military?) Dictatorship in Brazil in Historiographical Context," *Bulletin of Latin American Research* 37, no. 1 (January 2018): 33–42.

4 Vargas' first administration lasted from 1930 to 1945, the last eight years of which were a dictatorship. He was reelected democratically in 1951 and committed suicide in 1954.

5 This paragraph and the others in this section are drawn from Thomas E. Skidmore, *The Politics of Military Rule in Brazil, 1964–85* (New York: Oxford University Press, 1988); Maria Helena Moreira Alves, *State and Opposition in Military Brazil* (Austin: University of Texas Press, 1988); Daniel Aarão Reis Filho, *Ditadura e democracia no Brasil: do golpe de 1964 à Constituição de 1988* (Rio de Janeiro: Zahar, 2014); and Daniel Aarão Reis Filho et al., eds., *A ditadura que mudou o Brasil: 50 anos do golpe de 1964* (Rio de Janeiro: Zahar, 2014).

6 On the US support for the coup, see Carlos Fico, *O grande irmão da Operação Brother Sam aos anos de chumbo: O governo dos Estados Unidos e a ditadura militar brasileira* (Rio de Janeiro: Civilização Brasileira, 2008); Anthony Pereira, "The US Role in the 1964 Coup: A Reassessment," *Bulletin of Latin American Research* 37, no. 1 (January 2018): 5–17.

continued to support the dictatorship through its Office of Public Safety, a police training program, and the Alliance for Progress, a regional foreign aid program.[7]

The Brazilian officers in charge of the coup promoted the narrative that they were following a time-honored tradition in which the military helped overthrow unpopular governments and then stepped down shortly thereafter. The narrative failed to account for the continued influence of the military and authoritarianism in the civilian administrations that officers helped install, but it was nevertheless the military's popular account of its role in Brazilian political history.[8] That military officers used this narrative to justify their intervention in 1964 reflects the fact that those in charge of the coup cared much more about public opinion and approval than did other military dictatorships in the region.

However, the officers were divided into two camps on what should come next. Moderates argued that the takeover should swiftly purge the country of the radical leftist threat and then step down, whereas hardliners believed that more enduring and harsh measures were needed to restore order. Though the hardliners envisioned a longer mandate, even they did not plan for the military to retain control of the country indefinitely.

The military consolidated power and then held on. The first general to become president was Humberto Castelo Branco, a moderate. But thereafter, the hardline faction took control. What was originally planned as a short stint of military rule turned into more than two decades of dictatorship, which lasted from 1964 until 1985. The dictatorship was overseen by a procession of five generals, whose terms each ranged from three to five years.

Military officers and cabinet ministers chose the presidential successor, whom the military-controlled Congress then installed through indirect elections. Castelo Branco had envisioned that his own term would only last until 1965, the end date of Quadros' original presidential mandate, but his economic policymakers convinced him to extend his term for two

[7] See Stephen Rabe, *The Killing Zone: The United States Wages Cold War in Latin America* (New York: Oxford University Press, 2012), 108–113.

[8] Aarão Reis Filho, *Ditadura e democracia no Brasil*, 21. In 1889, the military intervened to dispose of the monarchy and then returned the country to civilian leadership. In 1930, the military responded to mounting opposition against elected candidate Júlio Prestes with a coup that removed the sitting president and installed opposition leader Getúlio Vargas as president. In 1945, the military forced Vargas – who had instituted a dictatorship in the 1930s – to step down and briefly took power before resuming elections.

more years so there would be sufficient time for their economic policies to produce results.[9] In 1965, he implemented a decree, Institutional Act No. 2, that made elections indirect and installed a strict two-party system consisting of the National Renewal Alliance [Aliança Renovadora Nacional, ARENA] and the Brazilian Democratic Movement [Movimento Democrático Brasileiro, MDB]. Both were right-wing and supported the dictatorship, though the latter was the more moderate of the two and was considered the opposition party.[10]

In 1967, Congress indirectly elected Artur da Costa e Silva as president, the first of two hardline presidents. The second was Emílio Garrastazu Médici, whose term lasted from 1969 to 1974. Both began their military careers at a young age and had become generals by March 1964, and both helped orchestrate the coup that overthrew João Goulart.[11] Their terms were the dictatorship's darkest years.

Both presidents relied on political repression to curb dissent. They justified their actions by citing the National Security Doctrine, which maintained that the military had the right to use extreme measures to protect itself from internal enemies. These hardliners widened the number of offenses considered threats to national security and increased the severity of punishments, even going so far as to reinstate the death penalty, which had long been illegal.[12] In December 1968, President Costa e Silva issued Institutional Act No. 5 (AI-5), which suspended both the Congress and habeas corpus rights. The latter meant that military officers could freely spy on and arrest anyone deemed a threat to national security and subject them to any form of punishment.

Officers did just that. During the military dictatorship, more than 350 officers tortured more than 6,000 people and killed or "disappeared" more than 430 people.[13] Since these figures are based on just the cases that can be traced, experts estimate that the true figures are much higher.

[9] Aarão Reis Filho, *Ditadura e democracia no Brasil*, 34–48.

[10] Colloquially, people called them the parties of Yes and Yes Sir. In 1979, ARENA and MDB were reorganized as the Social Democratic Party [Partido Democrático Social, PDS] and the Brazilian Democratic Movement Party [Partido do Movimento Democrático Brasileiro, PMDB], respectively.

[11] See Renato Lemos, "Biografia de Artur da Costa e Silva," and Sônia Dias, "Biografia de Emílio Médici," General Collection, CPDOC-FGV.

[12] Mariana Joffily, "O aparato repressivo: Da arquitetura ao desmantelamento," in Aarão Reis Filho et al., eds., *A ditadura que mudou o Brasil*, 158–171; Skidmore, *The Politics of Military Rule in Brazil*, 83–84.

[13] These figures come from Brazil's National Truth Commission [Comissão Nacional de Verdade, CNV], a government agency created in 2011 to report on the human rights

Institutional Act No. 5 also introduced censorship for all forms of media, which the military regime used alongside a robust propaganda department to protect its image and promote its agenda. The military purged universities of left-leaning professors and monitored lectures. Under such political repression, anyone who disagreed with the military government's plans could not speak out without risking their safety, including those who objected to the dictatorship's big dams.[14]

The military regime restored political liberties gradually. In 1974, Congress elected Ernesto Geisel, a moderate, as president. Geisel subscribed to Castelo Branco's original vision that military intervention should be short, and after a decade in power, repression was deteriorating the regime's image and stirring broad mobilization for the return to civilian rule. In response, he set about reinstating democratic freedoms through a gradual process he labeled *distensão* (decompression).

But restoring political rights while also appeasing his fellow officers proved difficult. In November 1974, Geisel permitted direct congressional elections with uncensored television advertising for all candidates. The opposition party, MDB, won in a landslide. To ensure that the opposition did not undermine military rule further, thereafter Geisel used the powers that AI-5 granted him to close Congress, make gubernatorial elections indirect, and retire political adversaries. Decompression had stalled before it could even get started. In 1975, Geisel did lift censorship from some major newspapers, but he was unwilling to curb torture or other human rights abuses.[15]

A more genuine restoration of political rights accelerated at the end of Geisel's term. In 1978, he began the second phase of his redemocratization plan known as *abertura* (opening). He softened the hardline interpretation of the National Security Doctrine, reducing the number of offenses considered crimes against the state and reducing penalties for others. More importantly, he implemented Constitutional Amendment

violations committed during the dictatorship. For the number of those tortured, see CNV, *Relatório Vol. I* (Brasília: CNV, 2014), 349, and for the number of killed or disappeared, see CNV, *Relatório Vol. III: Mortos e Desparecidos Políticos* (Brasília: CNV, 2014), 7–14. Both reports are digitized and available at http://cnv.memoriasreveladas.gov.br/ (last accessed April 2023).

14 Joffily, "O aparato repressivo"; Marcelo Ridenti, "Censura e ditadura no Brasil, do golpe à transição democrática, 1964–1988," *Revista Concinnitas* 2, no. 33 (2018): 86–100.

15 Aarão Reis Filho, *Ditadura e democracia no Brasil*, 73–94; Skidmore, *The Politics of Military Rule in Brazil*, 160–210.

No. 11, which abolished AI-5, thereby eliminating the president's ability to shut down Congress and dismiss political opponents. Terminating AI-5 also restored both habeas corpus and freedom of the press to newspapers, radio, and television. In December, he granted political exiles the right to return without prosecution.

Abertura had its limits. For example, Constitutional Amendment No. 11 carried a clause that permitted the president to declare a state of emergency that granted them powers resembling those in AI-5. Moreover, Geisel's reforms had done nothing to dismantle the government's surveillance programs and security forces, and the law still permitted officers to hold prisoners without granting them access to communication for at least eight days. Lastly, in 1979, João Figueiredo – Geisel's successor and the last of the military presidents – passed an amnesty law that shielded officers from being prosecuted for human rights abuses.

Despite these limits, abertura was a giant step toward restoring democracy, and Figueiredo continued the program. In 1982, the military government permitted direct elections at the state level for the first time. By 1984, there were widespread demonstrations for direct elections at the federal level. Though Figueiredo did not concede to the protestors' demands, his administration did endorse, for the first time, a civilian politician as president. In 1985, Congress indirectly elected popular Brazilian Democratic Movement Party [Partido do Movimento Democrático Brasileiro, PMDB] moderate Tancredo Neves, marking the formal end to military rule.

He never got the chance to serve as president. He became ill on the eve of his inauguration and died shortly thereafter. The presidency fell to his unpopular vice president, José Sarney, a corrupt congressman who had aligned himself with the pro-dictatorship party during military rule. Sarney nevertheless pressed forward with democratic reforms, and in 1988, his administration oversaw the writing of a new constitution. In 1989, direct elections returned, shedding the last vestiges of military dictatorship.[16]

Abertura enabled activism that changed the political landscape. Beginning in 1978, repressed social movements regained momentum, and protests erupted in support of women, Indigenous people, Black people, LGBT people, labor unions, and landless farmers, among other groups. The Catholic Church was an umbrella organization for many of

[16] Aarão Reis Filho, *Ditadura e democracia no Brasil*, 114–127.

these movements (see Chapter 4), although it did not support LGBT and reproductive rights movements.

Social movements found common cause in protesting the dictatorship. Restoring democratic freedoms was a shared goal, and many movements set aside more specific campaigns to fight broader injustices, such as authoritarianism. Indeed, social movements were considered relatively safe spaces to launch criticism against the dictatorship because they had broad enough support that it was harder for the military government to label activists socialist or communist subversives. Protests culminated in 1984 with a massive campaign calling for direct elections that brought millions of people out onto the streets. The return of civilian rule in 1985 and direct elections in 1989 further emboldened these movements, which became potential political forces thereafter. It was in this context that the modern environmental movement was born.

Environmentalism has deep roots in Brazil. Beginning in the 1820s, prominent intellectuals, many of whom were abolitionists, debated the ills of slash-and-burn agriculture and deforestation, though their comments did not inspire remedial efforts.[17] Starting in the 1930s, some scientists and politicians began lamenting deforestation and implemented limited regulations to curb it.[18] In 1934, the National Museum hosted the First Brazilian Conference on the Protection of Nature, and that same year, the government implemented modest environmental safeguards through Forest, Mining, and Water Codes, as well as a Hunting and Fishing Code.[19] During this period, the government also began creating national parks. However, meager enforcement did little to mitigate the damage

[17] José Augusto Pádua, *Um sopro de destruição: Pensamento político e crítica ambiental no Brasil escravista (1786–1888)* (Rio de Janeiro: Zahar, 2002).

[18] José Augusto Pádua, "Environmentalism in Brazil: An Historical Perspective," in J. R. McNeill and Erin Mauldin, eds., *A Companion to Global Environmental History* (Oxford: Wiley-Blackwell, 2012), 455–473.

[19] For more on environmental protection during this period, see José Luiz Franco and José Augusto Drummond, *Proteção à natureza e identidade nacional no Brasil, anos 1920–1940* (Rio de Janeiro: Editora Fiocruz, 2009); Regina Horta Duarte, *A biologia militante: O museu nacional, especialização científica, divulgação do conhecimento e práticas políticas no Brasil, 1926–1945* (Belo Horizonte: Universidade Federal de Minas Gerais, 2010), translated into English by Diane Grosklaus Whitty as *Activist Biology: The National Museum, Politics, and Nation Building in Brazil* (Tucson: University of Arizona Press, 2016).

from deforestation or other economic activities, and, after 1945, industrialization accelerated, unleashing a new set of environmental problems.[20]

The modern environmental movement began as a series of disconnected local campaigns during decompression and abertura. Its earliest rumblings began in 1971, when retired chemical engineer-turned-activist José Lutzenberger created the Gaúcha Association for the Protection of the Natural Environment [Associação Gaúcha de Proteção ao Ambiente Natural, AGAPAN] in Rio Grande do Sul.[21] But the organization's campaigns were limited. Popular environmentalism did not gain momentum until the end of the decade, when people began mobilizing against urban pollution and big infrastructure projects in the Southeast, such as airports and nuclear plants.[22]

In the late 1980s and 1990s, issues outside the industrial Southeast became more prominent, and disparate campaigns coalesced into national, and sometimes international, networks. For example, in the Amazon Rainforest, rubber tappers began promoting extractive reserves that would preserve both the forest and the livelihood of those living in it. Starting in the 1990s, national and international networks of environmentalists strengthened and helped local activists pressure the government to curb deforestation rates, which it did during the 2000s and 2010s.[23]

[20] For more on nature protection in the 1940s–1960s, see Marluza Marques Harres and Fabiano Quadros Rückert, *A Natureza, o tempo, e as marcas da ação humana: Políticas públicas e ambiente em perspectiva histórica, São Leopoldo, RS* (São Leopoldo: Editora Oikos, 2011); Daniel Porciuncula Prado, *A figueira e o machado: Uma história das raízes do ambientalismo no Sul do Brasil e a crítica ambiental de Henrique Roessler* (Carreiros: FURG, 2011); Elenita Malta Pereira, *Roessler: O homem que amava a natureza* (São Leopoldo: Oikos, 2013); Filipe Oliveira da Silva, "O Conselho Florestal Federal: Um parecer de sua configuração institucional (1934–1967)," *Historia Ambiental Latinoamericana y Caribeña (HALAC)* 7, no. 2 (2017): 101–129.

[21] For more on Lutzenberger, see Elenita Malta Pereira et al., *José Lutzenberger: Um ambientalista global* (São José: Editora Sobre o Tempo, 2022).

[22] See Kathryn Hochstetler and Margaret Keck, *Greening Brazil: Environmental Activism in State and Society* (Durham: Duke University Press, 2007), 75–83, 85, 189–202.

[23] For more on environmentalism in the Amazon Rainforest, see Chico Mendes and Tony Gross, *Fight for the Forest: Chico Mendes in His Own Words* (London: Latin American Bureau, 1989); Andrew Revkin, *The Burning Season: The Murder of Chico Mendes and the Fight for the Amazon Rainforest* (Boston: Houghton Mifflin, 1990); Susanna Hecht and Alexander Cockburn, *The Fate of the Forest: Developers, Destroyers and Defenders of the Amazon* (New York: HarperCollins, 1990); and Lise Sedrez, "Rubber, Trees and Communities: Rubber Tappers in the Brazilian Amazon in the Twentieth Century," in Marco Armiero and Lise Sedrez, eds., *A History of Environmentalism: Local Struggles, Global Histories* (London: Bloomsbury Academic, 2014), 147–166.

Environmental activism during the dictatorship generated a political response. The military government had already created some infrastructure to deal with environmental problems before the environmental movement gained momentum – it had rewritten the Forest, Hunting, and Mining Codes and created a National Sanitation Policy – but went much further thereafter.[24] In response to the 1972 Stockholm Conference, discussed in Chapter 3, the military government created a Special Secretariat of the Environment [Secretaria Especial do Meio Ambiente, SEMA].[25] It was housed within the Ministry of the Interior, then under the leadership of dam builder Costa Cavalcanti, who was uninterested in environmental protection. The government appointed environmentalist Paulo Nogueira Neto as SEMA's leader, who remained in charge until 1986. Nogueira Neto had been a well-known environmentalist since the 1950s, when he cofounded the Association in Defense of the Environment [Associação em Defesa do Meio Ambiente], one of Brazil's first conservation associations. Despite SEMA's high-profile leadership, it was understaffed and did little to protect the environment.[26]

During the 1980s, environmentalists pressured the government to implement more effective legislation and enforce it. Environmental licensing was particularly important. In 1981, the government promulgated legislation that required companies to obtain an environmental operating license – which was contingent upon completing an environmental impact study – before building big infrastructure projects.[27] Yet, specific resolutions from the newly created National Council for the Environment [Conselho Nacional do Meio Ambiente, CONAMA] were needed to enable these laws, and the council dragged its feet during the final years of the dictatorship.[28] The return to civilian rule emboldened the council, and in January 1986, it passed the enabling resolutions that required environmental licensing and impact studies for projects with potential to

[24] Hochstetler and Keck, *Greening Brazil*, 26.

[25] Médici, *Decreto No. 73.030, de 30 de outubro de 1973, Cria no âmbito do Ministério do Interior, a Secretaria Especial do Meio Ambiente – SEMA, e da outras providências*, October 30, 1973, ACD, laws and decrees accessible online. For more on SEMA's policies, see Roberto Guimarães, *The Ecopolitics of Development in the Third World: Politics and Environment in Brazil* (Boulder: Lynne Rienner, 1991).

[26] See Hochstetler and Keck, *Greening Brazil*, 27.

[27] João Figueiredo, *Lei No. 6.938, de 31 de agosto de 1981, Dispõe sobre a Política Nacional do Meio Ambiente, seus fins e mecanismos de formulação e aplicação, e dá outras providências*, August 31, 1981, ACD.

[28] Hochstetler and Keck, *Greening Brazil*, 35–40.

damage the environment.[29] Two years later, the federal government wrote these and other protections into the Constitution.[30] This legislation reflected the political impact of the environmental movement and its growth from a movement based on direct-action campaigns to one that counted on regulatory policies to forestall damaging projects and curb pollution.[31]

Dam critics were among those emboldened by abertura, and their efforts to mitigate the deleterious impacts of big dams were an integral part of the environmental movement. The immediate results of such activism were mixed, as the following chapters illustrate. In some cases, displaced populations won some victories, but many did not, and no group succeeded in modifying or halting a project altogether. Environmentalism was not strong enough to interfere with the military regime's plans for unrestrained industrial growth.

INDUSTRIAL GROWTH

Industrialization in Brazil dates to the 1930s, when then president Getúlio Vargas promoted domestic industrialization to reduce dependence on imported industrial goods, which became unaffordable after the Great Depression slashed demand for coffee exports.[32] Vargas believed that a strong federal government would be the engine for this growth, and he created a series of agencies and companies charged with orchestrating this industrial program. His plan was popular, and it continued to be the organizing principle of subsequent administrations.

Industrialization and urbanization began in earnest after the Second World War. Both trends created a positive feedback loop: Manufacturing jobs encouraged more people to migrate to cities, and urban living in turn demanded more industrial goods. In the 1940s, just 15 percent of

29 *Resolução CONAMA No. 001, de 23 de janeiro de 1986*, published in the Diário Oficial do União on February 17, 1986, ASF.

30 Hochstetler and Keck, *Greening Brazil*, 46–51.

31 In 1989, the government created the Brazilian Institute of the Environment and Natural Resources [Instituto Brasileiro do Meio Ambiente e dos Recursos Naturais, IBAMA] to replace SEMA. IBAMA continues to oversee the country's environmental protection today. Nelson Carneiro, *Lei No. 7.735, de 22 de fevereiro de 1989, Dispõe sobre a extinção de órgão e de entidade autárquica, cria o Instituto Brasileiro do Meio Ambiente e dos Recursos Naturais Renováveis e das outras providências*, February 22, 1989, ACD.

32 Stanley E. Hilton, "Vargas and Brazilian Economic Development, 1930–1945: A Reappraisal of His Attitude toward Industrialization and Planning," *The Journal of Economic History* 35, no. 4 (1975): 754–778.

Brazilians lived in cities, but thereafter urban growth exploded. In the 1990s, 75 percent of Brazil's population was urban (a trend underway throughout Latin America).[33] Industrial growth boomed alongside urbanization, and manufacturers built giant industrial centers.

The country's commitment to industrial and urban growth was neither partisan nor tied to a singular form of governance. However, the character of these trends was markedly shaped by the fact that one of the greatest periods of growth was orchestrated by a military dictatorship.

The military saw its intervention as necessary to forestall a communist threat and it believed that sustained growth could raise the standard of living and boost the country's international standing, which would help justify its illegal intervention. Its efforts were successful in the short term. Between 1967 and 1973, the hardline generals and their administrators orchestrated a big economic boom, later dubbed the Brazilian economic miracle, during which economic growth averaged about 11 percent annually. The industrial sector led the way, and the most prosperous industries, such as automobile manufacturing, electronics, and construction, saw annual growth rates upwards of 20 percent during this period.[34] The military regime looked toward big, government-built infrastructure projects such as dams as a means of stimulating such growth and providing much-needed energy.

The Brazilian military regime was part of a global cohort of military dictatorships that sought validity through their dedication to industrial and economic growth, a phenomenon scholars have called "military modernization."[35] Brazil's South American neighbors were some of the most eager of these military modernizers: There, the right-wing generals seized power and installed repressive dictatorships that shared a

[33] US-LOC, Federal Research Division, *Brazil: A Country Study* (Washington, DC: LOC, 1998), General Collection, US-LOC, accessible online.

[34] Aarão Reis Filho, *Ditadura e democracia no Brasil*, 58–60; Francisco Vidal Luna and Herbert S. Klein, "Transformações econômicas no período militar (1964–1985)," in Aarão Reis Filho et al. eds., *A ditadura que mudou o Brasil*, 92–111; Skidmore, *The Politics of Military Rule in Brazil*, 138–144.

[35] See Bradley R. Simpson, *Economists with Guns: Authoritarian Development and U.S.–Indonesian Relations, 1960–1968* (Stanford: Stanford University Press, 2008) and Thomas C. Field Jr., "Ideology as Strategy: Military-Led Modernization and the Origins of the Alliance for Progress in Bolivia," *Diplomatic History* 36, no. 1 (2012): 147–183.

commitment to industrial and economic growth through bureaucratic organization and technocratic decision-making.[36]

Military modernization was also popular in newly independent countries rebuilding their economies and political institutions. Some of these regimes embraced socialism. In 1952, Egyptian military officer Gamal Nasser staged a coup that brought the military to power. Thereafter, he carried out a modernization program based on socialism and Arab nationalism, which included building the gigantic Aswan High Dam on the Nile River.

Other military regimes were more ideologically aligned with Brazil. In 1965, right-wing Indonesian General Suharto led a coup that brought the military to power under his leadership, which ruled the country for more than thirty years thereafter. Suharto looked to industrial and economic growth as a means of legitimizing his rule and built many big dams.[37]

Energy was central to the Brazilian military regime's goal of accelerating economic and industrial growth. Energy is the most basic ingredient of all growth, whether the source is organic energy, such as human or animal labor, or inorganic energy, such as fossil fuels. Food production and mining, the two industries upon which all others depend, require inputs of energy and water, and big waterworks often require massive inputs of cheap energy to construct. Thus, the military regime's economic goals had a basic material requirement: The generals needed to expand the country's energy infrastructure if its cities and industries were to grow.

Energy projects were also central to the military's related plans for ensuring national security. Brazil had become independent with a gigantic territory, most of which was inhabited only by Indigenous communities.

[36] The most infamous were the Argentine, Chilean, and Uruguayan dictatorships. See Guillermo O'Donnell, *Modernization and Bureaucratic-Authoritarianism: Studies in South American Politics* (Berkeley: University of California Institute of International Studies, 1973); James Malloy, ed., *Authoritarianism and Corporatism in Latin America* (Pittsburgh: University of Pittsburgh Press, 1977); David Collier, ed., *The New Authoritarianism in Latin America* (Princeton: Princeton University Press, 1979); Guillermo O'Donnell, *El Estado burocrático: Triunfos, derrotas y crisis* (Buenos Aires: Editorial Belgrano, 1982), translated in English by James McGuire in collaboration with Rae Flory as *Bureaucratic Authoritarianism: Argentina 1966–1973 in Comparative Perspective* (Berkeley: University of California Press, 1988).

[37] George Aditjondro and David Kowalewski, "Damning the Dams in Indonesia: A Test of Competing Perspectives," *Asian Survey* 34, no. 4 (April 1994): 381–395; Sanjeev Khagram, *Dams and Development: Transnational Struggles for Water and Power* (Ithaca: Cornell University Press, 2004), 159–165.

Politicians and military officers worried that distant and underpopulated areas could be lost to border disputes or foreigners' imperial greed.[38] This fear was most pronounced in the Amazon Rainforest, which was rich in resources and sparsely populated compared to the industrialized Southeast.

The Cold War raised the specter of a second source of potential political disintegration and territorial loss: the appeal of socialism. Brazil has long-standing social and racial inequalities that the dictatorship put off addressing in favor of increasing the country's overall wealth.[39] Thus, during the dictatorship, large segments of the country lived in poverty, and political ideologies that promoted a more just distribution of resources were likely to appeal to much of the country's population. For example, in the late 1960s, communists gained the sympathy of locals in the Araguaia region of the Amazon and staged a rebellion. Between 1972 and 1974, the Brazilian military stamped out the uprising, but the incident reaffirmed the generals' fears that socialism and communism could gain traction among groups in the interior and cause political instability that could lead to territorial loss.

To forestall these threats, previous administrations had orchestrated campaigns to fill these underpopulated regions and stimulate economic and industrial growth, which they believed would discourage foreign influence. The dictatorship accelerated this program. In 1970, President Médici promulgated legislation that launched the National Integration Program, which built upon more targeted legislation from the previous administrations.[40] The military regime embellished its campaign with the

[38] This idea is presented in José Augusto Pádua, "The Dilemma of the 'Splendid Cradle': Nature and Territory in the Construction of Brazil," in John Soluri et al., eds., *A Living Past: Environmental Histories of Modern Latin America* (New York: Berghahn, 2018), 91–114.

[39] Regina Horta Duarte, "'Turn to Pollute': Poluição atmosférica e modelo de desenvolvimento no 'milagre' brasileiro (1967–1973)" *Revista Tempo* 21, no. 37 (2015): 64–87, 65.

[40] Médici, *Decreto-Lei No. 1.106, Cria o Programa de Integração Nacional, altera a legislação do imposto de renda das pessoas jurídicas na parte referente a incentivos fiscais e dá outras providências*, June 16, 1970; Médici, *Decreto-Lei No. 1.164, Declara indispensáveis à segurança e ao desenvolvimento nacionais terras devolutas situadas na faixa de cem quilômetros de largura em cada lado do eixo do rodovias na Amazônia Legal, e dá outras providências*, April 1, 1971; Médici, *Decreto-Lei No. 1.243, Eleva a dotação do Programa de Integração Nacional (PIN) criado pelo Decreto-Lei No. 1.106, de 16 de junho de 1970, altera o Decreto-Lei No. 1.164, de 1 de abril de 1971, e dá outras providências*, October 30, 1972, ACD.

slogan "integrate to not hand over," an alliteration in the original Portuguese, "integrar para não entregar."[41]

The program designated funds for projects in the North and Northeast to raise the standard of living there. It emphasized road building in the Amazon Rainforest and settling farmers from the Northeast and Southeast along these highways. It also encouraged industrial development in designated regional hubs in order to stimulate demographic and economic growth in underpopulated regions. But the program's importance transcended specific projects: It was the dictatorship's guiding manifesto for all its infrastructure projects and a core component of its two broad National Development Plans [Planos Nacionais de Desenvolvimento, PNDs], which were the design plans for the economic boom that the military hoped would legitimize its coup.[42]

As with the goal of accelerating industrial growth generally, expanding energy production was central to the National Integration Program. Highway construction and agricultural colonization overshadowed energy in the program's earliest legislation, but electricity was always a principal underpinning of its industrial component.[43] Indeed, the dictatorship slated some of its biggest power plants for the Northeast and the Amazon precisely because it thought these areas were most vulnerable to foreign influence or socialist subversion. For the military regime, both highways and transmission lines connecting power plants to consumption centers were threads that wove these strategic hinterlands into the national fabric, simultaneously making them more accessible to both military officers and landless farmers, and showcasing the benefits of government-led capitalism.

Geography limited the energy resources that the dictatorship could choose from. Brazil had modest coal seams in its southernmost states,

41 Castelo Branco is credited with popularizing the phrase in 1966 in reference to *Lei No. 5.173, de 27 de outubro de 1966, Dispõe sobre o Plano de Valorização Econômica da Amazônia; extingue a Superintendência do Plano de Valorização Econômica da Amazônia (SPVEA), cria a Superintendência do Desenvolvimento da Amazônia (SUDAM), e dá outras providências*, October 27, 1966, ACD.

42 On the PNDs, see Médici, *Lei No. 5.727, Dispõe sobre o primeiro Plano Nacional de Desenvolvimento (PND) para o período de 1972 a 1974*, November 4, 1971; Médici, *I Plano Nacional de Desenvolvimento*, 13–31; Geisel, *Lei No. 6.151, Dispõe sobre o Segundo Plano Nacional de Desenvolvimento (PND), para o período de 1975 a 1979*, December 4, 1974. ACD; and Ernesto Geisel, *II Plano Nacional de Desenvolvimento (1975–1979)* (Brasília: Congresso Nacional, 1974), 62–65, General Collection, BCN.

43 Médici, *I Plano Nacional de Desenvolvimento*, 13–31; Geisel, *II Plano Nacional de Desenvolvimento*, 62–65.

but these reserves have never been big enough to sustain the country's industrial ambitions.[44] Domestic petroleum was also scarce. Though Brazil is now among the global petroleum superpowers, it was not during the dictatorship (geologists discovered big offshore deposits in the late 2000s). Throughout the twentieth century, domestic oil reserves were too meager to fuel industrial growth.[45]

Foreign oil, however, did have a major role in Brazil's industrialization. Petroleum was the fuel that undergirded industrial development and economic growth in most countries during the postwar period, and Brazil was no exception. The government used its foreign exchange to buy petroleum through its state-owned company, Petrobras, and foreign oil companies set up downstream operations in the country. However, dependence on foreign oil inspired a widespread backlash from nationalists, which put pressure on the military government to substitute imported oil with domestic alternatives.[46]

Two oil crises during the 1970s exacerbated the problem. During the preceding decades, power balances among oil producers shifted as the United States depleted much of its domestic reserves fueling unprecedented demographic and industrial growth, while, at the same time, geologists discovered massive oil deposits in the Middle East. As the industry's epicenter moved to the Middle East, oil-producing countries there gained the power to influence the price of petroleum. In October 1973, Israel and Palestine were at war, and the Organization of Arab Petroleum Exporting Countries (OAPEC) – a Palestinian ally – decided to use its newfound economic power to protest the United States and other countries that were supporting Israel. The organization embargoed petroleum sales to the United States and a handful of other countries and cut production across the board, which caused a shortage that more than tripled the price of oil.

44 Dias Leite, *Energia do Brasil*, 51, 55–56, 67–68. See also Christian Brannstrom, "Was Brazilian Industrialisation Fuelled by Wood? Evaluating the Wood Hypothesis, 1900–1960," *Environment and History* 11 (2005): 395–430.

45 Natascha Otoya, "Petroleum and Science: The National Petroleum Council and the Development of Oil and Geology in Brazil," in Silvia Fernanda Figueirôa et al., eds., *History, Exploration & Exploitation of Oil and Gas* (New York: Springer, 2019), 37–49. Geologists discovered the first offshore field in 1968, and the first giant one, Tupi, in the Santos Basin, in 2007. See Tyler Priest, "Petrobras in the History of Offshore Oil," in Ben Ross Schneider, ed., *New Order and Progress: Development and Democracy in Brazil* (New York: Oxford University Press, 2016), 53–77.

46 Dias Leite, "Perspectivas da energia no Brasil," 1974, 8–9, File ADJ.0.0.38, FADL, AN-RDJ.

Brazil and other countries that were dependent on foreign oil were hit hard.

Just as the global economy began to recover, a second oil crisis erupted. In 1979, revolutionaries in Iran toppled the government in a violent confrontation that halted oil production. The revolution also transferred power to a government hostile to the United States and its allies, and businessmen worried that the government would withhold oil sales and cause another shortage. Once again, the price of petroleum shot up and caused a financial crisis in countries dependent on importing petroleum. Both oil crises put more pressure on the military regime to divest from foreign oil.

The military government implemented a multifaceted program to slash foreign oil consumption. The top priority was finding domestic oil reserves to substitute for imports.[47] Second, the government invested in developing a fleet of cars that would run on ethanol produced from sugarcane, already widely planted.[48] Third, it set out to replace petroleum-fired power plants with both nuclear and hydropower, and to substitute petroleum-powered engines with electricity wherever possible. Thermoelectric power plants that burned petroleum were not the norm in Southeast Brazil, but some remote cities, such as Manaus, generated electricity using oil (see Chapter 6). Of course, the military regime valued nuclear and hydropower for more than just mitigating the oil crises; they were important sources of electricity in their own right, which was essential to the dictatorship's plans for industrial development and national integration.

Nuclear power was popular within the military government, but it never approached the scale of hydropower, which was the cornerstone of the regime's electrification plans. During the 1950s and 1960s, optimism abounded around the world that nuclear fission, in addition to fueling bombs, could be used to generate cheap electricity. Brazil shared this enthusiasm for both atomic bombs and nuclear power, and in the 1970s, the dictatorship began planning a series of ambitious nuclear plants.[49] Both financial difficulties and opposition from environmentalists

[47] Ministério de Minas e Energia, *Balanços Energéticos Nacionais*, 1977–1980, Biblioteca Roberto Simonsen, São Paulo. I am grateful to my colleague Bruno Biasetto for generously sharing his copies of these documents with me.

[48] On Brazilian ethanol, see Eaglin, *Sweet Fuel*, and Rogers, *Agriculture's Energy*.

[49] Dias Leite, "Opções da energia nuclear," 1974, File ADJ.0.0.38, FADL, AN-RDJ.

derailed most of the projects, some of which never even got off the ground. Even before these setbacks, nuclear power investment was modest, as were the results of the nuclear program: In 1990, nuclear power accounted for just 1 percent of Brazil's electricity production.[50] For the Brazilian dictatorship, nuclear reactors were novel and unreliable, and building them would take time and money. Building hydropower plants also took time, but dams were familiar ground for the country's engineers and, according to the military government's economic advisors, comparatively cheap.

Hydropower dams have a long history in Brazil.[51] What the country lacked in accessible fossil fuels, it made up for in good dam sites. Brazil has many giant rivers with reliable flow, many of which run near population centers. Notably, the headwaters of the eastern tributaries of the mighty Paraná River begin in the coastal mountain range near São Paulo and Rio de Janeiro, flowing inland toward the main stem. These rivers all had ample dam sites. In the Northeast, the powerful São Francisco River plummets through a series of cascades in its lowest stretches that were ideal sites for power plants. The Amazon River had even more potential. It is the world's most powerful river, and its biggest tributaries all have propitious dam sites. These prodigious waterfalls, or "white coal," as boosters called them, made dams an appealing source of power.

Industrialists had long recognized the country's hydropower potential. Manufacturers built Brazil's first hydropower plant in 1883, and during the first three decades of the twentieth century, multinational companies built a handful of dams on the rivers near São Paulo and Rio de Janeiro. During this first dam-building campaign, power plants and reservoirs were small and disconnected from one another. Much has changed since these initial dams, but the position of hydropower in the country's electricity profile has remained remarkably consistent. Hydropower has been Brazil's principal source of electricity since 1900, always accounting for at least 60 percent of installed capacity and often much more, upwards of 80 percent (Figure 1.1).[52]

50 International Energy Agency (IEA)'s webpage on Brazil's energy profile, www.iea.org/countries/brazil (last accessed April 2023).

51 See Nathalia Capellini, "A grande aceleração e a construção de barragens hidrelétricas no Brasil," *Vária História* 34, no. 65 (2018): 315–346.

52 For installed capacity from 1883 to 1920, see Dias Leite, *Energia do Brasil*, Appendix 2-E.

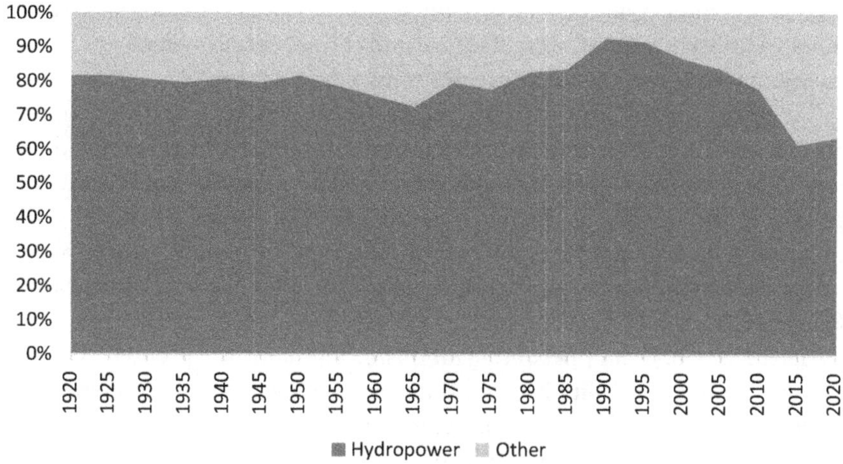

FIGURE 1.1. Installed electricity capacity by source, Brazil, 1920–2020. Area chart created by author using data from Antônio Dias Leite, *Energia do Brasil, 3ª edição* (Rio de Janeiro: Editora Nova Fronteira, 2014), Appendix 2-E for the 1920–1930 period; Memória da Eletricidade, *Panorama do setor de energia elétrica no Brasil* (Rio de Janeiro: Memória da Eletricidade, 1988), 99, 149, and 232, for the 1930–1970 period; the Ministério de Minas e Energia's online *Sistema de Informações Energéticos* for the 1970–1990 period; and the International Energy Agency's online database for the 1990–2020 period.

Between the 1940s and the 1960s, the basic characteristics and scale of dam building changed immensely. In the 1930s, President Vargas strengthened the federal government, and he and his successors created state-owned companies at both the state and federal levels to oversee dam construction. The Great Depression and World War II stalled the government's hydropower ambitions, but thereafter the government supplanted private companies as the country's principal dam builders.

Four additional features distinguished these dams from their predecessors. First, they were not confined to the coastal Southeast. During the late 1940s and 1950s, the government began building dams further inland from Southeastern population centers and in the Northeast. For example, the federal government built its first big dam, Paulo Afonso (1948–1955), on the São Francisco in the Northeast to electrify the region's coastal cities and industrial centers. The government also began using transmission lines to connect power plants with one another and create fledgling national grids. Third, these hydropower dams were much more powerful than their predecessors. For example, Furnas, the most powerful dam built during this period, had an installed capacity of more than 1,000 megawatts after upgrades. Lastly, these dams also had much larger reservoirs. The two biggest were Furnas and Três Marias, which each had reservoirs with surface areas greater than 1,000 sq. km.[53]

In March 1964, the military regime seized control of the country, taking over this government-orchestrated effort to build powerful dams with big reservoirs in distant hinterlands. The generals floored the gas pedal on the campaign. Dams were an obvious means of meeting the material requirements for economic growth, and hinterland power plants and transmission lines were useful tools for the National Integration Program.

There was one final benefit: symbolism. Big dams are a potent symbol of government capability because they are a striking visual representation of engineering on a herculean scale, which requires tremendous expertise and coordination. The symbolism surrounding the dictatorship's big dams was especially strong because the rivers it set out to tame were some of the world's most powerful.

For all these reasons, worldwide, building big dams was remarkably nonpartisan until the environmental movement gained momentum in the late twentieth century – capitalists, communists, democracies, dictatorships, and both imperial and revolutionary governments throughout the world all invested in big dams as a means of powering industrial growth and legitimizing their administrations.[54] Most did so with little to no attention to environmental safeguards until the rise of popular

[53] See Warren Dean, *With Broadax and Firebrand: The Destruction of the Brazilian Atlantic Forest* (Berkeley: University of California Press, 1995), 294–298.

[54] For capitalists, see Richard P. Tucker, "Containing Communism by Impounding Rivers: American Strategic Interests and the Global Spread of High Dams in the Early Cold War," in J.R. McNeill and Corinna R. Unger, eds., *Environmental Histories of the Cold War* (New York: Cambridge University Press, 2010), 139–164, and Christopher Sneddon, *Concrete Revolution: Large Dams, Cold War Geopolitics, and the US Bureau of Reclamation* (Chicago: University of Chicago Press, 2015). For communist and socialist regimes, see Paul Josephson et al., *An Environmental History of Russia* (New York: Cambridge University Press, 2013), 162–172, and Shapiro, *Mao's War against Nature*, 2001, 48–65. For imperial governments, see chapters in Deepak Kumar et al., *The British Empire and the Natural World: Environmental Encounters in South Asia* (New York: Oxford University Press, 2011), and Matthew P. Johnson, "'Thirsty Sugar Lands': Environmental Impacts of Dams and Empire in Puerto Rico since 1898," *Environment & History* 27, no. 3 (August 2021): 337–365. For revolutionary and newly independent governments, see Stephan Miescher, *A Dam for Africa: Akosombo Stories from Ghana* (Bloomington: Indiana University Press, 2022), and Nancy Reynolds, "Building the Past: Rockscapes and the Aswan High Dam in Egypt," in Alan Mikhail, ed., *Water on Sand: Environmental Histories of the Middle East and North Africa* (New York: Oxford University Press, 2012), 181–205.

environmentalism. There is no better example than the Brazilian military dictatorship, which built some of the world's biggest and most controversial dams.

MODERN PHARAOHS

During the dictatorship, many Brazilians referred to big dams as pharaonic projects, a term meant to highlight their grandiose nature by comparing them to the pyramids of ancient Egypt. Project proponents sometimes used the term to laud grand projects, but more often critics used the term to discredit projects they deemed economically senseless and thus having alternative motives, such as the symbolic benefits of carrying out an impressive engineering feat, or the private gains for construction firms and politicians.

The term could easily be applied to other dams throughout the world. Indeed, on at least one occasion an Egyptian politician made celebratory associations between a big dam and the monuments of Pharaonic Egypt.[55] On the critical side, one North American journalist used the term to describe the US congressmen funding economically senseless multipurpose dams in the American West.[56] But such references are rare. Although the Brazilian military regime was not alone in building dams that could be labeled pharaonic, its citizens appear to be unparalleled in using the term so widely and consistently.

The term's modern usage began in reference to Brasília. In 1956, the federal government began building a new capital in the middle of the country, about 900 kilometers northwest of Rio de Janeiro, the capital at that time. Politicians designed Brasília to represent their effort to better integrate the country's territory by filling its underpopulated areas, a program that the dictatorship later accelerated. During construction, senators critical of the ambitious scheme began protesting, labeling it a pharaonic project.[57]

The term spread widely during the dictatorship, and dams became one of the biggest targets. In the late 1980s, the popular magazine *Veja* ran an exposé on corruption and engineering projects during the dictatorship

55 Elizabeth Bishop, "Talking Shop: Egyptian Engineers and Soviet Specialists at the Aswan High Dam" (PhD Diss., University of Chicago, 1997), 125.

56 Reisner, *Cadillac Desert*, 511.

57 See Senator Mem de Sá, "Discurso no Senado Federal," February 5, 1960, and Senator Jefferson de Aguiar, "Discurso no Senado Federal," February 9, 1960, ASF.

titled "a game of pharaohs and contractors." Of the ten projects the article identified, seven were dams – four of which are featured in this book – two were nuclear plants, and one was a railroad.[58]

The pharaonic phrasing dovetails with another reference that historians have used to highlight the symbolism surrounding big dams. In the 1950s and 1960s, India's prime minister Jawaharlal Nehru gave two speeches at the Bhakra Dam, in North India: the first when construction began and the second when it finished. During the speeches, he called the dam a "temple," and in subsequent years, he continued to refer to dams and other big infrastructure projects as the "temples of modern India," a phrase that has become somewhat of a cliché among historians writing about dams.[59]

Nehru's phrase is powerful because it captures both an unwavering faith in science and the symbolism associated with big dams. Since the Enlightenment and the Industrial Revolution, governments have looked toward science and technology to overcome disease and hunger and to improve the standard of living of their constituents. Faith in science reached its apogee in the 1950s and 1960s, when governments around the world and across the ideological spectrum witnessed remarkable gains in health and prosperity through modern science and became convinced that research and big engineering projects – such as mega dams – could overcome most natural limitations. Nehru's comment captures this worldview, as it imbues secular science with religious reverence. His phrase also highlights the potent symbolism surrounding big dams, which showcase the capabilities of the countries and administrations that build them.

Such symbolic benefits were especially important during the era of decolonization in the second half of the twentieth century. During this period, newly independent countries such as India worked to overcome legacies of racism and imperialism. One of the reasons governments in

58 "Um jogo de faraós e empreiteiros," *Veja*, May 20, 1987, 30. The seven dams were Itaipu, Tucuruí, Itaparica, Balbina, Xingó, Pedra do Cavalo, and Salto Segredo. North American historian Lewis Mumford also compared big dams to Egyptian pyramids in "The Architecture of Power," *New Yorker* (June 1941), 58, though he did not use the phrase "pharaonic projects," and it is unclear if his writings influenced Brazilians who coined the term.

59 See Jawaharlal Nehru, "Temples of the New Age Speech, July 8, 1954," in *Jawaharlal Nehru's Speeches Vol. 3, March 1953–August 1957* (New Delhi: Ministry of Information and Broadcasting, 1958), 1–4, and Nehru, "Bhakra Project Inauguration Speech," October 22, 1963, available on the website of the Bhakra Beas Management Board, https://bbmb.gov.in/speech.htm (last accessed April 2023).

both India and Egypt pressed ahead with environmentally harmful dams was because building them would refute racist imperial stereotypes that both countries were inferior to Europeans in science and engineering.[60]

Brazilians shared this faith in science and also looked toward engineering feats to overcome such stereotypes. Though Brazil's independence had come much earlier than India's, it remained a predominantly agrarian country with little industrial growth through World War II. Thereafter, Brazil set out to industrialize and thereby overcome its peripheral position in the global economy and prove that its engineering prowess matched that of the industrial world.

Energy projects were essential to this ambition. The Brazilian government championed both domestic petroleum and ethanol as sources of economic independence and liberation, though both remained mostly latent until the 1980s.[61] Both industries were also venues for Brazil to showcase its technological genius. Starting in the 1970s, Brazilian scientists and engineers became world leaders in designing and building equipment for offshore oil production, and they developed an ethanol-fueled car that was successful and popular in the 1980s.[62]

Hydroelectricity was also an important arena for developing domestic energy and engineering expertise, and mega dams were an even more grandiose symbol of Brazil's industrial capabilities. Like India's, Brazil's giant hydroelectric dams were its modern temples. The biggest and most iconic were temples of the military dictatorship, a cadre of generals and technocrats – "modern pharaohs" – who saw dams as a means to gain political legitimacy.

The first group worthy of this designation might be the presidents. Three presidents in particular played an important role in the

[60] Christopher Hill, "Case Study C: Narmada Bachao Andolan (Save the Narmada Movement)," in Christopher Hill, *South Asia: An Environmental History* (Santa Barbara: ABC-CLIO, 2008), 234–248; J. Donald Hughes, "Case Study C: The Aswan Dams and Their Environmental Results," in J. Donald Hughes, *The Mediterranean: An Environmental History* (Santa Barbara: ABC-CLIO, 2005), 213–232.

[61] See Antoine Acker, "A Different Story in the Anthropocene: Brazil's Post-Colonial Quest for Oil (1930–1975)," *Past and Present* 249 (November 2020): 167–211; Eaglin, *Sweet Fuel*; and Rogers, *Agriculture's Energy*.

[62] See Edmilson Moutinho dos Santos and Peyerl, "The Incredible Transforming History of a Former Oil Refiner into a Major Deepwater Offshore Operator: Blending Audacity, Technology, Policy, and Luck from the 1970s Oil Crisis up to the 2000s Pre-salt Discoveries," in Silvia Fernanda Figueirôa et al., eds., *History, Exploration & Exploitation of Oil and Gas* (New York: Springer, 2019), 87–100; Priest, "Petrobras in the History of Offshore Oil"; Eaglin, *Sweet Fuel*; and Rogers, *Agriculture's Energy*.

dictatorship's hydropower program: generals Costa e Silva, Médici, and Geisel. But the generals in charge of the government were not trained as economists or engineers themselves and thus designed their plans based on the guidance of experts, such as economic ministers Antônio Delfim Netto and Roberto Campos. For the dictatorship's big dam-building boom, two energy ministers in particular stand out for having played an outsized role in orchestrating the campaign.

The first was José Costa Cavalcanti, a military officer and politician. Born in 1918 in Fortaleza, the capital of the northeastern state of Ceará, Costa Cavalcanti began his military career in 1935, when he enrolled in military school in Rio de Janeiro. In 1950 and 1951, he studied at the United States Army Infantry School, at Fort Benning, Georgia, now infamous for training many of the officers responsible for installing repressive dictatorships across Latin America during the Cold War. Indeed, Costa Cavalcanti supported the coup that toppled Goulart in March 1964, and once the military was in power, he sided with the hardline faction that advocated for strict censorship, torture, and other forms of violence to curb dissent. Costa Cavalcanti championed AI-5 and was close with Costa e Silva, the first hardline president who promulgated the act.[63]

Costa Cavalcanti played a big role in the energy sector during the dictatorship. In 1967, President Costa e Silva appointed him to the position of Minister of Mines and Energy, where he oversaw the passage of legislation designed to stimulate research for building dams in the Northeast and the Amazon (see Chapter 2). In 1969, the president transferred him to the Department of the Interior, but his hiatus from the energy sector was brief. In 1974, President Geisel appointed Costa Cavalcanti to head Itaipu Binacional, the binational state company in charge of building the colossal Itaipu Dam. In 1981, President Figueiredo made him president of Eletrobras, the state company whose regional subsidiaries were in charge of building dams across the country. Costa Cavalcanti remained in charge of both Itaipu and Eletrobras until 1985, when the military stepped down. He died in 1991.[64]

The second important energy minister was Antônio Dias Leite. Trained as an engineer and economist, he was one of Brazil's most respected energy experts, which set him apart from Costa Cavalcanti, who had no

63 Fátima Valença, "Biografia de José Costa Cavalcanti," General Collection, CPDOC-FGV.
64 Ibid.

formal training in engineering. Dias Leite was born in 1920 in Rio de Janeiro, and in the late 1930s, he enrolled in the National Engineering School at the Universidade do Brasil (now the Universidade Federal do Rio de Janeiro). During the 1940s, he was an economic consultant for the government, and in 1946, he became a professor of economics at the National Engineering School and, soon thereafter, at the Universidade do Brasil's Economic Sciences Department. He earned doctorates in both engineering and economics while teaching at the university. In 1963, Dias Leite began his career in government as an undersecretary of economic affairs in the Ministry of Finance.[65]

After the coup, the generals recognized his expertise and appointed him to key positions in the military government. In 1965, President Castelo Branco put Dias Leite in charge of an economic planning commission that both Delfim Netto and Campos participated in. In 1967, the president nominated Dias Leite to head the state mining company, Companhia Vale do Rio Doce, and, in 1969, transferred him to the position of Minister of Mines and Energy, where he replaced Costa Cavalcanti.

Dias Leite remained energy minister until 1974.[66] During his tenure, he oversaw the planning stages of the dictatorship's most iconic dams. For example, he was on the three-member commission that decided Itaipu was the best solution to a diplomatic rift with Paraguay, and he championed Tucuruí as a tool of the National Integration Program (see Chapter 2). After stepping down as energy minister, he remained a consultant on energy matters. Dias Leite's expertise as an engineer and economist, rather than as a military officer, made him exemplary of the technocratic elite that guided the generals who oversaw the big dam-building boom.[67]

Costa Cavalcanti and Dias Leite believed strongly in building big dams at all costs. Like their superiors, they maintained that hydropower had appealing economic and political benefits. They also shared their superiors' uncompromising attitude regarding environmental protection. Both contended that economic growth trumped environmental preservation and that there was no middle ground until the country had advanced further economically (see Chapter 3). These values, shared by others in

65 "Biografia de Antônio Dias Leite Júnior," General Collection, CPDOC-FGV. 66 Ibid.

67 After the dictatorship, Dias Leite returned to his position teaching at the Universidade Federal do Rio de Janeiro and later published the most comprehensive overview of energy in Brazil to date. See Dias Leite, *Energia do Brasil*.

the military regime, reflected the logic of the generals and energy ministers in charge of building the dictatorship's big dams.

These modern pharaohs had collaborators. Generals and ministers relied on a small cohort of domestic construction companies to build their dams.[68] International companies had been major protagonists in earlier construction booms, but the nationalistic dictatorship designed policies to strengthen domestic engineering and construction companies. President Costa e Silva's tenure was the starting point for this policy. In 1969, he promulgated legislation that required state companies to hire domestic firms to build public works projects. The law was regularly upheld, except on rare occasions, and came just in time to ensure that domestic construction companies got most of the contracts for the government's big dams.[69]

Dam construction became concentrated in the hands of six domestic companies, the biggest of which were Camargo Corrêa and Mendes Júnior.[70] These six companies were responsible for building the dams featured as case studies in this book, as well as the dictatorship's other big reservoirs.[71] Of the twenty-five big dams identified in the Introduction, 80 percent were built by these six companies.[72]

[68] The following paragraphs draw heavily on Pedro Henrique Pedreira Campos, *Estranhas Catedrais: As empreiteiras brasileiras e a ditadura civil-militar, 1964–1988* (Rio de Janeiro: Editora da Universidade Federal Fluminense, 2014).

[69] Costa e Silva, *Decreto No. 64.345, de 10 de abril de 1969, Institui normas para a contratação de serviços, objetivando o desenvolvimento da engenharia nacional*, ACD, and Campos, *Estranhas Catedrais*, 67.

[70] The other four companies are Andrade Gutierrez, Companhia Brasileira de Projetos e Obras (CBPO), Cetenco, and Servix. Camargo Corrêa was the most prolific dam-building company during the dictatorship in terms of both the number of dams and installed capacity. Mendes Júnior was the second, though it built more of the 25 powerful dams with big reservoirs identified in the Introduction than Camargo Corrêa. For more on these companies, see Campos, *Estranhas Catedrais*, 65–112, 320–324.

[71] Servix built Sobradinho, Camargo Corrêa built Tucuruí, Mendes Júnior built Itaparica, and Andrade Gutierrez built Balbina. Two consortiums, one Brazilian and one Paraguayan, built Itaipu. The Brazilian consortium was responsible for 90 percent of the construction, although the workforce was divided evenly between the two countries. The Brazilian consortium consisted of 5/6 of the companies listed in the preceding footnote: Camargo Corrêa, Mendes Júnior, Andrade Gutierrez, CBPO, and Cetenco. Ibid., 389.

[72] The only dams not built by these companies were São Simão, Samuel, Paulo Afonso IV, Salto Santiago, and Estreito. Camargo Corrêa built Itaipu, Tucuruí, Jupiá, Ilha Solteira, and Água Vermelha. Mendes Júnior built Itaipu, Itaparica, Itumbiara, Marimbondo, Volta Grande, and Boa Esperança. Andrade Gutierrez built Itaipu, Emborcação, Salto Osório, and Balbina. CBPO built Nova Avanhandava, Capivara, Chavantes, and Foz de Areia. Cetenco built Itaipu and Promissão, and Servix built Sobradinho and Rosana.

These firms were close with the generals and their civilian ministers. In some cases, ministers worked for construction companies before or after their tenure in government. Both Costa Cavalcanti and Dias Leite spent parts of their careers working for private engineering firms, though neither worked for the six big dam-building companies. In 1985, Shigeaki Ueki – who succeeded Dias Leite as energy minister from 1974 to 1979 – became the executive director of Camargo Corrêa.[73] In other cases, ministers and construction firms were close without formal ties. César Cals, energy minister from 1979 to 1985, was close friends with the presidents of big construction companies, such as Cetenco's president Eduardo Celestino Rodrigues.

During the dictatorship, such relationships were also common in other departments and ministries, and at different levels of government. For example, Mário Penna Bhering, who was president of Eletrobras from 1967 to 1975 and 1985 to 1990, had close ties with private construction firms, including Servix, which built Sobradinho. In São Paulo, two engineers with ties to Camargo and other São Paulo-based firms were in charge of the Companhia Energética de São Paulo (CESP), the government-owned company then overseeing dam building in the state.[74]

To be sure, friendships between government officials and construction firms were not an exclusive feature of the dictatorship. Similar relationships permeated the Kubitschek administration's construction projects and continued long after the military stepped down from power. For example, the president of Eletrobras during Luiz Inácio Lula da Silva's administration (2003–2010) was a former Camargo Corrêa employee.

Corruption was rampant during the dictatorship. Some of this corruption originated from within the companies, which often proposed cheap blueprints to secure the contract and then added additional features during construction to raise the cost and generate more revenue.[75]

Companies also collaborated with politicians to further increase their profit margins. The most common form of such collusion was rigged bids: Government dam-building companies ensured that most of the contracts went to a small cohort of big firms. The presidents of these firms acknowledged and defended the practice. In 1980, Andrade Gutierrez's president Sérgio Andrade argued that over the previous decade, the government had begun "building big projects that demanded big companies. To build such projects, a company needed equipment ..., specialized technical

[73] Campos, *Estranhas Catedrais*, 318. [74] Ibid., 316–325. [75] Ibid., 399–410.

teams, unique construction methods, lots of capital, versatility, and a series of other factors that only big companies can provide."[76] But even within this small stratum of elite businesses, state companies rigged bids to favor certain firms, often those from the same state as the politicians in charge. For example, Camargo Corrêa's close ties to São Paulo's political elites helped it win dam contracts there. Sometimes the practice caused tensions among these construction firms, as was the case at Itaparica, São Simão, and Nova Avanhandava. In the case of the latter, CESP cancelled the bidding auction outright and simply gave the contract to Camargo Corrêa.[77]

In return for securing contracts, politicians got kickbacks. In one high-profile case presented to Congress in 1978, a whistleblower exposed Delfim Netto's illicit charges to companies for helping secure contracts for equipment for Água Vermelha and Tucuruí.[78] Though such corruption was not always well documented, it was common practice during the dictatorship. Historian Warren Dean argues that during the 1970s–1980s, Brazilians were well aware that corruption accounted for much of Itaipu's inflated US$20 billion price tag, which was double its original cost estimate.[79]

Overpriced dams drained public coffers. Starting in the late 1970s, the US Federal Reserve Board raised interest rates on foreign loans to fight inflation, which, by the early 1980s, caused a recession in Brazil and other borrowing countries (see Chapter 3). Most of Brazil's loans had financed public works projects, which accounted for 70 percent of the country's debt during the 1980s.[80] The ensuing debt crisis crippled the economy and compelled the government to curtail public spending. Politicians and the businessmen in charge of big construction firms, on the other hand,

[76] Ibid., 126. [77] Ibid., 322. [78] Ibid., 403–404.

[79] Dean, *With Broadax and Firebrand*, 298. Itaipu Binacional's official figures for the dam's final cost are slightly lower, at US$ 17.6 billion. See www.itaipu.gov.br/en/press-office/faq (last accessed April 2023). The final cost of the dictatorship's other big dams was also often much greater than original estimates. Tucuruí's earliest studies estimated that the dam would cost US$4.2 billion, taking into account interest. The World Commission on Dams estimated the final cost to be US$7.5 billion, and prominent Brazilian journalist Lúcio Flávio Pinto argues the figure is closer to US$10 billion. See World Commission on Dams (WCD), *WCD Case Study Tucuruí Hydropower Complex Brazil, Final Report* (Cape Town: WCD, 2000), viii; Lúcio Flávio Pinto, "De Tucuruí a Belo Monte: A história avança mesmo?" *Boletim Museu Paraense Emílio Goeldi*, 7 no. 3 (2012): 777–782.

[80] "Um jogo de faraós e empreiteiros," 28.

made a fortune from these contracts and became some of the country's wealthiest people.

Consider the case of Camargo Corrêa, which won the contracts for some of the dictatorship's biggest and most prestigious dams.[81] São Paulo businessman Sebastião Ferraz Camargo founded the company in 1938 and shortly thereafter teamed up with Silvio Brand Corrêa – the brother-in-law of São Paulo's governor – who had capital and important political connections. Its earliest construction projects were railroads, cityscapes, and highways, mostly in the state of São Paulo. In the 1960s, the company began building dams, and it turned into such a prolific dam builder that hydropower plants became its trademark. Camargo Corrêa was building so many dams during the 1970s that it became the single biggest consumer of cement in Brazil. The profits from these dam contracts enabled the company to grow tremendously. In the 1980s, Camargo Corrêa was the company that owned the most Caterpillar construction equipment in the world, and the machinery retailer judged it to be the biggest construction company on the planet.

Camargo shared the military dictatorship's ideology and agenda. In 1967, the Escola Superior de Guerra awarded him the title of honorary diplomat, and he developed a close friendship with Paraguayan dictator Alfredo Stroessner (introduced in Chapter 2). Camargo also helped finance Operação Bandeirantes, an intelligence gathering center established in June 1969 – shortly after AI-5 was implemented – that sought to locate and capture dissidents, who were often tortured and disappeared. Such ideological ties further cemented bonds between the construction giants and the government agencies awarding contracts for dams.

These contracts made Camargo and his family rich. He accumulated a fortune of more than US$1.3 billion before he died in 1994. Upon his death, his wife, Dirce Navarro de Camargo, took over the company and increased its total revenue tenfold. By 2012, she had amassed a personal fortune of roughly US$13 billion, which made her the second-richest

[81] The following passages on Camargo Corrêa are drawn from Campos, *Estranhas Catedrais*, 85–89, 118–119, and 128–129. For more on the company and its dams, see Wilson Quintella, *Memórias do Brasil Grande: A história das maiores obras do país e dos homens que as fizeram* (São Paulo: Editora Saraiva, 2008). Quintella worked for Camargo Corrêa between 1949 and 1984. He was a close associate of Sebastião Camargo and became a manager and then president of the company during the dictatorship. Former finance minister Delfim Netto wrote the preface to his memoirs, further illustrating the close ties between construction companies and government elites.

person in Brazil. When she died in 2013, her three daughters inherited the estate.

The profits that flowed to both construction firms and politicians were a powerful supplementary motivation for the government's commitment to big dams. The degree to which such personal gains influenced decision-making varied on a case-by-case basis. Some authors have suggested that pressure from construction firms and the lure of profit were the main drivers behind the dictatorship's construction boom,[82] but the military government had sufficient motivation to build big dams without such enticements. The need for cheap electricity and the political appeal of dams were reasons enough for the generals to invest in building dams. But creating jobs for domestic engineering firms and profiting from kickbacks sweetened the deal and was a secondary motivation that shaped the character of the dam-building campaign. The generals and their ministers all agreed on the economic and political reasons for building dams, and for some – it is impossible to know how many – profits for their friends and kickbacks for themselves provided additional incentive to prioritize construction at all costs.

CONCLUSION

This chapter provides essential context on repression and the gradual restoration of democratic freedoms, environmentalism, the dictatorship's plans for industrial growth and energy production, the symbolism that surrounds big dams, and the influential generals and engineers responsible for orchestrating the military regime's dam-building campaign. The dictatorship of 1964–1985 was a period of severe repression during which military officers tortured and disappeared critics they deemed subversive. Censorship was also rife, and such repression enabled the military regime to begin building its dams without public debate. The gradual redemocratization process (1978–1989) emboldened the regime's critics and enabled a diverse set of social movements to flourish. Among them was the environmental movement, which was part of the global reaction to the deleterious ecological impacts that unrestrained industrial growth had unleashed. It was in this context in the late 1970s and 1980s that Indigenous communities, their allies, and environmentalists

[82] "Um jogo de faraós e empreiteiros," 28; Philip M. Fearnside, "Brazil's Balbina Dam: Environment and the Legacy of the Pharaohs in Amazônia," *Environmental Management* 13, no. 4 (1989): 401–423; and Campos, *Estranhas Catedrais*.

organized to protest dams, although outpourings of opposition came much too late to modify or halt the projects.

This chapter also lays the foundation for this book's first argument, that politics influenced the decision-making processes undergirding the dictatorship's dam-building boom. The chapter argues that the military regime accelerated the government's dam-building program, expanded its geographic reach, and scaled up dams and reservoirs because the material benefits of hydropower dovetailed with the goals of its National Integration Program and offered potent symbolism of aptitude that would help legitimize the regime, which had come to power in a coup and installed an enduring and increasingly unpopular dictatorship. In short, dams had appealing political characteristics in addition to their basic technical merits as power plants.

Political pressure also influenced the military regime's dam-building campaign in a second and more profound manner: It often competed with technical and environmental considerations in debates about where to build the biggest and most controversial dams. It is to these stories about choosing dam sites that the next chapter turns.

2

Building "the Big Dam"

Economic and Political Considerations during Planning, 1960s–1970s

INTRODUCTION

On October 26, 1982, the town of Guaíra awoke to some very strange sights and sounds. The first was the bats – thousands upon thousands of bats. They began invading the town a few nights prior and made homes anywhere they could.[1] Stranger still was the silence. For as long as most residents could remember, they fell asleep and awoke to the sound of the neighboring Guaíra Falls thundering in the background.[2]

Guaíra Falls was a massive grouping of waterfalls on the stretch of the Paraná River that forms the Brazilian–Paraguayan border. The falls supported a burgeoning tourist economy, which expanded after 1961 when the Brazilian government made the site a national park (see Chapter 5). Locals and visitors alike frequented the park, both for the sights and to bathe in the pools that dotted the islands between the falls. The cluster of waterfalls also supported a rich ecological web of fish, birds, bats, and other flora and fauna.

The bats and the disturbing silence that shrouded Guaíra were signs that the residents' worst nightmare had come to pass: Guaíra Falls was gone, submerged under the reservoir of the giant Itaipu Dam. From the beginning, the project's blueprints had entailed flooding Guaíra Falls – as well as the land of Indigenous and non-Indigenous farmers – but the falls

[1] "Morcegos que fugiram de Sete Quedas atacam Guaíra," *O Globo*, November 27, 1982, 8.

[2] Interview with Guaíra writer Edson Galvão, Guaíra, September 12, 2018.

were far enough upstream that many held out hope that the reservoir would not reach them.[3]

During the two weeks it took for the reservoir to fill, some residents went to the river only to find an ecological disaster unfolding.[4] Confused fish, birds, bats, and other animals searched desperately for their homes. The cries of the slaty-breasted wood rails brought one group of onlookers to tears and compelled them to leave.[5] Local naturalist Ernst Mann remembered watching the reservoir cover the falls. Years later, he reflected in an interview, "we sat there crying ... watching millions and millions of fish swimming upstream ... it was horrible. I do not want to pass through that again. I spent five years in the war, saw so much, and what shocked me the most, was this." The memory brought him to tears and prevented him from finishing the interview.[6] Guaíra Falls was a natural wonder, and those who knew it best watched in horror as Itaipu's reservoir flooded it.

This chapter is about the Brazilian military government's decision to commit to building big dams and reservoirs regardless of their environmental footprints. It covers the period between 1967 and 1975, when military generals and their energy ministers chose to build Sobradinho, Itaipu, Tucuruí, Itaparica, and a spate of other large reservoirs. These four dams alone flooded a combined area of more than 9,000 sq. km, a surface area greater than the island of Puerto Rico. The rising floodwaters covered Indigenous land, fertile farmland, and the homes of hundreds of

[3] Interview with Guaíra historian Ana Menél, Guaíra, September 13, 2018.

[4] Itaipu's reservoir filled between October 13 and October 26, 1982.

[5] Ernst Mann, "O inferno molhado para os pássaros das Sete Quedas," undated, Folder "Artigos originais escritos por Ernest Mann," FEM, CEPEDAL; "Lago de Itaipu avança sobre as Sete Quedas," *O Estado de Paraná*, October 19, 1982; "A los manotazos sacaban los peces," *ABC Color*, October 25, 1982.

[6] "Entrevista com Sr. Ernesto Mann, feita pelo repórter Lincoln Leduc, para o programa personalidade da semana da Rádio Difusora do Paraná, apresentado no radio no dia 08 de dezembro de 1990," 11, Folder "Files 001-035," FEM. Mann was a Central European-born naturalist and local history enthusiast, who moved with his wife Charlotte to Southern Brazil in 1954 and to Guaíra in 1963. He ran the town's only tourist agency and collected as much information as he could about the falls. His records are a unique and invaluable source of information about the history of Guaíra Falls. They also contain some racist remarks, as this quote attests. Mann fought for the German army during World War II, and the fact that he was more troubled by the loss of Guaíra Falls than the horrors of that war is a disturbing reflection. For more on Ernst and Charlotte, see Vilma Fiorotti Lima, "Guaíra nos tempos de Ernst Mann," undergraduate thesis, Universidade Estadual do Oeste de Paraná, Marechal Cândido Rondon, 1994.

thousands of non-Indigenous Brazilians (see Chapter 4), as well as cherished natural waterscapes such as Guaíra Falls. The dams also changed the local ecology and unleashed a flood of environmental consequences that affected fishermen, lakeside residents, and others whose livelihoods were tied to soils, waters, and biota of these newly formed lakes (see Chapter 5).

This chapter argues that political pressures encouraged the military government to build dams with giant reservoirs and to do so quickly and without regard to their social and environmental footprints. The military regime looked to big dams as a means to secure political legitimacy, and it wasted no time in starting construction because it takes a long time to build such projects, often the better part of a decade and sometimes longer. The 1973 oil crisis added urgency, as imported petroleum became prohibitively expensive and pushed the government to invest in alternative sources of energy. The crisis encouraged the military regime to double down on the big dams already under construction and to plan a host of new ones.

Political pressures also made their way into debates about specific dam sites. The most prominent case was Itaipu, where the military government had to factor in geopolitical considerations. The combined result of all these political pressures was a firm commitment to building large reservoirs in environmentally sensitive areas without public debate and without completing thorough environmental impact studies.

THE DESERT BECOMES THE SEA

In October 1967, Senator Manoel Novaes stood in front of the National Congress and applauded the federal government for finally taking initiatives to build a large reservoir on the São Francisco River. He was worried, however, that it would not build the dam that he wanted. Novaes supported building a large dam at the Sobradinho rapids, in the state of Bahia, which would create a reservoir that would cover 4,120 sq. km of semidesert. This ample water supply could be used to irrigate fields and generate hydroelectricity onsite, but its most important role would be regularizing the river to ensure a steady supply of water for the downstream hydroelectric dams that were powering the region's industrial and urban growth. The government, however, was considering building its large storage dam at Itaparica Falls, about 300 kilometers downstream from Sobradinho in the neighboring state of Pernambuco, which had greater energetic potential than Sobradinho. Novaes had been dreaming of Sobradinho for decades and had been fighting in Congress for the funds to build the dam since 1963. He was not ready to see his project lose out to a competing dam site.[7]

Luckily for Novaes, he had an important supporter on his side. Minister of Mines and Energy José Costa Cavalcanti also supported building Sobradinho. Costa Cavalcanti was an important architect of the legislation that had directed the federal government to build the dam in question. On July 26, 1967, President Costa e Silva signed into law a decree that created an interministerial subcommission to study potential sites for multiple-use engineering projects along the São Francisco River, including the big storage reservoir.[8] Engineers had long proposed and carried out some minor engineering projects on the lower-middle section of the river to facilitate transportation, but the idea of creating a big storage reservoir was comparatively new. President Juscelino Kubitschek (1956–1961) was the first to seriously consider the idea, and he sent the São Francisco Valley Commission to assess the Sobradinho site. Its studies, carried out between 1960 and 1962, became the basis for the interministerial subcommission's subsequent plans.

Though the subcommission was nominally charged with studying the multiple uses of the river, the decree made it clear that energy was the priority. It gave the commission about four months to come up with the next project that the state-owned Companhia Hidrelétrica do Rio São Francisco (CHESF) would build along the river. Costa Cavalcanti favored Sobradinho because it would be a huge boon to the river's hydropower production and also offer greater possibilities for multiple-use than competing designs.[9] The interministerial subcommission was inconclusive, however, and Costa Cavalcanti lobbied to create another commission, this time entirely within the Ministry of Mines and Energy [Ministério de Minas e Energia, MME], to carry out more determinative studies. On September 26, 1969, the interim president signed into law a decree that created this committee, which set about studying the hydroelectric potential of the lower-middle stretches of the São Francisco.[10]

[7] Manoel Novaes, "Discurso na Câmara de Deputados," October 18, 1967, ACD.

[8] Artur da Costa e Silva, *Decreto No. 61.076, de 26 de julho de 1967, Institui Subcomissão Interministerial de Estudos para a Utilização Múltipla do Rio São Francisco*, ACD.

[9] José Costa Cavalcanti, "Inaugural Speech as Minister of Mines and Energy," 1967, quoted in Novaes, "Discurso na Câmara de Deputados."

[10] Augusto Hamann Grünewald, *Decreto No. 65.237, de 26 de setembro de 1969, Cria no Ministério das Minas e Energia o Comitê Coordenador dos Estudos Energéticos da Região Nordeste do Brasil*, ACD.

The committee was charged with building what engineers referred to as "the big dam." Since the 1940s, when the federal government began studying the hydroelectric potential of the São Francisco River, engineers had dreamed of a giant dam on the lower-middle stretches of the river that would create an immense reservoir. Such a dam was the key to unlocking the river's full energetic potential because it would regulate water flow. The São Francisco River flows through the driest region of the country and generally runs low except during the few months of the year when it rains in the upper watershed and floods rush downstream. Without regulation, the output of any hydroelectric dams built on the river is unreliable and subject to the vagaries of the region's climate.

In 1948, the government took the first step toward taming the river when the newly created CHESF began building a big hydroelectric dam at the renowned Paulo Afonso Falls – the Brazilian government's first big dam.[11] These waterfalls were an obvious site for a hydroelectric dam (Figure 2.1). They had captured the imagination of poets, painters, engineers, and even the Emperor Dom Pedro II, who made the treacherous trek to visit the falls in the 1860s. In 1855, Dom Pedro II had commissioned German engineer Henrique Halfeld to study potential engineering projects along the São Francisco that would improve navigation. In his 1860 report, whose publication had inspired the emperor's visit, Halfeld marveled at Paulo Afonso. He reflected that the thundering sound made it impossible to hear the people standing next to him, and he boasted about the falls' power.[12] In 1906, Brazilian engineer Teodoro Sampaio, who had participated in a second government-commissioned engineering expedition between 1879 and 1880, concluded: "Paulo Afonso Falls, the famous haunt of old chroniclers and travelers, is indeed one of the most stupendous spectacles imaginable."[13] The falls were so grand that these engineers simply deemed that stretch of the river unnavigable and focused their attention on reengineering smaller rapids upstream. But in the first decades of the twentieth century, hydroelectric dams were

[11] CHESF completed the first phase of Paulo Afonso between 1948 and 1955 and continued adding turbines thereafter. See Luiz Fernando Motta Nascimento, *Paulo Afonso: Luz e força movendo o Nordeste* (Salvador: Associação Recreativa dos Servidores da CHESF, 1998).

[12] Henrique Guilherme Fernando Halfeld, *Atlas e relatório concernente a exploração do Rio de São Francisco desde a Cachoeira da Pirapora até ao Oceano Atlântico, 1852–1854* (Rio de Janeiro: 1860), 44.

[13] Teodoro Sampaio, *O Rio São Francisco e a Chapada Diamantina, 2ª edição* (Salvador: Progresso Editora, 1955), 51 (first edition published in 1906).

FIGURE 2.1. Paulo Afonso Falls, circa 1860s. This is a cropped version of an image that comes from the Brazilian National Library's Brasiliana Fotográfica Collection.

becoming increasingly popular, and engineers began considering the falls' energetic potential.

The fight to domesticate the river at Paulo Afonso was not easy. Building the dam was an incredible engineering feat that involved excavating large caverns for underground powerhouses and diverting the giant and tempestuous São Francisco River. The project was headed by Brazilian engineer Octávio Marcondes Ferraz, who famously protested the original plans that North American engineers had laid out and built the dam according to his own design, which offered greater generating capacity. His protests won over his superiors.[14] In 1955, his team completed Paulo Afonso, successfully damming the mighty São Francisco River. The project put the Brazilian government on the map as a force capable of building big dams and brought Marcondes Ferraz into the annals of Brazil's national heroes.[15]

But one problem remained. Engineers had deemed Paulo Afonso an unsuitable site for a large storage reservoir and had instead built a run-of-the-river dam adjacent to the falls, which had unreliable power output because it could not store water to produce power during periods of

[14] The engineering challenges of Paulo Afonso's construction and Marcondes Ferraz's battle with US technicians are covered in rich detail in Folder 1, Series 45.02.10; Folders 1–10, Series 45.05.11; Folders 1–2, Series 48.06.22; and Folder 1, Series 48.07.07, FOMF, CPDOC-FGV.

[15] For more on Marcondes Ferraz, see Memória da Eletricidade, *Octávio Marcondes Ferraz: Um pioneiro da engenharia nacional* (Rio de Janeiro: Memória da Eletricidade, 1993).

reduced streamflow. The solution was to build a storage dam upstream where conditions were more favorable for a massive reservoir. With more dependable water flow, engineers could install more powerhouses at Paulo Afonso and turn it into a prolific hydroelectric complex that would take full advantage of the powerful falls. The idea became so celebrated among Brazilian engineers that they gave the project a special nickname, *barrajão*, which is an uncommon, colloquial way of saying "big dam."[16] The big storage dam was the dream, and the military government was determined to make it a reality.

The military government was interested in building the big dam on the lower-middle stretch of the São Francisco River because it would help its National Integration Program and earn its regime political legitimacy in a region that previous administrations had neglected.[17] One of the military government's top priorities was stimulating industrial and urban growth in relatively unpopulated and economically underdeveloped regions of the country. The idea was that raising the standard of living in these areas would better integrate them into the national economy, which would showcase the benefits of government-led capitalism and diminish the appeal of socialist policies designed to redistribute wealth or topple the military government (see Chapter 1). The Northeast was the country's second-most populated region, but its economy lagged far behind that of the Southeast. It was, and remains, one of the country's poorest regions, and reoccurring droughts and their associated economic crises periodically encouraged Northeasterners to move south in search of better economic opportunities. The military government hoped that investment in large infrastructure projects like big dams could settle the population and raise the standard of living without having to resort to land reform or other wealth redistribution policies.

Moreover, a dam the projected size of Sobradinho or Itaparica – when completed Sobradinho became the largest artificial lake in South America – would have tremendous symbolic value and earn the fledgling military regime high prestige for having reengineered the São Francisco River in such an audacious manner. Promoting the dam in the Senate on

[16] Apolônio Sales' remarks in CHESF, *Relatório Anual 1970*, 3, BCHESF. Portuguese speakers frequently use diminutives and augmentatives for emphasis. In this case, the augmentative-*ão* is added to *barragem*, the Portuguese word for dam, along with a small spelling change to maintain pronunciation. The word is very unusual. The conventional way of saying big dam is *grande barragem*, using the Portuguese word for big, *grande*.

[17] "Antônio Dias Leite to Presidente Emílio Médici," November 17, 1971, File BR RJANRIO ADJ.0.0.7, 127–131, FADL, AN-RDJ.

November 21, 1963, Senator Wilson Roriz exclaimed: "Sobradinho will really be that large dam capable of immortalizing whichever government [builds it] because it will be the grand dam in a specifically northeastern sense, because only with it can we combine the great binomial of salvation in the Northeast, energy and irrigation."[18] The military regime was looking for both prestige and to save the Northeast from poverty without social reform, and it believed that building the big dam on the São Francisco River would accomplish both these goals.

Though Senator Novaes seemed to have found the support he needed to realize his dreams of building Sobradinho, there was one problem: All the studies for a dam on the lower-middle stretch of the river showed that an alternative dam site was better. If the government built its large storage dam at Itaparica, the reservoir could be much deeper and therefore have greater storage benefits to regularize the river. Moreover, the power potential at Itaparica Falls was much greater, so it would be more cost-effective. Lastly, it would slow siltation at downstream dams, whereas Sobradinho would accelerate it. Large dams trap silt, leaving the water they release free of sediment and with greater capacity to erode shorelines downstream and collect sediment. Scientists call such rivers "sediment hungry." Thus, Sobradinho would create a long stretch of sediment-hungry river that would deposit its silt load at the Paulo Afonso dams, whereas Itaparica was close enough that it would not. Indeed, it would have the opposite effect – its giant reservoir would collect upstream sediment and prevent them from gumming up the dam infrastructure at Paulo Afonso. The site did not offer much in the way of irrigation, but that did not matter much to the military regime, who, despite the protests of some senators, had made it clear that its priority was hydroelectric generation.[19]

The two sites were not mutually exclusive. Indeed, the government ultimately planned to build both dams, but there was only room for one giant reservoir on that stretch of the São Francisco River, so building a large reservoir at Sobradinho meant limiting the size and power output at Itaparica (Figure 2.2). Likewise, erecting a high dam at Itaparica – as the engineering studies had recommended – meant building a much smaller dam at Sobradinho. Moreover, the military government wanted to build

[18] Wilson Roriz, "Discurso na Câmara de Deputados," November 23, 1963, ACD.

[19] Ministério das Minas e Energia, Comitê Coordenador dos Estudos Energéticos da Região Nordeste (CCEERN), *Estudos Energéticos do Nordeste, Vol. 1*, April 1972, 11.16, Box T2052, ACHESF.

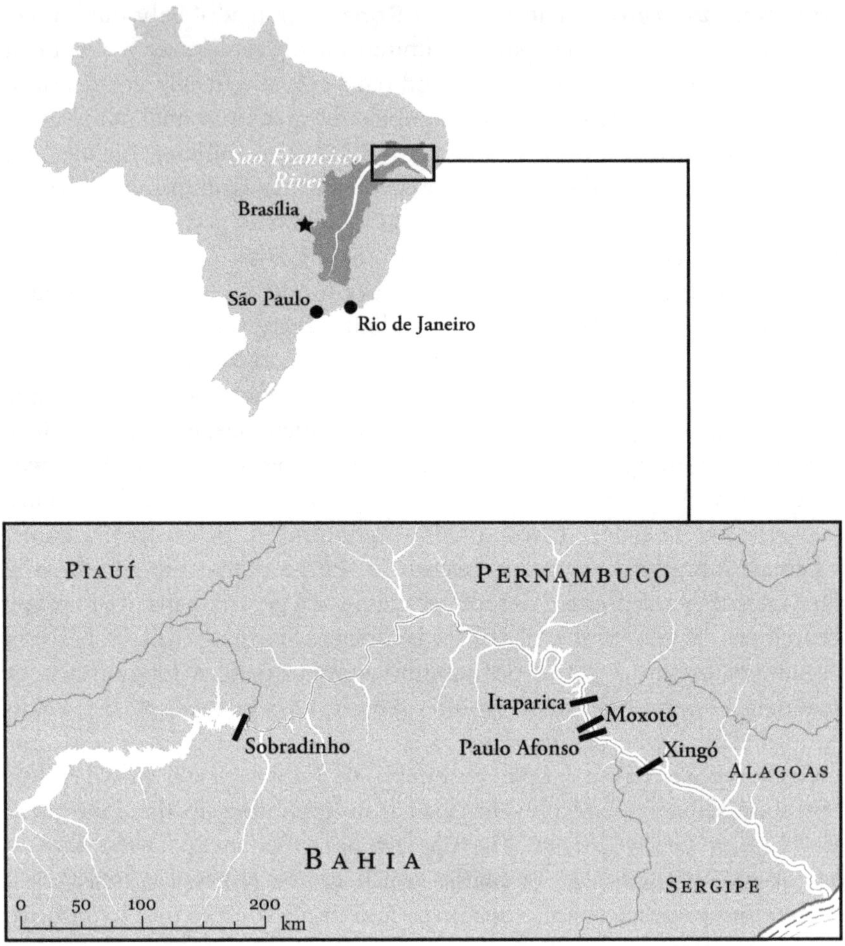

FIGURE 2.2. Hydropower dams on the lower-middle stretch of the São Francisco River, 1940s–1990s. Map by Geoffrey Wallace (G. Wallace Cartography & GIS).

its showcase project immediately, and in the short term, demand in the Northeast only warranted building one dam. So for the time being, the choice was Sobradinho or Itaparica. To the delight of Novaes, in June 1971, the MME chose to build its large storage dam at Sobradinho, despite the technical superiority of Itaparica.[20]

Two factors discouraged the government from choosing Itaparica. First, Itaparica had a huge social and environmental footprint that the

[20] CHESF, "Síntese do projeto Sobradinho," September 1972, Box T2372, ACHESF.

government deemed unacceptable.[21] The dam would flood valuable agricultural land and displace about 65,000 people. It was not the footprint itself that the government cared about, but the political repercussions. Sobradinho too would flood valuable agricultural land and displace the exact same number of people. Yet, the military government considered the land around Itaparica to be much more valuable because wealthy landowners – including CHESF president Apolônio Sales – had already invested in building infrastructure there.[22] To support these wealthy landowners, the local government had already begun investing in transportation networks and infrastructure in nearby towns. Displacing wealthy agriculturalists was sure to be unpopular and bring bad press to the military regime. At Sobradinho, in contrast, the people who lived there were poor and the land was less developed. Most of the people living alongside the river practiced floodplain agriculture, and many did not have the title to their land. It would be much easier to expel them without protest than it would be for those at Itaparica.[23]

Sobradinho was also cheaper. Itaparica was a bigger and more expensive dam that was only worth its cost if the government installed hydroelectric turbines to generate energy. Sobradinho, on the other hand, was affordable without turbines, which the military government deemed unnecessary at this stage.[24] Its big dam did not need to generate power; it just needed to have a reservoir large enough to control the river's water flow. That would support the planned expansion at Paulo Afonso downstream, which engineers had already begun – once completed in the late 1970s, the Paulo Afonso power plant would boast three additional powerhouses and a diversion canal and small storage reservoir for the fourth powerhouse.[25]

Between the two dam sites, Sobradinho won out: It was "the big dam" capable of immortalizing the military government, and it could be built quickly, cheaply, and without a huge political backlash. The military regime began construction on its then turbine-less storage dam in July 1973, and within five years the reservoir began filling, creating a

21 "Dias Leite to Médici" (November 17, 1971).

22 "Entrevista com o engenheiro Mario Santos, outubro 2010," in CBDB, *A História das Barragens no Brasil*, 495–496.

23 CCEERN, *Estudos Energéticos do Nordeste, Vol. 1*, 11.15.

24 "Dias Leite to Médici" (November 17, 1971).

25 For an excellent social and labor history of this campaign, see Jamile Silva Silveira, "'Braços cruzados, máquinas paradas': trabalhadores, experiências, e conflitos na capital de energia (1960–1980)," *Rios Eletrônica* 16 (2018): 9–22.

body of water large enough to stand out in satellite images. Just as Antônio Conselheiro – leader of the Canudos settlement – had prophesized, the desert was becoming the sea.

Though the military government was willing to choose a less efficient dam in order to shift the social costs onto a poorer and less politically powerful group, it was unbending in its decision to build a giant reservoir that would have a large socioenvironmental footprint; when alternatives were proposed, it discounted them. On June 20, 1972, Senator Manoel de Almeida took to the chamber floor to express the unpopular position that the government should build a series of smaller dams instead of Sobradinho.[26] Not long ago, Senator Almeida had advocated for big dams on the São Francisco River – in the late 1950s, he had fought vehemently in the chamber to get funding to build a large dam at Três Marias (far upstream near the river's headwaters), and he had supported Senator Novaes in 1967 when he had advocated for Sobradinho.[27] But he had since been in touch with an ecologist at the Universidade Federal de Pernambuco, and he was starting to change his mind.

Big dams made less sense to him the more he considered potential evaporation rates. The Northeast was so hot and dry that the water of such a large reservoir was sure to evaporate quickly and diminish the dam's ability to perform its function. He exclaimed, "mega dams like Sobradinho are inconvenient given the environmental conditions of the semidesert region, where high rates of evaporation plague bodies of water with large surface areas."[28] CHESF's engineers believed that the dam was economic despite endemic drought and high evaporation rates, but Almeida worried they underestimated the threat and that drought would all but erase the reservoir's multipurpose functions and limit its ability to produce energy.

Moreover, the experience of Três Marias convinced him that the social costs of displacing people were much greater than the government anticipated. He argued that the government did a poor job of compensating the people whose land the reservoir had flooded and that many now lived in poverty as a result. Sobradinho threatened to displace an even larger number of people, and he was not sure the

[26] Manoel de Almeida, "Discurso na Câmara de Deputados," June 20, 1972, ACD.
[27] Novaes, "Discurso na Câmara de Deputados."
[28] Almeida, "Discurso na Câmara de Deputados."

government was up to the task of properly relocating and compensating them.

Rather than building a large storage dam, he suggested that the government build a series of smaller dams along the tributaries of the São Francisco River that together could do a better job of controlling floods, regulating the river, and providing multipurpose benefits for the people who lived near them. Such a solution would also have a much more benign environmental footprint and would not involve relocating tens of thousands of people. Some other senators interrupted his speech to commend him for speaking up, but no one – in the chamber or elsewhere – supported his position against the dam. Though criticism of the project's social and environmental impacts ballooned during and after construction, there is almost no record of protest against the dam on environmental grounds during the planning stages. Almeida's opinion was not popular, and his protests seem to have never left the chamber floor.

But even had Almeida's protests garnered more support, it is unlikely they would have done much to change the government's plans. During the late 1960s and early 1970s, the Brazilian military government committed to building giant reservoirs because they promised short-term economic growth that would serve the goals of its National Integration Program and because such grand dams had the symbolic benefits of showcasing its regime's efficacy in controlling nature for the betterment of its constituents. In its calculations, the political benefits of such projects far outweighed any backlash that might emerge in response to its dams' environmental costs. Such a backlash did emerge, most prominently at Itaipu, a large dam on the Paraná River in Southern Brazil, but it did little to impede the dam-building ambitions of the military government and its engineers.

SÃO PAULO CANNOT STOP

In the 1950s, São Paulo was outgrowing its electricity supply, and the local utility began rationing consumption. At that time, the city got most of its electricity from small hydropower and thermal plants managed by a Canadian-owned company called São Paulo Light, commonly referred to as "Light." The crisis worsened between 1952 and 1955, when drought hit Southeastern Brazil and dried up the rivers and reservoirs that powered the region's small hydroelectric plants. Some cried malpractice on the part of Light, but the company had not orchestrated the

Shortage.[29] The city was simply growing too fast. True to the popular saying of the time, São Paulo could not stop.

People were pouring into the city at an unprecedented rate, attracted by its booming industrial growth.[30] Between 1940 and 1960, the population of the state of São Paulo grew from roughly seven to thirteen million people, and much of that growth was concentrated in urban centers.[31] In 1958, the city had grown to roughly three million people and had become Latin America's largest industrial center; its growth slowed only by the periodic electricity rationing that gripped the region.[32]

São Paulo Light looked toward Guaíra Falls, about 800 kilometers west, as the solution. There, the great Paraná River cut through the Maracaju Mountains, dropping 115 meters and compressing its five-kilometer breadth into a narrow canyon.[33] This powerful set of waterfalls consisted of hundreds of cataracts – about twenty of which were substantial in their own right – that together grouped into seven prominent clusters. It was the largest waterfall in the world measured by volume, with incredible potential to generate power.[34]

Damming Guaíra Falls was an old dream among Brazilian and Paraguayan engineers. Early investigations were carried out by engineers in both countries in the 1890s and 1900s, and Brazilian engineers published the first studies on the site's energetic potential in 1908.[35] Over the

[29] Memória da Eletricidade, *Notas sobre racionamento de energia elétrica no Brasil (1940–1980)* (Rio de Janeiro: Memória da Eletricidade, 1996), 44–66.

[30] See Werner Baer, *Industrialization and Economic Development in Brazil* (Homewood: Irwin, 1965), and Paulo Fontes, *Migration and the Making of Industrial São Paulo* (Durham: Duke University Press, 2016).

[31] Instituto Brasileiro de Geografia e Estatística (IBGE), *Censo demográfico de 1940* (Rio de Janeiro: Serviço Gráfico do IBGE, 1940), 67; IBGE, *Censo demográfico de 1960* (Rio de Janeiro: Serviço Gráfico do IBGE, 1960), 80.

[32] Furnas, "The Furnas Hydroelectric Project Development," 1958, 1, Folder 36, FJC, ACME.

[33] Mann, "As 'Sete Quedas' do Rio Paraná," Folder "Livro," FEM.

[34] Marcondes Ferraz, "O sacrifício de Sete Quedas," October 8, 1982, Folder 6, Series 61.11.23, FOMF. For more on how interest in the site's energetic potential eclipsed earlier interest in the falls as a natural wonder, see Richard Niland, "Death by Water: The Rise and Fall of Los Saltos del Guairá," *Environmental History* 23, no. 1 (2018): 56–81, and Richard Niland, *La muerte del agua: El ascenso y el descenso de Los Saltos del Guairá* (Asunción: El Lector, 2021).

[35] On the Paraguayan investigations, the earliest of which were carried out between 1893 and 1896, see Efraim Cardozo, *Los derechos del Paraguay sobre los Saltos del Guairá* (Asunción, 1965), 138–141. On the Brazilian expeditions and studies, see Ivone Teresinha Carletto Lima, *Itaipu: As faces de um megaprojeto de desenvolvimento* (Marechal Cândido Rondon: Editora Germânica, 2006), 117–119.

next several decades, specialists in both countries continued to study the hydroelectric potential of the famous waterfalls. Brazilian engineers took the lead, as their country had a greater demand for electricity and more resources to carry out research.

These studies, however, produced scant concrete data. They were more ideas and aspirations than practical blueprints, and at the time they attracted little interest from the Southeast's utility companies. Light surely recognized the potential of the falls, but distance diminished the appeal: Guaíra Falls was far from the county's large population centers, and transmission line technology was then unable to transport electricity over long distances without significant losses that would render the project unprofitable. Moreover, utility companies had access to other dam sites closer to Brazil's Southeastern cities.

The situation changed in the 1950s, and Light became interested in Guaíra Falls as the remedy for São Paulo's energy crisis. The company had exhausted all the best dam sites near the city, and improvements in transmission line technology were making it possible to transport electricity efficiently over greater distances.[36] In 1953, the company appealed to the Department of Agriculture for a concession to build the dam. The Department of Agriculture ignored Light's request and, on December 31, 1957, formally denied Light the concession, granting it instead to an interstate commission.

The government itself had become interested in the Guaíra Falls project. Federal and state governments were becoming more capable energy producers in the 1950s and, reflecting the influence of nationalism during that time, became increasingly antagonistic to foreign-based power companies such as Light. Despite the government's newfound interest in the Guaíra Falls project, it decided that in the short term, a comparatively smaller dam closer to the city would be more practical to address the energy crisis. Between 1958 and 1964, the federal government built the Furnas Dam on the Grande River, which resolved São Paulo's electricity shortages for the time being.[37]

During this same period, the federal government became the only entity responsible for building new power plants. In 1961, Light announced that it only had plans to expand its infrastructure until 1963, after which it would stop building new generating plants. The

[36] Montreal Engineering, "Presentation to Eletrobras: Sete Quedas Transmission Study Survey of the State of the Art," January 28, 1972, xvii, Folder 699, FJC.

[37] Lima, *Itaipu*, 119–123.

fourteen power plants under construction or expansion in 1964 belonged exclusively to the state or federal government, and that year, 55 percent of São Paulo Light's energy sales came from reselling electricity from the newly completed Furnas Dam.[38]

The success of Furnas inspired subsequent governments to build bigger dams. On January 31, 1961, President Jânio Quadros began his tenure and looked toward large dams to boost his popularity. He wanted to outdo Furnas, and he reportedly asked his aides to find the biggest project he could undertake. His advisors suggested Guaíra Falls. Quadros was enthusiastic about the dam, but he did not last long enough to oversee his showcase project. Seven months into his presidency, he quit, leaving the position, and his ambitious dam-building plans, to João Goulart.

The new president was not in such a hurry to build the dam as his predecessor, but he nevertheless agreed that reengineering Guaíra Falls to produce electricity would boost his administration's popularity.[39] Moreover, the federal government had recently taken steps to get more involved in planning and building large dams. In 1960, the federal government had created the Ministry of Mines and Energy, and in 1962, Goulart oversaw the final formation of Eletrobras, the federal government's electric utility company whose subsidiaries went on to build Brazil's largest dams.[40] The federal government was well positioned to take on a bigger role in building dams, and Goulart thought it should start with Guaíra Falls.

In March 1962, Minister of Mines and Energy Gabriel Passos contacted Marcondes Ferraz – the famous engineer who had designed Paulo Afonso – to carry out a viability study for a dam at Guaíra Falls.[41] This study stood out from previous surveys because it was carried out by the federal government and in the hands of one of Brazil's most respected

[38] Memória da Eletricidade, *Notas sobre racionamento de energia elétrica no Brasil*, 146–147.

[39] Memória da Eletricidade, *Notas sobre os antecedentes da criação de Itaipu Binacional* (Rio de Janeiro: Memória da Eletricidade, 1999), 26–30.

[40] Eletrobras was created on April 25, 1961, by Lei No. 3.890-A, and became operational on June 11, 1962. See www18.fgv.br/cpdoc/acervo/dicionarios/verbete-tematico/eletrobras-centrais-eletricas-brasileiras-s-a (last accessed April 2023).

[41] Marcondes Ferraz, "Comentários sobre a solução de Itaipu," palestra proferida no Clube de Engenharia do Rio de Janeiro, April 27, 1976. This speech was reproduced in the May 2, 1976 edition of *O Estado de São Paulo* and in the official transcript of Paulo Brossard's Discurso no Senado Federal, May 5, 1976, ASF. Correspondence between Gabriel Passos and Marcondes Ferraz concerning this study and the contract they signed can be found in Folder 1, Series 61.11.23, FOMF.

dam builders. The study's final report, published in 1962, did more than simply outline the technical parameters for the Sete Quedas Dam (the title he gave the project) – it set off a border dispute that raised the specter of Brazil and Paraguay's bellicose past and created a tense standoff that both countries thought best resolved with a different dam that would flood Guaíra Falls.[42]

SPECTER OF A BELLICOSE PAST

The Brazilian government's plans for damming Guaíra Falls stirred up trouble with Paraguay, then under the dictatorial rule of military officer Alfredo Stroessner (ruled 1954–1989), threatening to reignite old grievances. Between 1864 and 1870, Brazil, Argentina, and Uruguay fought against Paraguay in a notably bloody war (called the War of the Triple Alliance, or the Paraguayan War in Brazil), in which Paraguay lost 50 percent of its total population and as much as 90 percent of its male population.[43] The trauma of the conflict dominated Brazilian-Paraguayan relations thereafter and influenced Paraguay's reaction to Brazilian plans along their shared border, all the more so because the war's hundred-year anniversary was approaching.

Brazil wanted desperately to avoid raising the specter of the Paraguayan War because its international reputation had suffered heavily as a result of the conflict. Since Brazil had provided most of the troops on the side of the alliance, it bore the brunt of the criticism for the war and its destruction. International opprobrium accumulated after the conflict and has since been reaffirmed by foreign historians. Renowned North American historian Thomas Skidmore writes,

[42] Octávio Marcondes Ferraz, *Relatório preliminar sobre o aproveitamento do Salto de Sete Quedas (Guaíra) Rio Paraná* (São Paulo: Escritório Técnico O.M.F., December 1962). This report can be found in at least three locations: the BPP, DDP; FOMF; and FJC.

[43] This is the traditional estimate. The exact number of people who died in the Paraguayan War is unknown and the subject of debate. Estimates of the death toll range widely from 7 to 70 percent of Paraguay's population. See Vera Reber, "The Demographics of Paraguay: A Reinterpretation of the Great War, 1865–1870," *Hispanic American Historical Review* 68, no. 2 (1988): 289–319; Thomas L. Whigham and Barbara Potthast, "The Paraguayan Rosetta Stone: New Insights into the Demographics of the Paraguayan War, 1864–1970," *Latin American Research Review* 34, no. 1 (1999): 174–86; Jan Kleinpenning, "Strong Reservations about 'New Insights into the Demographics on the Paraguayan War'," *Latin American Research Review* 37, no. 3 (2002): 137–142.

"victory over such a small, poor, and desolate country hardly qualified Brazil for the annals of glorious warfare, despite the triumphalist rhetoric of some patriots in Rio. On the contrary, it raised fundamental questions about whether their own ill-integrated society was ready to join the race to modernity."[44] Brazil had emerged from the war with its reputation in tatters, and diplomatic cooperation between both countries ground to a halt.

In the 1950s, more than eighty years after the war, both countries began the process of rapprochement in an effort to further shared economic goals.[45] Brazil helped Paraguay build a highway from Asunción to the Paraná River (National Route 2) and a bridge crossing the river to the Brazilian town of Foz de Iguaçu, which Brazilians called the Friendship Bridge. The two countries' political interests aligned further after March 1964, when Brazil's right-wing military seized power (see Chapter 1).

Nevertheless, rapprochement was inchoate. As it went about planning its dam at Guaíra Falls, the Brazilian government wanted to avoid reopening old wounds that would set back its diplomatic efforts with Paraguay and sully its international standing.

But trouble erupted almost immediately. On February 13, 1962, *Jornal do Brasil* published a short article announcing the Brazilian government's decision to contract Marcondes Ferraz to carry out a viability study for a dam at Guaíra Falls.[46] One month later, on March 12, Paraguay's foreign minister wrote to Brazil's foreign minister inquiring about the article. The news was a problem, the foreign minister said, because the land on the western bank of the Paraná River above Guaíra Falls belonged to Paraguay, so the country would need to be consulted before surveyors could go about their work.[47] The Brazilian foreign minister replied that

[44] Thomas Skidmore, *Brazil: Five Centuries of Change, 2nd Edition* (New York: Oxford University Press, 2010), 70.

[45] See Peter Lambert, "The Myth of the Good Neighbour: Paraguay's Uneasy Relationship with Brazil," *Bulletin of Latin American Research* 35, no. 1 (2016): 34–48, 36.

[46] "Ministério das Minas nomeia Ferraz para saber como aproveitar Sete Quedas," *Jornal do Brasil*, February 13, 1962.

[47] "Raul Peña to Francisco Clementino, 'MRB No. 94'," March 12, 1962; "Afonso Arinos de Melo Franco to Raul Peña," September 19, 1962, Folder 6, FMT, ACME. The full series of diplomatic cables sent between both foreign ministers between 1962 and 1966 is preserved in its entirety in Folder 6 of FMT, and in Folder 2, Entry Pi1966.06.21, FJM, CPDOC-FGV.

Brazil was happy to consult with Paraguay about its dam-building plans, but that the land in question on the Paraná's western bank belonged to Brazil.

This exchange – triggered by the news of Marcondes Ferraz's dam proposal – set off a tense border dispute.[48] The conflict worsened in December 1962, when Marcondes Ferraz published his report on the Sete Quedas Dam, which proposed building a short dam above the falls that would divert water into a canal along the Brazilian side of the river that led to a powerhouse downstream.[49]

Diplomats in both countries failed to resolve the matter through telegrams, and as tensions built, the conflict threatened to turn violent. In March 1965, the Paraguayan government sent troops to occupy the disputed border region and held a flag-raising ceremony. In June, the Brazilian government responded by sending its military to occupy the same area and build roads, airstrips, buildings, and other infrastructure. To make matters worse, in October, Brazilian officers detained a group of Paraguayan government officials who had been sent to explore the Brazilian outpost. In response, in November, the youth sectors of two opposition parties organized a demonstration that turned riotous when participants began vandalizing the Brazilian embassy and various Brazilian-owned businesses.[50]

The border conflict revolved around differing interpretations of the 1872 Treaty of Loizaga-Cotegipe, which was signed in the aftermath of the Paraguayan War and established the border between the two countries (Figure 2.3). The treaty used ambiguous language in two key clauses to describe the border near Guaíra Falls. Its first clause stated that the border would run north along the Paraná River until it reached the

[48] This border dispute, which was not entirely resolved until April 26, 1973, is covered in great detail in former Paraguayan Itaipu Director Enzo Debernardi's *Apuntas para la historia política del Itaipu* and Memória da Eletricidade's *Notas sobre os antecedentes da criação de Itaipu Binacional.* John Cotrim, Itaipu's technical director between 1975 and 1986, was the lead author of the latter, and further information about his role in the book can be found in Folders 48, 92–97, and 807 of FJC. See also Tomaz Espósito Neto, *Itaipu e as relações brasileiro-paraguaias de 1962 a 1979: Fronteira, energia e poder* (Saarbrücken: Novas Edições Acadêmicas, 2017), and Jacob Blanc, *Before the Flood: The Itaipu Dam and the Visibility of Rural Brazil* (Durham: Duke University Press, 2019), 20–52.

[49] His design was similar to what Light engineers and Pedro Henrique Rupp had envisioned in the late 1950s. See "Aproveitamento de Sete Quedas," February 1963, Folder 93, FLL, ACME, and Folder 3, Series 61.11.23, FOMF.

[50] Jacob Blanc, *Before the Flood*, 34–42.

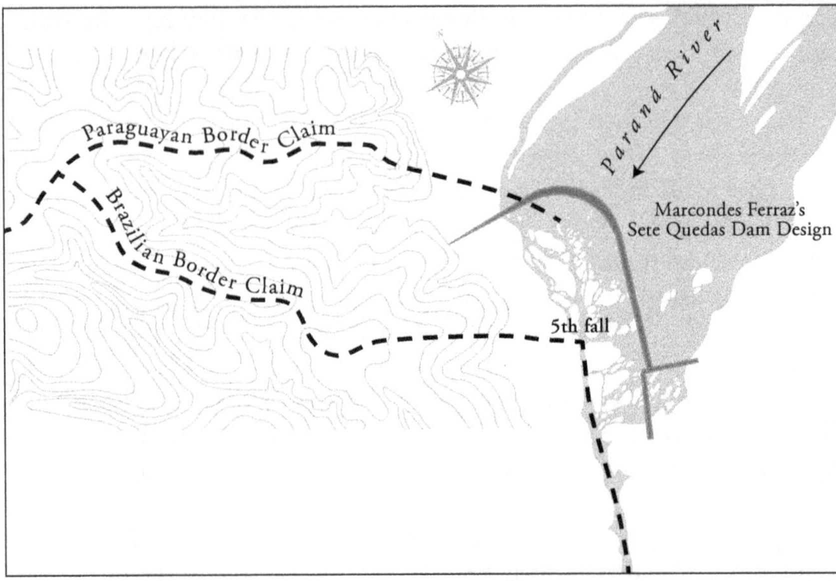

FIGURE 2.3. Octávio Marcondes Ferraz's Sete Quedas Dam design and the disputed Brazilian–Paraguayan border, 1962–1982. Map by Geoffrey Wallace (G. Wallace Cartography & GIS), based on the map in Comisión Mixta Demarcadora de Límites Paraguayo-Brasileña, "El límite entre Paraguay y Brasil en el Salto del Guairá," Folder 6, Fundo Mauro Thibau, Arquivo do Centro da Memória da Eletricidade, Rio de Janeiro, Brazil, and a map in Efraim Cardozo's *Los derechos del Paraguay sobre los Saltos del Guairá* (Asunción, 1965).

"Great Fall of the Seven Falls," and from there it would follow the "highest peak of the Maracaju Mountains."[51] The summit of the Maracaju Mountains splits just west of the Paraná River, creating a northern and southern peak. Paraguay contended that the treaty referred to the northern range, which terminated just above Guaíra Falls. This reading corresponded with its interpretation of the first clause, in which the term "Great Fall of the Seven Falls" referred to the entire set of

[51] The original text in Portuguese reads: "O territorio do Imperio do Brasil divide-se com o da Republica do Paraguay pelo alveo do rio Paraná, desde onde começão as possessões brasileiras na foz do Iguassú até o Salto Grande das Sete Quedas do mesmo rio Paraná; Do Salto Grande das Sete Quedas continúa a linha divisoria pelo mais alto da Serra de Maracajú até onde ella finda; D'ahi segue em linha recta, ou que mais se lhe aproxime, pelos terrenos mais elevados a encontrar a serra Amambahy." The full text of the treaty can be found in Folder 2, Entry Pi1966.06.21, FJM.

waterfalls. Under this reading, Paraguay owned the land north of the falls where Marcondes Ferraz planned to build his dam.[52]

The Brazilian government interpreted this language differently. The Maracaju Mountains' highest peak, it asserted, was its southern range, which led to the fifth and largest fall of the Guaíra Falls.[53] According to this reading, the "Great Fall of the Seven Falls" did not refer to the entire set of waterfalls, but rather the largest of them, which was widely regarded as the fifth fall. This interpretation would give Brazil complete jurisdiction over the land immediately north of Guaíra Falls and permit building a dam there without Paraguayan approval.

Though building on disputed land was the immediate point of contention, the conflict also concerned water diversion. Building a dam immediately upstream of the falls would give Brazil complete control of the water along the shared stretch of the Paraná. The legal institutions structuring water rights in the region were feeble and open to interpretation, and at the time Brazil argued it had no obligation to share water with downstream neighbors.[54] Thus, Paraguay was defending not just its claim to the disputed land, but its rights to Paraná River water as well.

The situation almost became intractable. In June 1966, the foreign ministers of both countries met to try and come to an agreement, but neither would concede. During these tense negotiations, Brazilian foreign minister Juracy Magalhães reportedly exclaimed that "a border treaty can only be altered by another treaty or a war in which we are defeated." Paraguay's foreign minister Raul Sapena Pastor responded, "is that a threat?"[55]

The two countries avoided hostilities by signing, on June 22, 1966, the Ata de Iguaçu, an agreement that both sides would work together on any dam planned in the disputed area, or elsewhere along their mutual border. The agreement was an important step toward easing tensions, but it did not erase the problem. The Ata de Iguaçu had not outlined plans for a specific project, and Paraguay remained suspicious of Brazil's engineering plans near the disputed area.

52 "¿Es nuestro el salto del Guaíra?" *Tribuna Liberal*, a ten-part series published between July 31 and August 10, 1962, consulted at BNP-AS.

53 Juracy Magalhães, Minister of Foreign Affairs, "Discurso no Senado Federal," May 19, 1966, ASF.

54 Blanc, *Before the Flood*, 25.

55 Enzo Debernardi, *Apuntas para la historia política del Itaipu* (Asunción: Editora Gráfica, 1996), 12.

Desperate to find a solution before tensions turned to hostilities, in November 1966, President Castelo Branco transferred diplomat Mario Gibson Barboza from Europe to Paraguay. According to Gibson Barboza, immense tumult greeted him in Asunción, where Paraguayans were burning Brazilian flags in the streets. On his first night, he claimed that blaring car horns outside his window kept him from sleeping, and when morning came, he found that someone had vandalized the Brazilian embassy with the phrase "get out Brazilian invader." When he went out to a restaurant or a store and spoke with a Brazilian accent, people refused to serve him. The degree of public resentment toward Brazil that he witnessed convinced him that Paraguay's authoritarian president Stroessner would not capitulate. He understood that Paraguay's military government could not retract from its stance without serious political repercussions. He was convinced, however, that Brazil's interpretation was correct, and he, following his superiors, refused to concede.[56]

What was at stake for the Paraguayan military regime was its political legitimacy. Having denounced Brazil for its imperialist actions, it would be difficult for Stroessner to backtrack without losing the support of important political elites. Moreover, because at that time Paraguay lacked the resources to harness the power of the Paraná River, the possibility of a binational dam offered an otherwise unavailable opportunity for economic growth that could improve living conditions and bolster popular support for Stroessner's regime.[57] These pressures encouraged Stroessner to be stubborn, but he treaded carefully because he had previously worked hard to cultivate Brazil as a political ally and did not want to lose its support.

What was at stake for the Brazilian military regime was not just the Guaíra border, but the precedent that capitulation would set. Castelo Branco feared that if the government conceded on this dispute, it would open all its borders up to further litigation. However, the fledgling military regime was also anxious about being the international bully once again, which could discredit its rule and jeopardize its access to loans from multinational banks. Gibson Barboza reflected, "the impasse was total and insurmountable. Of small proportions, but with extremely unfavorable publicity for Brazil. In Europe, government and diplomatic circles were saying that Brazil, a big country with huge national territory,

[56] Mario Gibson Barboza, *Na diplomacia, o traço todo da vida* (Rio de Janeiro: Editoria Record, 1992), 85–86, 90.

[57] Debernardi, *Apuntas para la historia política del Itaipu*, 65.

was once again wanting to oppress Paraguay, a small country without riches or a coastline, and take a piece of their territory ... we were the oppressors, imperialists who wanted, once again, to oppress Paraguay."[58] The Brazilian military government was determined not to concede and wanted desperately to find an alternative solution to a problem that threatened to undermine its political legitimacy.

With Brazil's international reputation on the line, Gibson Barboza began promoting a novel solution: a big dam downstream that would flood the disputed area. Gibson Barboza's memoirs suggest the idea was his own, but it is just as likely that the idea originated among Brazilian engineers. He was in touch with Brazilian dam builders around that time, and during the buildup to and aftermath of the Ata de Iguaçu, engineers floated various possibilities for dams along the shared stretch of the river. No specific studies had been made, but the idea of a dam further downstream than Marcondes Ferraz's original scheme was already in circulation around the time that Gibson Barboza arrived in Asunción.[59]

He soon conferred with Eletrobras officials and then met with Sapena Pastor and presented the idea. Gibson Barboza suggested that a large dam downstream would provide such immense benefits that they would forget about their dispute. Moreover, this dam could flood the contested area and make the problem disappear altogether. He reflects in his memoirs: "what we had to do was search for a project of mutual interest for Brazil and Paraguay of such stature that it would generate benefits for both countries and transform the territorial dispute into something secondary, or better, erase it." He continued, "I thought, if we build a hydroelectric dam on the border ... the territory in dispute would be forcibly covered by the reservoir's water."[60] Sapena Pastor liked the idea, and both diplomats presented it to their superiors.

Both presidents liked the idea too and, in February 1967, established a joint technical commission – composed of Italian and North American

[58] Gibson Barboza, *Na diplomacia*, 87.

[59] Ibid., 90–91, and Debernardi, *Apuntas para la historia política del Itaipu*, 62–77.

[60] Gibson Barboza, *Na diplomacia*, 90–91. Gibson Barboza's memoirs may exaggerate the threat of war and his role in resolving it, but other sources affirm that by 1966, the border dispute had become riotous, and diplomats in both countries agreed that building a big dam was the ideal solution. Energy ministers, engineers, and senators debating the merits of Marcondes Ferraz's scheme in the 1970s – whose statements are discussed at length in the following section – all reflected bluntly on Itaipu's geopolitical origins. Gibson Barboza's claims are also corroborated by two former Itaipu directors, Debernardi and Cotrim, whose books are cited in preceding footnotes.

engineers – to study the possibility of building a big dam. In 1967, Brazil's foreign ministry reflected in a confidential report that the dam "should flood the entire disputed zone, and as such, would finally resolve this problem."[61]

Between February 1971 and October 1972, the Mixed Technical Commission studied over fifty schemes across ten dam sites. Among these schemes were a handful of variations on the original Marcondes Ferraz proposal, all of which the engineers deemed viable but economically inferior to other schemes with big reservoirs.[62]

The Commission's report presented detailed studies of the two schemes it thought were economically superior to all others. One option was to build two small dams, one near Foz de Iguaçu, at a site known as Itaipu, and another upstream near Porto Mendes, at a site called Santa Maria. Or they could build one gigantic dam at Itaipu.

The engineers favored the latter option because it was the most cost-effective. Moreover, one giant reservoir, rather than two small ones, would make it easier to regulate the river. The commissioners deemed the Itaipu High Dam the best choice among the options they studied.[63] Reflecting on the study results one year later, Antônio Dias Leite, then energy minister and one of the architects of the 1973 Itaipu treaty, celebrated the happy coincidence that the most acceptable choice politically was also the most advantageous in terms of energy production.[64]

The dam's colossal size also meant it would generate lots of work for Brazilian construction companies that would benefit from the contracts. Corruption was rampant in both the Brazilian and Paraguayan military governments (see Chapter 1), and lobbying from construction companies surely provided additional incentive to choose Itaipu.

Furthermore, both governments could tout such economic benefits and the grandness of Itaipu as evidence of their effectiveness. Dams have symbolic value in demonstrating that governments can both control the natural environment and improve the lives of their people. Itaipu offered more powerful symbolism than most dams: When finished, it would be the largest hydroelectric plant in the world with tremendous capacity to power economic growth. Itaipu became a symbol of Brazilian engineering

[61] Quoted in Blanc, *Before the Flood*, 21.

[62] Comissão Mista Técnica Brasileiro-Paraguaia, *Estudo do Rio Paraná: Relatório preliminar*, October 1972, Section III, BIB-AS.

[63] Ibid., S-6.

[64] Dias Leite, quoted in Alexandre Costa, "Discurso no Senado Federal," May 15, 1976, ASF; Jarbas Passarinho, "Discurso no Senado Federal," May 15, 1976, ASF.

prowess and international greatness that the dictatorship used to bolster its legitimacy.[65] Itaipu was created to solve a geopolitical problem, but the Brazilian dictatorship was happy that it delivered other benefits that dovetailed with its political ambitions.

The only factor that might have encouraged the government to consider alternatives was Itaipu's environmental footprint. Building Itaipu meant creating a 1,350 sq. km reservoir that would displace thousands of people, inundate the iconic Guaíra Falls, and unleash a series of ecological changes in the region. Such environmental consequences hardly troubled the military regime in the late 1960s. But buildup to the United Nations Conference on the Environment in Stockholm in 1972 had convinced the military government that it would need to make some environmental concessions to stay in the favor of international lending agencies such as the World Bank, which was then greening its projects in response to the rising tide of environmentalism (see Chapter 3).

The Brazilian military regime understood that in order to maintain access to World Bank funding, its projects would have to pass new environmental assessments.[66] Though the World Bank was not funding Itaipu, the dictatorship was borrowing heavily from the bank and was in the middle of an unprecedented economic boom. With the economy growing, the generals wanted desperately to keep the spigot of international lending running.

But geopolitical pressures outweighed those of the burgeoning environmental movement. Since neither Brazil nor Paraguay had strong environmental lobbies, Itaipu's architects thought it sufficient to contract an ecologist to carry out a hasty environmental impact assessment that sidestepped the question of site selection and outlined some mitigation measures.[67] Choosing Itaipu meant flooding Guaíra Falls, but both governments considered the loss an acceptable sacrifice given all the benefits that the large dam would bring.[68] On April 26, 1973, the Brazilian and

[65] Maria de Fátima Bento Ribeiro, "Itaipu, a dança das águas: Histórias e memórias de 1966 a 1984" (PhD Diss., Universidade Estadual de Campinas, 2006), 9.

[66] "Anexo 4 – Filosofia ambiental do Banco Mundial," in Eletronorte Diretoria de Engenharia Usina Hidrelétrica de Tucuruí, "Diretrizes para os estudos dos problemas ambientais," July 1979, Document 02822, AELN.

[67] Comissão Mista Técnica Brasileiro-Paraguaia, "Estudo do Rio Paraná: Relatório especial no. 4. Reconhecimento dos efeitos ecológicos do projeto," April 1973, BIB-AS.

[68] Debernardi, *Apuntas para la historia política del Itaipu*, 23–24; Chefe do Gabinete da Casa Militar, "Estudo sucinto no. 37," May 11, 1973, AN-BSB.

Paraguayan military governments signed a treaty that created Itaipu Binacional and committed them to building the Itaipu High Dam.[69]

MARCONDES FERRAZ'S DEFENSE

Marcondes Ferraz had been following the project's evolution through short stories in local newspapers. What he was reading in April 1973 worried him: The government, it seemed, was about to choose Itaipu over his proposal and had yet to consult him. He was desperate for more information because the articles offered very little.[70]

The Mixed Technical Commission had carried out its studies in secret because both governments were worried that open debate about dam sites would give life to Marcondes Ferraz's proposal and reopen the geopolitical wound they were struggling to suture. In May 1973, Brazil's Chief of the Military responded to criticism about the manner in which the studies and negotiations were carried out. He wrote, "the advantages that could have been gained from widely spreading the studies for the Rio Paraná Dam ... seem to not compensate the risks of debates, where arguments could emerge that are contrary to the Brazilian position in the Rio de la Plata Basin, an extremely delicate issue in our foreign policy."[71] The military government had calculated the costs and benefits of publicity and chose secrecy. The Mixed Technical Commission had hired North American and Italian engineers to carry out the studies, and they had not even bothered to consult Brazil's own engineering experts, such as Marcondes Ferraz, who was left to follow the project's developments through the press.

Direct pressure from colleagues also prevented Marcondes Ferraz from promoting his alternative. In 1970, two Eletrobras directors – Amir Borges Fortes and Léo Amaral Penna – approached him and asked for his silence so as not to upset the delicate geopolitical situation.[72] Not wanting to cause trouble, he complied and remained silent.

However, when he learned that the government was about to make a decision without consulting him or the rest of Brazil's engineers, he changed his mind. There were multiple characteristics of his project that

69 The Brazilian Congress passed the treaty in less than a month without altering any of the original text. See Chaves, "Discurso no Senado Federal," September 6, 1977, ASF.

70 Marcondes Ferraz, "Comentários sobre a solução de Itaipu."

71 Chefe do Gabinete da Casa Militar, "Estudo sucinto no. 37."

72 Marcondes Ferraz, "Comentários sobre a solução de Itaipu."

he believed made it superior to Itaipu, and foremost among them was that it would preserve Guaíra Falls. In April 1976, he declared to an audience in the Club of Engineers in Rio de Janeiro: "At that time, the newspapers announced that the Itaipu solution was adopted, which among other inconveniences, would make the magnificent Sete Quedas disappear." He continued, "To permit the destruction of Sete Quedas greatly reduces the moral stance of the authorities in defense of ecology and nature … I confess that this fact scared and surprised me, because I never imagined that a project of this size would be adopted in such secrecy and without the proper studies."[73] The environmental impact of Itaipu alarmed Marcondes Ferraz, and as the government revealed more information about the project in the months preceding the April 26, 1973 Treaty, he decided he had to speak up.

Marcondes Ferraz first went to the President. In April 1973, a week before the treaty was signed, he met with President Emílio Médici, who heard him out but was unconvinced that his proposal was superior to Itaipu.[74] Over the following years, he wrote hundreds of letters to important politicians and engineers protesting Itaipu, including soon-to-be president Ernesto Geisel and then secretary of the environment, Paulo Nogueira Neto.[75] His plans resonated with at least one admiral, José Luiz Belart, a conservation-minded military man, but otherwise found little support within the military government and its associates.[76]

When construction began, he felt he was running out of time and decided he had to spread his argument to wider audiences. In April 1976, he went to the Club of Engineers in Rio de Janeiro to present his proposal and denounce the government for not having consulted him, other Brazilian engineers, or the public, before making its decision to build Itaipu and flood Guaíra Falls. His speech impressed Paulo Brossard, a Brazilian Democratic Movement [Movimento Democrático Brasileiro, MDB] senator and opposition leader from Rio Grande do Sul, who, along with Leite Chaves, an MDB congressman from Paraná, became staunch supporters of the engineer's plan. Both believed in restoring democracy, and, for them, Itaipu exemplified the shortcomings of dictatorship because it foreclosed debate to the detriment of the national

[73] Ibid.

[74] Marcondes Ferraz, "Observações sobre o projeto de aproveitamento no Rio Paraná," March 22, 1973, AN-BSB; Gibson Barboza, *Na diplomacia*, 96.

[75] Much of this correspondence can be found in Folders 4–6, Series 61.11.23, FOMF.

[76] Teresa Urban, *Missão (quase) impossível: Aventuras e desventuras do movimento ambientalista no Brasil* (São Paulo: Peirópolis, 2001), 34–36.

interest. Brossard brought Marcondes Ferraz's cause to the attention of the National Congress, which, after some debate, agreed to invite him to share his argument.[77]

Marcondes Ferraz argued that his scheme was superior to Itaipu in three respects. First, his dam would produce more energy for Brazil. At the time, Itaipu's installed capacity was projected to be around 12,000 megawatts, which would be split evenly between Brazil and Paraguay. His Guaíra Falls proposal, he contended, would have the capacity to produce about 10,000 megawatts, nearly as much as Itaipu, but Brazil would not have to divide this energy with Paraguay because the entire dam would be on Brazilian soil. Building its own dam also meant that Brazil had more administrative control than in a binational scheme. Secondly, his plan was cheaper. Itaipu entailed building a giant concrete dam over 150 meters tall that required lots of labor and materials. His project, in contrast, entailed a short 10-meter dam above the falls, a 60-kilometer canal, and an underground powerhouse, all of which he claimed would be much cheaper to build than Itaipu. He also accounted for construction time in this calculation. He believed that the rate of inflation – which was then about 34 percent – would continue to rise over the next decade and make any large infrastructure project increasingly expensive as the years progressed. He argued that his dam could be built faster and thus avoid the long-term costs of inflation that would incur with Itaipu, which would take nearly a decade to build.[78] Lastly, his project would have a comparatively benign environmental footprint. The dam he proposed had a much smaller reservoir. Its main purpose was to divert water into a canal, and it would flood neither Indigenous land nor

[77] During the dictatorship, there were two dominant political parties, the National Renewal Alliance [Aliança Renovadora Nacional, ARENA] and the MDB. Both were right-wing, though the latter was considered the opposition party (see Chapter 1). Almost all of Marcondes Ferraz's sympathizers in Congress – such as Lazaro Barbosa, Amaral Peixoto, and Benjamin Farah – were from the MDB, though at least one ARENA congressman from Paraná, Túlio Vargas, expressed interest in building a scheme that would spare Guaíra Falls. Many of the sympathetic MDB politicians changed their mind after June 2, 1976, when Itaipu director Costa Cavalcanti gave them a tour of Itaipu. See Arnon de Mello, "Discurso no Senado Federal," June 10, 1976, ASF.

[78] Inflation rates mushroomed during the 1970s and 1980s. In 1970, the annual rate of inflation was 16.4 percent, and it rose to 33.9 percent in 1975. It shot up to 110 percent in 1980 and 235.1 percent in 1985. It continued to rise thereafter, reaching an apogee of 2708.2 percent in 1993, before falling to 14.8 percent in 1995. Skidmore, *Brazil*, 172, 214.

large areas of farmland. Most importantly for Marcondes Ferraz, it would spare Guaíra Falls.[79]

The senators who supported Itaipu paid their respects to Marcondes Ferraz but were not persuaded by his arguments. Some touted the technical benefits that Itaipu would bring. They argued that Brazil was desperate for affordable energy – even more so after the 1973 oil crisis – and that if geopolitical exigencies required Brazil to split either dam's output evenly, Itaipu would promise more electricity for Brazil than Marcondes Ferraz's scheme.[80] These opponents also highlighted that the group of Italian and North American engineers had studied Marcondes Ferraz's scheme as part of their survey and favored Itaipu.[81]

But more important than technical considerations to these senators was that Marcondes Ferraz's project was unacceptable politically. In the debate about whether to invite Marcondes Ferraz to speak, Senator Alexandre Costa referenced earlier testimony from energy minister Dias Leite, one of the three Brazilian architects of the 1973 Itaipu treaty. In this speech, Dias Leite had remarked, "there was an alternative in which the dam would remain in Brazilian territory. This, however, was politically unviable, because of Paraguay's total refusal. So, the alternative was not even considered."[82] Most senators agreed with Dias Leite and stated bluntly that the disappearance of Guaíra Falls was lamentable, but that the political benefits of Itaipu were simply more important.[83]

Marcondes Ferraz and his supporters defended their position. He clarified that he never intended for his project to prevent Paraguay from tapping the river's energetic potential and argued that there were other ways in which Brazil and Paraguay could cooperate without flooding Guaíra Falls. He pointed to examples from other parts of the world where two countries built dams along a shared border without resorting to warfare or giant dams. The Canadian and United States governments both built their own hydropower plants along the Niagara

79 Marcondes Ferraz, "Discurso no Senado Federal," September 6, 1977, ASF.

80 Alexandre Costa, Roberto Saturnino, Herbert Levy, Virgílio Távora, and Dirceu Cardoso, "Discursos no Senado Federal," September 6, 1977, ASF.

81 Cotrim, "Itaipu-Sete Quedas: Projeto Marcondes Ferraz," March 25, 1994, Folder 810, FJC.

82 Dias Leite, quoted in Costa, "Discurso no Senado Federal" (May 15, 1976). The other two treaty architects were ambassador Gibson Barboza and President Médici. See Leão, "Discurso no Senado Federal," September 4, 1972, ASF, and Gibson Barboza, *Na diplomacia*, 95.

83 Saturnino, "Discurso no Senado Federal," September 6, 1977, ASF.

River, and Spain and Portugal shared dams along the Douro River. He argued that the Paraguayan and Brazilian governments could cooperate on building similar projects along their shared border that would spare Guaíra Falls.[84]

These arguments resonated with a handful of sympathetic congressmen who argued that no other country in the world would flood a natural wonder as grand as Guaíra Falls.[85] These senators were influenced by the rise of international environmentalism in the 1960s and 1970s, and they were right to suggest that other countries might have tried harder to spare a natural wonder as monumental as Guaíra Falls. Government-built dams have submerged many scenic landscapes, but some of the world's most treasured natural sites have been spared from floodwaters, such as Niagara Falls and the Grand Canyon in North America (see Chapter 5). Paulo Brossard pleaded with his colleagues to take up Marcondes Ferraz's proposal: "I ask God, Mr. President, that our current generation does not come to be remembered by history as the generation that ... erased from the universe the wonder of Sete Quedas and created grave problems for future generations."[86] Itaipu would indeed flood the famous waterfalls and cause grave environmental problems for future generations, but these considerations were not priorities for most of the senators who considered, and discounted, Marcondes Ferraz's alternative.

That Marcondes Ferraz made it as far as he did was due to the fact that he was an engineering legend in Brazil. From its beginning, the Brazilian dictatorship harshly censored any criticism of Itaipu. Skeptical senators such as Leite Chaves were openly ridiculed and encouraged to keep quiet. Those who had no political or engineering clout suffered graver consequences. In June 1982, Brazilian journalist Juvêncio Mazzarollo was jailed for criticizing Itaipu and remained incarcerated for nearly two years.[87]

Marcondes Ferraz's engineering prowess may have earned him an audience with Congress, but his reputation alone was not enough to convince the government to abandon Itaipu. The 1966 Ata de Iguaçu had obliged both countries to develop their shared stretch of the Paraná River together, and there were few in the Ministry of Foreign Relations or the Ministry of Mines and Energy that shared Marcondes Ferraz's optimism

84 Marcondes Ferraz, "Discurso no Senado Federal."

85 Chaves, "Discurso no Senado Federal," October 10, 1982, ASF.

86 Brossard, "Discurso no Senado Federal."

87 See Jacob Blanc, "The Last Political Prisoner: Juvêncio Mazzarollo and the Twilight of Brazil's Dictatorship," *Luso-Brazilian Review* 53, no. 1 (2016): 153–178.

that his plan could be developed with Paraguayan approval. Ambassador Gibson Barboza, one of the 1973 treaty's architects, recounted: "I judged it [Marcondes Ferraz's project], above all, suicide. There was not an international court that would not have sided with Paraguay in a case like that ... if a small border problem had already almost brought us to war, imagine if we would have went ahead and diverted the Paraná River to build a hydroelectric dam in our territory."[88] Dias Leite, energy minister and fellow treaty architect, agreed with Gibson Barboza. He reflected, "I believe, with complete frankness, that we did not have many options, since any hypothesis that involved a canal was not approved by the Republic of Paraguay. So, there was no discussion."[89] Marcondes Ferraz's proposal had some merits, but it was geopolitically unfeasible.

This conclusion was the principal refrain among all in Congress who supported Itaipu. Senator Arnon de Mello admitted that he judged Marcondes Ferraz's scheme to be superior in technical terms, and clarified that the choice to build Itaipu was primarily political.[90] Senator Luiz Viana remarked, "politically, there was no way for Brazil to build the eminent engineer's dam ... because if Brazil wanted to force the construction of a dam completely within its own territory, we would have naturally encountered extreme difficulty and we cannot say to what point that would have taken us."[91] The Mixed Technical Commission had discarded Marcondes Ferraz's proposal because of its geopolitical shortcomings, and the Senate was doing the same.

Furthermore, Itaipu's supporters argued that the country already had a large waterfall in the region that they could showcase as a natural wonder. About 30 kilometers southeast of Itaipu, the binational stretch of the Iguaçu River (on the Brazil–Argentina border) creates a giant waterfall before reaching the Paraná. In the 1930s, both Brazil and Argentina created national parks at their respective sides of the iconic Iguaçu Falls and encouraged tourism there.[92] Further promoting tourism at Iguaçu Falls was an excellent strategy for softening the opportunity cost of flooding Guaíra Falls. Dias Leite pointed this out, remarking that "next to Guaíra Falls we have Iguaçu Falls, which in contrast to Guaíra Falls, have perhaps an even greater value in terms of tourism if preserved

88 Gibson Barboza, *Na diplomacia*, 95.

89 Dias Leite, quoted in Costa, "Discurso no Senado Federal" (May 15, 1976).

90 Mello, "Discurso no Senado Federal."

91 Luiz Viana, "Discurso no Senado Federal," May 15, 1976, ASF.

92 See Frederico Freitas, *Nationalizing Nature: Iguazu Falls and National Parks at the Brazil-Argentina Border* (New York: Cambridge University Press, 2021).

as a natural park, rather than exploited for their energetic value."[93] Other Itaipu proponents made the same argument: Since Western Paraná already had a large waterfall, there was little need for a second.[94]

The decision had been made in April 1973, and it was only a matter of time before Itaipu's reservoir flooded the falls. Environmentalism was not a priority of either dictatorship, and the political and economic benefits of Itaipu simply overshadowed any concern for the dam's environmental footprint. Marcondes Ferraz's protests fell on deaf ears, and Itaipu's construction proceeded apace.

INDUSTRIALIZING THE RAINFOREST

Sobradinho and Itaipu were not the only large dams that Congress was discussing in the 1970s. On August 8, 1975, Congressman Gabriel Hermes gave a speech in the chamber thanking the government for making a financial commitment to building Tucuruí, a giant dam on the Tocantins River whose reservoir would flood 2,850 sq. km of the Amazon Rainforest. Just three days prior, Minister of Mines and Energy Shingeaki Ueki had approved funds for detailed studies and initial construction of the dam, and Hermes, who had long championed the project, was excited to inform Congress about this monumental step. He opened his speech by going over the history of the government's involvement in the Amazon, which had been minimal in the decades preceding 1964, when the military came to power and rekindled interest in the region. Other senators interrupted his speech to register their agreement. The generals, they said, had rediscovered the rainforest and its economic potential after years of neglect.[95]

Indeed, they had. As with the Northeast, the Amazon fit the bill for the government's plan to increase population growth and economic development in relatively unpopulated and economically underdeveloped areas of the country in order to diminish the appeal of socialist policies designed to redistribute wealth or topple the government. Since the collapse of the once-booming rubber industry in the 1910s, the Amazon had become an economic backwater and the government had very little control in the

[93] Dias Leite, quoted in Costa, "Discurso no Senado Federal" (May 15, 1976).
[94] Urban, *Missão (quase) impossível*, 35.
[95] Gabriel Hermes, "Discurso na Câmara de Deputados," August 8, 1975, ACD.

region.[96] After seizing power, the military government immediately accelerated the Amazonian highway-building program that Kubitschek had started and began passing legislation to encourage industrial growth in the region.

Belém was one of the population centers that the government hoped to industrialize. Belém is a port city at the mouth of the Amazon River whose fortunes had risen and fallen with the late-nineteenth- and early-twentieth-century rubber boom. By 1964, it had fallen far behind the Southeast in terms of demographic and economic growth. Electricity consumption statistics for Belém during this period are not readily available, but the remarks of energy minister Dias Leite give an idea of just how slow industrial and urban growth was in Belém compared to Southeastern cities. In 1973, Dias Leite estimated that in 1980, the demand for energy in Belém would be a mere 230 megawatts.[97] In comparison, the Southeast's industrial belt was already consuming 2,200 megawatts annually in the late 1950s, and the government expected demand there to grow at an exponential rate in the coming decades.[98] Urban and industrial growth was even slower in the region surrounding Belém, where most people lived off the land or earned an income from extractive activities. The military government believed that it could turn Belém and its hinterland into an industrial center, and it became the starting point for its program to integrate the Amazon into the rest of the nation.

Senator Hermes was a dedicated industrialist. He was born in 1909, in the small town of Castanhal, which was about 65 kilometers east of Belém. After finishing school, he entered the world of commerce and soon became a champion of industry. In 1950, he was elected president of the Industrial Federation of the State of Pará, a position he held for seventeen years, and in 1954, he was elected to the Chamber of Deputies as a representative of Pará.[99]

Hermes was above all else a technocrat, and he believed that with the appropriate application of science and technology, the Amazon could be

[96] For more on the rubber boom, see Barbara Weinstein, *The Amazonian Rubber Boom, 1850–1920* (Stanford: Stanford University Press, 1983), and Warren Dean, *Brazil and the Struggle for Rubber: A Study in Environmental History* (New York: Cambridge University Press, 1987).

[97] "Dias Leite to Médici," September 5, 1973, File BR RJANRIO ADJ.0.0.9, 269–286, FADL, AN-RDJ.

[98] World Bank, "Appraisal of Furnas Hydroelectric Project Brazil," September 9, 1958, 3, WBGA.

[99] FGV, "Biografia de Gabriel Hermes Filho," General Collection, CPDOC-FGV.

turned into a large industrial center that would raise the standard of living for everyone. He had long argued that large hydroelectric dams were the key to the region's industrial growth, a position his experiences abroad strengthened.[100] In 1952, Hermes traveled to the United States, and he spent most of his time touring the dams of the Tennessee Valley Authority, bragging afterward that he visited each of the thirty dams between the river valley's headwaters and its mouth.[101] Subsequent trips to Eastern Europe further reaffirmed his conviction that multipurpose dams could provide the Amazon with what it needed most: cheap energy and shipping routes.

Hermes believed that he had found the perfect place to build such a dam. About 300 kilometers southwest of Belém, there was a series of large waterfalls called Itaboca Falls, near a small town called Tucuruí. The falls were known for impeding navigation along the lower Tocantins and showed tremendous energetic potential. In 1966, Hermes wrote a book about the hydraulic resources of the lower Amazon basin in which he compared Itaboca to Guaíra Falls.[102] He believed that a dam there could resolve the problems of transportation at the same time it would electrify Belém and provide energy for an industrial center. He wrote: "Those who have utopian visions about the immediate approval of Sete Quedas are not wrong; even happier and more certain are those who believe in the Itaboca hydroelectric dam."[103] For Hermes, the rapids were the key to realizing his technocratic dreams.

Like Senator Novaes in the Northeast, Hermes had found an ally in the military government, which also had ambitious plans to industrialize the Amazon. Shortly before passing the legislation that funded the Energetic Studies of the Northeast, energy minister Costa Cavalcanti promulgated legislation for a similar set of studies for the Amazon basin, which Congress and President Costa e Silva signed into law on December 31, 1968.[104] The initial set of studies was wide-reaching and included surveys of dam sites at all the development centers throughout the Amazon. The initial studies were finished in 1972 and were inconclusive. The

100 Gabriel Hermes, *No País dos Dólares* (Belém, 1952).
101 Hermes, "Discurso na Câmara de Deputados," November 26, 1971, ACD.
102 Gabriel Hermes, *O Baixo Amazonas* (Rio de Janeiro: Editora Leitura, 1966).
103 Ibid., 94.
104 Costa e Silva, *Decreto No. 63.952, de 31 dezembro de 1968, Cria no Ministério das Minas e Energia o Comitê Coordenador dos Estudos Energéticos da Amazônia*, ASF.

commission struggled to find good dam sites. Some were simply too flat, while others had insufficient demand to warrant building a large dam.[105]

Such pessimism did not deter Dias Leite, who had taken over the position of Minister of Mines and Energy from Costa Cavalcanti in 1970. He began a new series of Amazonian energy studies in 1972, claiming that the priorities of the National Integration Program justified projects that would not have been considered feasible before.[106] The National Integration Program sought to encourage the economic integration of distant hinterlands through government-stimulated industrial growth, as well as the physical integration of the country through infrastructure, such as highways and transmission lines. Dias Leite believed that building a large dam in the Amazon could engender industrial growth there and that its transmission lines could also connect through Sobradinho to the electricity grid in the Northeast.

He also believed that the plan required a certain sense of urgency, and he was unwilling to wait until the commission finished its studies of the entire basin to proceed. He shared the generals' belief that Belém was an obvious starting point for Amazonian industrialization and Hermes' belief that Itaboca was the region's best dam site. In November 1972, Dias Leite attached a map to a letter he wrote to President Médici that depicted a large dam at the town of Tucuruí and transmission lines connecting it to the Northeast. This was the first time Dias Leite had identified the starting point for the government's energetic policies in the Amazon and the moment that Tucuruí moved from an abstract dream to a concrete reality.[107]

There was just one problem. Few industrial giants were interested in investing in the region without reliable supplies of energy, and Dias Leite had a hard time convincing his superiors to build large power plants without guarantees that industrialists would utilize their electricity.[108] By 1972, the situation had come to a complete standstill. Dias Leite himself admitted, "Amazônia is characterized, at the present moment, by the availability of large hydraulic potential, but demand for energy is scarce. The current market does not justify the construction of large hydroelectric dams." Yet, he was optimistic and believed that the government should take the first step in order to attract industrialists.

[105] "Dias Leite to Médici," November 23, 1972, Box 1, FADL, ACME.
[106] "Dias Leite to Médici" (September 5, 1973).
[107] Dias Leite, "Parecer sobre o estudo de caso de Tucuruí," no date, Box 1, FADL, ACME.
[108] Hermes, "Discurso na Câmara de Deputados" (August 8, 1975).

He continued, "However, there is the prospect of building them, still in this decade, together with large industrial plants, which cannot be installed without a previous guarantee of electric energy."[109] The National Integration Program compelled Dias Leite to try to create demand where there was none.

The 1973 oil crisis bolstered Dias Leite's position and made it easier for him to convince his superiors that it was a good idea to build a dam at Tucuruí. Its impact was twofold. First, it encouraged the federal government to divest from fossil fuels and to invest instead in alternative energy sources such as hydroelectric dams.[110] The original plans for Tucuruí estimated that it could produce almost 4,860 megawatts in its first stage and had the potential to double that amount if it were later upgraded.[111] It would be one of Brazil's largest dams, second only to Itaipu, which was then under construction.

Second, the oil crisis encouraged aluminum producers to consider building refineries in the Amazon.[112] The aluminum industry is one of the most energy-intensive industries in the world. The average total amount of electricity consumed in primary aluminum production is around 15 MWh per ton.[113] Eletronorte, which the government created in 1973 and immediately put in charge of the Tucuruí project, expected Amazonian aluminum production to use about 19 MWh per ton of metallic aluminum.[114] The Amazon had significant reserves of bauxite – the base material of aluminum – but companies had only just started to map and mine the region in the early 1970s.[115] The rainforest's two

[109] "Dias Leite to Médici," September 11, 1973, File BR RJANRIO ADJ.0.0.9, 269–286, FADL, AN-RDJ.

[110] Cotrim, "Brazilian Energy Policy 1973/84," June 23, 1986, Folder 122, FJC.

[111] Eletronorte, *Aproveitamento Hidrelétrico de Tucuruí: Estudos de viabilidade, Vol. 1 – Texto*, December 1974, File 03613_B, AELN.

[112] Dias Leite, "Parecer sobre o estudo de caso de Tucuruí."

[113] Glenn Switkes, *Foiling the Aluminum Industry: A Toolkit for Communities, Activists, Consumers, and Workers* (Berkeley: International Rivers Network, 2005), 19. The Australian government's Commonwealth Scientific and Industrial Research Organization estimates that the total energy used to produce aluminum is 211 gigajoules (about 60 MWh) per metric ton compared to 22.7 gigajoules (about 6 MWh) per metric ton for the steel industry. See Geoffrey Brooks, "The Trouble with Aluminum," *The Conversation*, May 24, 2012.

[114] Eletronorte, *Aproveitamento Hidrelétrico de Tucuruí.*

[115] For histories of the aluminum industry, see Robin S. Gendron et al., eds., *Aluminum Ore: The Political Economy of the Global Bauxite Industry* (Vancouver: University of British Columbia Press, 2013), and Mimi Sheller, *Aluminum Dreams: The Making of Light Modernity* (Cambridge: MIT Press, 2014).

largest deposits were located near Porto Trombetas and Paragominas, both in the state of Pará. Under the prevailing economic conditions before the oil crisis hit, these deposits could be mined and moved cheaply along the Amazon River, in the case of the Porto Trombetas mines, and along highways or railways, in the case of the Paragominas mines, to Belém, where ore would be shipped to refineries elsewhere in the world.

The oil crisis made shipping bauxite long distances on petroleum-fueled tankers prohibitively expensive, which encouraged aluminum companies to consider building refineries near the mines. It also made refining bauxite in fossil fuel-powered plants prohibitively expensive and thus further incentivized companies to continue a long-held tradition of working with local governments to build large hydroelectric dams to power their refineries.[116]

In November 1973, just weeks after the oil crisis hit, a group of Japanese industrialists reached out to Dias Leite about building an aluminum refinery and a large power plant in the Amazon. Dias Leite met with them in Tokyo and established a tentative agreement to build a large hydroelectric dam that would ensure a reliable power supply to aluminum refineries that Japanese companies would build.[117] The Japanese industrial group sent two technicians to work with Eletronorte as it carried out its studies. When the studies concluded in 1975, the Japanese industrialists decided the venture would not be profitable and terminated their agreement with the Brazilian government. Nevertheless, experience had convinced Dias Leite and his successor Shingeaki Ueki that there would be interest in building aluminum refineries near the bauxite deposits if there was a dependable supply of electricity readily available.[118]

Dias Leite was so sure that aluminum companies could be enticed to build their refineries near the region's bauxite mines that he even suggested building Tucuruí before the entire Tocantins River was surveyed. In 1973, he wrote: "in order for the dam that began as part of the National Integration Program to help establish a great energy center in the region ... it will be of fundamental importance ... to start the engineering projects now, even before the conclusion of the basin-wide

116 Some examples include a handful of dams built in the United States and Canada during the 1930s and 1940s; Krasnoyarsk Dam, Russia, built between 1956 and 1972; Afobaka Dam, Suriname, built between 1961 and 1965; Akosombo Dam, Ghana, built between 1961 and 1965; Manapouri Dam, New Zealand, built between 1964 and 1971; and Búrfell Dam, Iceland, built between 1966 and 1969.

117 "Toshiwo Doko to Dias Leite," November 14, 1973, Box 1, FADL, ACME.

118 Dias Leite, "Parecer sobre o estudo de caso de Tucuruí."

studies."[119] Such haste might seem out of character for a technocratic regime that was otherwise concerned with applying rational scientific principles to the development of the country's natural resources. But political pressures trumped patient scientific planning. Dias Leite felt tremendous pressure to meet the goals of the National Integration Program, and his priority was to build a huge power plant quickly so that he could entice companies to build refineries and factories in the area. The potential for demand alone was sufficient justification to plan and build a large dam. His successor Ueki agreed, and in August 1975, he signed an agreement that committed the government to building the Tucuruí Dam.[120]

To be sure, Dias Leite and Ueki were not planning to build a dam that was uneconomic. Itaboca Falls was the region's most propitious dam site, and enough studies had been made to know that building a dam there was feasible. But unlike the São Francisco or Paraná rivers, there would be little discussions about competing dam sites, the logic of building one dam before another, or the social and environmental impacts of alternative sites. Over the preceding decades, engineers had produced a wealth of surveys and studies about the São Francisco and the Paraná rivers, which provided ample knowledge about both basins and their potential dam sites. The Tocantins Basin, in contrast, had fewer surveys, and Dias Leite and Ueki were not going to wait for engineers to produce these time-consuming studies before making their decision.[121]

Therefore, there were no political or environmental factors that limited the size of the reservoir. The rainforest that the reservoir would flood was inhabited by Indigenous communities, small farmers, loggers, and other resource extractors, but the government did not care about flooding their land. Neither the Indigenous communities nor the small farmers had a large lobby, so the government was not worried about protests as it was at Itaparica, where powerful agribusiness and industrial interests had installed infrastructure in the area to be flooded. In contrast, those living near Tucuruí had not built any infrastructure that the government believed was worth sparing. Moreover, although Hermes had compared Itaboca to Guaíra Falls, the Amazonian falls had attracted little attention,

[119] "Dias Leite to Médici" (September 5, 1973).
[120] Hermes, "Discurso na Câmara de Deputados" (August 8, 1975).
[121] The US Bureau of Reclamation and the US Agency for International Development (USAID) carried out the first hydraulic studies of the Tocantins River in 1964. Eletronorte, *Aproveitamento Hidrelétrico de Tucuruí*, 1.

and the government had never considered turning them into a national park. The falls likely meant something to the people who lived there, but they were not as dramatic as either Guaíra or Iguaçu Falls and were mostly unknown to domestic and international audiences.[122]

The only factor that might have limited reservoir size was mining deposits. Eletronorte's viability report noted the irony that mining was both the principal reason for building a large dam and the largest constraint on the size of the dam because the government would not readily flood rich mineral deposits. However, studies found no such deposits in the area, and the government approved Tucuruí.[123]

During the mid-1970s, the country's energy ministers also decided to revive old plans to build two smaller dams in the region. The first was Coaracy Nunes, on the Araguari River in Amapá, and the second was Curuá-Una, on the Curuá-Una River in Pará near the city of Santarem. Both are considered large dams in technical terms, but are much smaller than the dictatorship's showcase projects: Both have less than 80 megawatts of installed capacity and reservoirs that are less than 150 sq. km. Due to their comparatively small size, Eletronorte finished both before Tucuruí.

The energy crisis, which had raised interest in Amazonian dams, also encouraged the military government to accelerate its plans for damming the São Francisco River. In May 1974, it decided to install turbines at Sobradinho, which was originally designed as a storage dam.[124] Then, in 1975, the military regime decided to build Itaparica, despite its environmental impact.[125] This version of Itaparica was not as big as the one that engineers had proposed as an alternative to Sobradinho a few years prior, but it still had a large social and environmental footprint that included flooding Indigenous and agricultural land and displacing tens of thousands of people. The military government had planned to build Itaparica at some point, and the rising price of oil pushed it to build the dam immediately.

[122] Robert Goodland, *Environmental Assessment of the Tucuruí Hydroelectric Project, Rio Tocantins, Amazonia* (Brasília: Eletronorte, 1977), 61, General Collection, US-LOC.

[123] Eletronorte, *Aproveitamento Hidrelétrico de Tucuruí.* For more on the planning and construction of Tucuruí, see Nathalia Capellini, "Historiciser les barrages en Amazonie Brésilienne: Environnement, conflit et politique dans la planification et la construction de Tucuruí" (PhD Diss., Université Paris-Saclay, 2019).

[124] CHESF, *Relatório da Diretoria*, 1974, 3, BCHESF.

[125] CHESF, *Relatório da Diretoria*, 1975, 4, BCHESF.

Political considerations have shaped almost every dam-building program in the world in some manner, most often as an impetus for the initiative itself. In some cases, politics have matched or surpassed economic criteria in influencing which rivers and projects are developed first. For example, during the 1950s and 1960s, the Lebanese government's political priorities encouraged it to build the Qaraoun Dam on the upper stretches of the Litani River rather than a different dam on the river's lower stretches despite the fact that the latter site was superior both geologically and economically. The southern project was near the Israeli border, and the Lebanese government worried that building the dam might entice its bellicose neighbor to sequester the reservoir's water.[126]

But it is rare that political priorities intrude in debates about site selection or design once a particular project is chosen. In China during the 1950s and 1960s, political goals compelled the government to press ahead with a single high dam at Sanmenxia on the Yellow River despite the fact that engineers had warned that siltation would immediately render the dam obsolete and advised building smaller dams on tributaries instead.[127] But analogous stories are uncommon. Sobradinho and Itaipu are exceptional for the prominent role of political considerations in the site selection and design process.

The impact on future planning was also distinct. In the Litani and other similar cases, the possibility remained open to build the competing project. However, for both Sobradinho and Itaipu, the final choice foreclosed the competing option. Building a big reservoir at Sobradinho made the Itaparica scheme (at its original scale) obsolete, and Itaipu's reservoir inundated the area where Marcondes Ferraz had proposed to build his dam.

CONCLUSION

Between 1967 and 1975, the Brazilian military government planned some of the world's largest and most controversial hydroelectric dams. At every

[126] Sneddon, *Concrete Revolution*, 2015, 63–73. Other examples from the 1950s and 1960s include the debates surrounding dam construction in the Zambezi River basin in Southern Africa and the Fraser River basin in Western Canada. See Julia Tischler, *Light and Power for a Multiracial Nation: The Kariba Dam Scheme in the Central African Federation* (London: Palgrave Macmillan, 2013), and Matthew Evenden, *Fish versus Power: An Environmental History of the Fraser River* (New York: Cambridge University Press, 2004).

[127] Shapiro, *Mao's War against Nature*, 48–65.

step of the way, the government had choices. In most cases, there were alternative schemes, such as a series of small dams on tributaries or upstream diversion dams, that would not require large reservoirs. In understudied areas such as the Amazon, the Brazilian government could have waited to complete a thorough survey of possible dam sites and consider more environmentally benign options. It could have waited to study the full environmental impact of its dams before committing to large reservoirs, and it could have made its decision-making transparent so that those people affected by the dams and the rest of the public could voice their opinions.

Instead, it chose to build big dams quickly and without debate. The government did so because it believed that the political benefits of these large dams would far outweigh any backlash from any fledgling environmental movement. Murmurs of an environmental lobby did start to emerge as the military regime went about planning its big dams, but it ignored these protests and pressed ahead regardless.

3

Pharaonic Environmentalism

Mitigation Efforts during Planning and Reservoir Filling, 1970s–1980s

INTRODUCTION

Rupert Fothergill was getting better at wrestling rhinos. At the end of 1959, Fothergill and his team had been working tirelessly for almost a year saving as many animals as they could from the rising floodwaters of the Kariba Dam, which the Northern and Southern Rhodesian colonial governments built on the Zambezi River between 1955 and 1959. The dam was the largest in the world at the time, and it had a massive social and environmental footprint. Notably, it flooded the land of the people and animals that lived in the Kariba Gorge. The rescue team spent most of its time saving small and medium-sized animals, such as antelopes and monkeys, but they occasionally had to rescue big ones. Rhinos were the most troublesome, and all of Fothergill's efforts to save them had failed. But he was convinced that his newest scheme – trapping rhinos in foot snares and towing them to shore on rafts of oil drums – would have higher success rates.[1] A global team of reporters covered the rescue operations, making Fothergill and his team international stars.

Operation Noah, as the Kariba animal rescue came to be called, was unprecedented. The Northern and Southern Rhodesian Game Departments had been anticipating the impact that the rising floodwaters would have on the region's animals and had planned some preliminary measures to drive them away from the floodwaters and to capture some of

[1] Eric Robins and Ronald Legge, *Animal Dunkirk: The Story of Lake Kariba and "Operation Noah" the Greatest Animal Rescue Since the Ark* (London: Herbert Jenkins, 1959), 170–171.

those stranded. But the operation's evolution into a full-fledged animal rescue mission with international coverage was unexpected and largely thanks to the spontaneous efforts of many unnamed volunteers and a few high-profile rangers such as Fothergill, who devoted tremendous amounts of their energy and courage to saving animals.[2]

Those following the story abroad became enamored with the rescue efforts and clamored to help when called on. When reporters publicized that hemp ropes chaffed animals, sympathetic onlookers sent in heaps of nylon stockings; thousands arrived within the first twenty-four hours. Two British reporters who published an account of Operation Noah in 1959 reflected, "seldom has any appeal so captured the public imagination."[3]

The whole episode was rife with racism. The well-intentioned people sending stockings to help rescue animals had seemingly forgotten that people were affected by the dam too. The Kariba Dam had displaced tens of thousands of Gwembe Tonga who lived there. The Gwembe Tonga had both an economic and spiritual attachment to their land, were ill-informed about the project, and ill-prepared to move. The imperial government in Rhodesia merely wrote off their displacement as part of the march toward progress, and its plans met no resistance from the public. The Gwembe Tonga's plight generated very little attention in the international press at the time compared to the animal rescue missions.

The enterprise was also of dubious effectiveness. In such scenarios, most animals are able to flee floodwaters of their own accord, and many of the animals rescued have a hard time surviving in their new environments because they suffer trauma from capture and are released into environments where they have to compete with existing animals for food and shelter. Moreover, between 1957 and 1961, the Northern and Southern Rhodesian governments culled 750,000 animals in the area surrounding Kariba as part of a tsetse fly control operation, which was more than one hundred times the roughly 7,000 animals that Operation Noah saved.[4]

2 For more on Operation Noah, see Robins and Legge, *Animal Dunkirk*; Charles Lagus, *Operation Noah* (New York: Coward-McCann, 1960); JoAnn McGregor, *Crossing the Zambezi: The Politics of Landscape on a Central African Frontier* (Martlesham: James Curry, 2009), 117–120. For more on the Kariba Dam, see Frank Clements, *Kariba: The Struggle with the River God* (London, Methuen and Co., 1959); David Howarth, *The Shadow of the Dam* (London: Macmillan Press, 1961); Tischler, *Light and Power for a Multiracial Nation.*

3 Robins and Legge, *Animal Dunkirk*, 58.

4 McGregor, *Crossing the Zambezi*, 117–120.

But such considerations did not trouble sympathetic onlookers. As far as the international community was concerned, large dams could proceed apace regardless of their social and environmental costs, all the more so if there were courageous people willing to rescue animals trapped by rising floodwaters. Twenty years later, the Brazilian military government hoped that the international community would feel the same way about its large dams and animal rescue missions.

This chapter is about the Brazilian military regime's efforts to ameliorate the social and environmental impacts of its large dams. It covers the period between 1972 and 1986, when the rise of international environmentalism pressured Eletrobras and its subsidiaries to undertake environmental impact studies and design environmental mitigation programs. This series of studies produced copious information about nearly all the social and environmental impacts that these dams would generate, and mitigation programs paid lip service to addressing them. But these measures were far from effective environmental protection, and they failed to win over the international community.

This chapter argues that the environmentalism that the Brazilian military regime's energy sector practiced was selective and designed to protect power plant infrastructure and showcase environmental care without fundamentally altering project designs or slowing down construction. Eletrobras and its subsidiaries contracted ecologist Robert Goodland to study their dams, but did so only after already committing to particular dam sites. Furthermore, they gave him a very short window to undertake his studies, put pressure on him to censor his conclusions, and applied his findings selectively. The mitigation projects that the government implemented were those designed to protect or prolong the life of the power plants themselves, such as lakeside forest reserves intended to curb erosion and sedimentation, and those that would dramatically showcase its environmental consciousness, such as animal rescue missions. These measures were "pharaonic environmentalism": environmental protection designed to bolster the image of the military dictatorship as a regime that could build durable mega dams and simultaneously protect the environment.[5] The generals largely succeeded at the former but failed at the latter.

[5] As suggested by my colleague Gabriel Lopes, such measures might also be called "ambientalismo para inglês ver" ["environmentalism for the English to see"] in reference to a famous Brazilian saying popularized after the 1830s, when Brazil began ostensibly subscribing to English laws prohibiting the slave trade but widely ignoring them in practice.

ENVIRONMENT AND INTERNATIONAL LOANS

The Brazilian delegation to the 1972 Stockholm Conference on the Environment caused quite the uproar. The world had met in Stockholm to commit to reducing its environmental footprint, and the Brazilians had politely refused. It was not that they did not believe that pollution was regrettable; they had joined their colleagues in lamenting the destruction that industrialization had wrought. But the Brazilians argued that economic stagnation was worse than pollution, and they had no intention of slowing their industrial growth, which they maintained would rid their country of poverty. They contended that environmentalism was only relevant in industrial countries, and their priority was industrial growth, no matter its environmental cost. Indeed, some even remember the delegation as having gone as far as to celebrate pollution because it was a sign of progress.[6] The Brazilian position found some support among other industrializing countries such as India and Egypt, but most other participating countries – including Latin American neighbors Colombia, Chile, and Mexico – expressed an interest in both development and environmental protection.[7] Brazil's intransigent position alarmed most conference participants, and the Swedish media claimed that the Brazilians had outright sabotaged the conference.[8]

The leader of the Brazilian delegation to the Stockholm Conference was none other than General José Costa Cavalcanti, one of the important generals and energy ministers referred to as "modern pharaoh," in Chapter 1. Costa Cavalcanti was the Minister of Mines and Energy between 1967 and 1969, during which time he designed and oversaw legislation that created the regional energetic studies for the Northeast and Amazônia that resulted in Sobradinho and Tucuruí, respectively, as discussed in Chapter 2. In 1974, President Ernesto Geisel appointed Costa Cavalcanti to be the president of Itaipu Binacional, a position he held for over a decade, where he oversaw dam construction and the flooding of Guaíra Falls.

Costa Cavalcanti was anything but an environmentalist. He believed that environmental destruction was necessary for economic growth,

6 The World Bank Oral History Program (WBOHP), "Transcript of Interview with Robert Goodland," January 26, 2006, 36, WBGA. See also Duarte, "'Turn to Pollute'," 64–87.

7 Duarte, "'Turn to Pollute'," 72–75, 83.

8 Antonio Teixeira de Barros, "Brazil's Discourse on the Environment in the International Arena, 1972–1992," *Contexto Internacional* 39, no. 2 (May–August 2017): 421–442, 424.

which he prioritized above all else. His statement at the Stockholm Conference reflected his values and summarized the Brazilian delegation's position perfectly. He argued, "it was precisely economic growth that allowed industrial countries to make great progress in eliminating mass poverty, ignorance, and poor health, making it possible to prioritize environmental considerations ... a country that has not reached a satisfactory level in providing basic essentials is not in a condition to divert considerable resources to environmental protection."[9] Costa Cavalcanti agreed to abstain from the most predatory forms of environmental destruction but was otherwise uninterested in altering his plans for economic growth.[10]

The Brazilian military government held its ground at the Stockholm Conference, but it left knowing that its position had made it tremendously unpopular. The conference had made it clear that environmental protection was becoming a big part of the industrial world's agenda, even though – indeed because – these countries were themselves imperfect environmental stewards. Brazil understood it would need to soften its hardline position on industrial growth if it were to be in the good graces of the international community. There were many advantages to being regarded favorably by the global community, and none more immediate than continued access to loans from multinational lending agencies, such as the Inter-American Development Bank (IDB) and the International Bank for Reconstruction and Development, or World Bank.

The IDB lent a lot of money to the Brazilian power sector during the dictatorship. It helped finance Sobradinho and five other of the dictatorship's powerful dams identified in the Introduction and lent more than US$1 billion cumulatively to the power sector in the form of general and transmission line loans during the military regime.[11]

Nevertheless, of the two big banks, the World Bank was the more involved. It was founded in 1944 to help rebuild Europe after World War II, and in

[9] "Anexo C, Discurso do Ministro José Costa Cavalcanti, Chefe da Delegação do Brasil (Estocolmo, 6 de junho de 1972)," in Ministério do Interior, *Relatório da Delegação do Brasil à Conferência das Nações Unidas sobre Meio Ambiente Humano, Estocolmo, 1972, Vol. 2* (Brasília: Ministério do Interior, 1972), 13, accessed through the website of the Companhia Ambiental do Estado de São Paulo (CETESB). https://cetesb.sp.gov.br/proclima/wp-content/uploads/sites/36/2013/12/estocolmo_72_Volume_II.pdf (last accessed April 2023).

[10] For more on Costa Cavalcanti's disinterest in environmentalism, see WBOHP, "Transcript of Interview with Robert Goodland," 36.

[11] See https://www.iadb.org/en/projects-search?country=BR&sector=EN&status=&query= (last accessed April 2023).

the 1950s and 1960s, it shifted its resources toward big infrastructure projects in Latin American, Africa, and Asia.

In the lead-up to the 1972 Stockholm Conference, the World Bank was becoming increasingly concerned with the negative publicity surrounding the social and environmental impacts of some of the large projects it funded.[12] Growth in industrial countries during the preceding decades had put tremendous strain on the environment. By the early 1960s, pollution, suburban sprawl, and disappearing wilderness were becoming more and more visible in the industrial world and causing alarm. Specialists and groups of concerned citizens began publicizing the consequences of this environmental degradation and lobbying their governments to enact legislation to protect the environment. This was the birth of modern environmentalism, and it resulted in a flood of new laws and government agencies designed to ameliorate the environmental impact of industrial growth. Since pollution and other forms of degradation do not respect political borders, many countries understood that effective protection was only possible through international cooperation. In 1968, the United Nations Economic and Social Council passed a resolution to design a conference to address the issue of environmental protection globally and decided the following year that it would be held in Stockholm in 1972.[13]

In anticipation of this event, the World Bank began to invest in greening its projects. In 1970, the World Bank created the position of environmental advisor – the first multinational lending agency to do so – who was charged with assessing the social and environmental impacts of potential projects and making recommendations.[14]

World Bank president Robert McNamara doubled down at the conference itself. Ecologist Robert Goodland, whom the World Bank hired in 1978, reflected in a later interview, "well, at the 1972 Stockholm conference on the environment, Dr. Lee and President McNamara made a big splash. They promised much [...] Lee and McNamara publicly stated that all Bank projects were environmentally scrutinized early in the design phase such that no project with unacceptable environmental impacts

[12] Robert McNamara, "Address to the UN Economic and Social Council," November 13, 1970, WBGA.

[13] Ministério do Interior, *Relatório da Delegação do Brasil à Conferência das Nações Unidas sobre Meio Ambiente Humano, Estocolmo, 1972, Vol. 1* (Brasília: Ministério do Interior, 1972), 4, accessed through the CETESB website.

[14] McNamara, "Address to the UN Economic and Social Council."

would go ahead."[15] Lee and McNamara's grand claims were an effort to improve the World Bank's image, which had become tarnished because the bank was then funding large engineering projects without any environmental protection measures at all.

Three years after the Stockholm Conference, the World Bank expanded the position of environmental advisor into the Office of Environmental Affairs.[16] In its early years, the Office had little practical impact because it was short staffed and underfunded, and much of the environmental impact studies were done on an ad hoc basis and met stiff resistance from superiors.[17] Nevertheless, the World Bank's fledgling environmental consciousness meant that in the early 1970s it was at least starting to be more selective when it came to funding environmentally controversial projects. The Brazilian military government understood that in order to maintain its access to World Bank funding, its projects would have to pass these new environmental assessments.[18]

The dictatorship was then borrowing heavily from the World Bank to help finance an unprecedented boom that was later dubbed the Brazilian economic miracle.[19] The World Bank had been lending to Brazil since the late 1940s. It cut off funding during the left-leaning João Goulart administration (1961–1964), but resumed lending to Brazil on an even larger scale after the military coup on March 31, 1964. For the rest of the 1960s, World Bank lending to Brazil rose from zero to US$73 million a year, and by the mid-1970s, it reached roughly US$500 million a year.[20] Brazil was borrowing so much that it became one of the World Bank's biggest clients – the country was responsible for nearly 10 percent of the bank's accumulated lending worldwide through 1991. Brazil was then the World Bank's third-biggest borrower behind India and Indonesia in terms of the number of projects and third behind India and Mexico in terms of the total money spent.[21]

15 WBOHP, "Transcript of Interview with Robert Goodland," 2.

16 World Bank, "World Bank Approaches to the Environment in Brazil: A Review of Selected Projects, Report No. 10039, Vol. 1: Overview," April 30, 1992, 8, WBGA.

17 WBOHP, "Transcript of Interview with Robert Goodland," 5.

18 "Anexo 4 – Filosofia ambiental do Banco Mundial," in Eletronorte Diretoria de Engenharia Usina Hidrelétrica de Tucuruí, "Diretrizes para os estudos dos problemas ambientais," July 1979, Document 02822, AELN.

19 See Skidmore, *The Politics of Military Rule in Brazil, 1964–85*, 89–93, 138–144.

20 Bruce Rich, *Mortgaging the Earth: The World Bank, Environmental Impoverishment, and the Crisis of Development* (Boston: Beacon Press, 1994), 99.

21 World Bank, "World Bank Approaches to the Environment in Brazil, Vol. 1," 10.

This steady stream of money from the World Bank was part of a wave of lending from the United States – the largest shareholder in the Bank – and its allies to Latin America in an effort to fight communist influence in the region. Latin America was one of the battlegrounds of the Cold War, and after the Cuban Revolution of 1959, the United States and its allies invested lots of money in anti-communist regimes in the region to ensure that anything remotely resembling communism did not spread.

Much of this lending went to the Brazilian power sector. Of the first seventeen projects that the World Bank funded in Brazil, thirteen of them were to the power sector, starting in 1949, with a US$90 million loan to the Canadian-owned Brazilian Light and Power Company. When state companies came to control the electricity sector in the late 1950s and 1960s, the World Bank began lending directly to the Brazilian government for its power projects, most of which were hydropower dams and their associated transmission facilities. Between 1948 and 1990, the World Bank lent the Brazilian power sector more than US$2 billion.[22] The Brazilian military government wanted desperately to keep the door open to international loans so that it could continue to expand its electricity sector.

It saw rising environmentalism as a potential threat to its access to power sector loans. The military government had tried to argue at Stockholm that expanding the electricity grid was compatible with the conference's goals, as one of its major findings was the importance of electricity for socioeconomic growth.[23] The Brazilian government used electricity as a metric of socioeconomic well-being and argued that it was well behind the industrial world. During the 1970s, industrial countries in North America and Europe consumed between 3,300 and 10,000 kilowatt hours/inhabitant/year, whereas Brazil consumed only 771 kwh/inhabitant/year. There were also internal inequalities, as the Southeast consumed 1,313 kwh/inhabitant/year, whereas the North consumed only 289 kwh/inhabitant/year.[24] This argument, however, was not enough. The Brazilian delegation left the conference with the understanding that it could not get away with simply arguing that industrial growth was more important than environmental care. Much to its displeasure, it would have to make some allowances for environmental protection.

[22] World Bank, "Project Performance Audit Report, Brazil Power Sector Loan (Loan 2720-BR), Report No. 9267" December 31, 1990, 2, WBGA.

[23] Eletronorte, "UHE Tucuruí: Estudos e atividades de proteção ambiental realizados," undated, 3, Document 426_1, AELN.

[24] Ibid.

But the military regime was not about to give up its large dams. In response to protests in the chamber against Itaipu in 1972 (see Chapter 2), energy minister Antônio Dias Leite responded: "with relation to … the preservation of the falls, from the conservationist point of view, I want to repeat the Brazilian position at Helsinki [sic]: we are not rich enough to permit ourselves the luxury of preserving the environment at a high economic cost."[25] He contended that cheap energy was fundamental to economic development and that growth was simply more important than environmental protection. The hardline industrial position persisted long after Stockholm when it came to choosing sites for large hydroelectric dams, and the Brazilian government instead sought other ways in which to showcase its burgeoning environmentalism.

The key for the military dictatorship was to find ways to improve the environmental image of its big dams without fundamentally altering their designs. The first step was to assess the social and environmental impact of the dams it was planning, which would generate the basic information that it needed to plan environmental programs. In 1972, the Brazilian and Paraguayan military governments contracted ecologist Robert Goodland to study the environmental impact of their giant Itaipu Dam. The following year, the World Bank and IDB contracted Goodland to study Sobradinho, and in 1977, Eletronorte contracted Goodland to study Tucuruí. The reports did not end up having much impact on mitigating each project's environmental footprint, but they generated useful data and were an excellent tool to show the world that the military government was paying attention to the social and environmental consequences of its giant dams.[26]

GOODLAND'S STAMP OF APPROVAL

In 1973, the World Bank and IDB began the pre-appraisal process for loans to the Brazilian government for its large storage dam at Sobradinho. The loans included funding for more than just the one dam. It was to finance a project known as Paulo Afonso IV, which, in addition to Sobradinho, included a fourth installment to the downstream Paulo Afonso dams that required a canal and a small storage reservoir. The World Bank had financed the original Paulo Afonso Dam, and the IDB

25 Quoted in Alexandre Costa, "Discurso no Senado Federal," May 15, 1976, ASF.

26 I found Goodland's reports in three locations: the libraries of the respective Eletrobras subsidiaries that contracted him, the BCME, and the US-LOC.

had financed its two expansions during the 1950s and 1960s, as well as the smaller Moxotó Dam upstream. The Brazilian government approached both agencies in mid-1972, and they agreed to fund the project the following July-August.[27]

Sobradinho was not only the project's principal feature, but it was also the feature most likely to be environmentally controversial because its giant reservoir would displace tens of thousands of people. The World Bank and the IDB considered the project viable from an economic perspective, but, as environmentalism swelled, both banks became anxious that the dam's social and environmental impact might make the project unpopular. In response, they thought it prudent to conduct an environmental impact study.[28]

The World Bank and IDB chose Goodland because of his experience with Brazil. Goodland was born in Montreal, Canada, and studied biology as an undergraduate at McGill University.[29] He stayed at McGill to complete his master's degree and decided to study tropical ecology in Guyana because McGill had a research connection there. Despite having family in Guyana, he found living conditions there to be very difficult and was reluctant to choose it as a site for further research. He believed that he could address the same questions more easily in neighboring Brazil, and the Canadian government awarded him a scholarship for PhD research on semi-Amazonian ecosystems in the Cerrado – Central Brazil's tropical savanna – which he completed at McGill in 1969.[30] In 1972, Goodland began working as a consultant for the Cary Arboretum of the New York Botanical Gardens, a position he held until 1978, when the World Bank

[27] Both banks approved and signed their respective loans between December 1973 and June 1974. See World Bank, "Report No. 396a-BR, Appraisal of the Paulo Afonso IV Hydroelectric Power Project," May 15, 1974, and World Bank, "Report No. P-1414-BR, Report and Recommendation of the President to the Executive Directors on a Proposed Loan to Companhia Hidro Elétrica do São Francisco (CHESF) with the Guarantee of the Federative Republic of Brazil for the Paulo Afonso IV Hydroelectric Power Project," May 22, 1974, WBGA.

[28] Robert Goodland, *Sobradinho Hydroelectric Project Environmental Impact Reconnaissance* (Millbrook: Cary Arboretum of the New York Botanical Garden, September 1973), General Collections, US-LOC.

[29] The following profile of Goodland is largely drawn from a 2006 interview that the World Bank conducted as part of its oral history program. WBOHP, "Transcript of Interview with Robert Goodland," 1, 7. Biographical information on Goodland can also be found in his obituaries in the *Guardian* and *Washington Post* and in his memorial fund website, https://goodlandrobert.com/?page_id=2 (last accessed April 2023).

[30] Goodland, "An Ecological Study of Cerrado Vegetation of South-Central Brasil" (PhD Diss., McGill University, 1969).

hired him to work in its Office of Environmental Affairs. In 1974 – while still working at the Cary Arboretum – he returned to Brazil to teach tropical ecology and environmental assessment, first at the Universidade de Brasília and then at the National Institute for Amazonian Research [Instituto Nacional de Pesquisas da Amazônia, INPA] in Manaus.[31]

Goodland also had experience with dams. He had cut his teeth in 1971, when the World Bank hired him to carry out an environmental impact assessment of Brazil's São Simão Dam. The following year, the Brazilian-Paraguayan Mixed Commission hired him to carry out an environmental impact assessment of its Itaipu Dam,[32] and the Brazilian government hired him to do the same for Furnas' Itumbiara Dam. In 1973, the Frankfurt-based dam-building consortium Consorcio LAMI hired Goodland and Richard Pollard to carry out an environmental impact assessment of its Chixoy Dam in Guatemala, which they did over the course of ten days sometime between January and March.[33] Goodland's experience assessing dams in Central and South America marked the beginning of one of the most important careers in the history of large dams and international lending.[34]

Goodland arguably did more than any single person to highlight the environmental consequences of large dams and persuade international lending communities to adopt policies to soften their impacts. In 1978, Goodland left the Cary Arboretum to work for the World Bank, where he stayed until he retired in 2001. There he earned the reputation of being the World Bank's environmental conscience and was one of a handful of people responsible for its environmental safeguards. He began working in the understaffed Office of Environmental Affairs where he and two others sorted all of the World Bank's projects based on their social and

[31] See "Robert Goodland: Ecologist in Motion," *Newsletter of the Cary Arboretum of the New York Botanical Garden* 1, no. 3, September–October 1975; "Goodland in Brazil and Suriname," *Newsletter of the Cary Arboretum of the New York Botanical Garden* 3, no. 5, September–October 1977; and "Goodland to World Bank," *Newsletter of the Cary Arboretum of the New York Botanical Garden* 4, no. 1, January–February 1978. These Cary Arboretum newsletters are digitized and available on its website.

[32] Comissão Mista Técnica, "Estudo do Rio Paraná: Relatório especial no. 4.

[33] Robert Goodland and Richard Pollard, *Chixoy Development Project: Environmental Impact Reconnaissance* (Millbrook: Cary Arboretum of the New York Botanical Garden, 1974), General Collection, US-LOC.

[34] In addition to the studies discussed in-text, Goodland also conducted environmental impact assessments for El Salvador's Cerrón Grande Dam in 1973, Brazil's Salto Santiago Dam in 1974, and Brazil's Foz de Areia Dam in 1975.

environmental impact and flagged the ones they believed would be the most disastrous.

Goodland perceived almost immediately that ad hoc assessments were inefficient and that he could be more effective if he drafted comprehensive environmental policies that the Bank and its borrowers had to adhere to. The World Bank had no such policies then and was reluctant to adopt them. But Goodland lobbied persistently, and by the end of the decade, he and his team had convinced the World Bank to pass eight wide-reaching environmental policies. Goodland reflected, "I had to produce dozens of workshops, conferences, publications, brown-bag lunch colloquia ... and campaigns before the Bank would adopt each policy [and] ... as soon as the Bank adopted a policy, I had to offer a lot of training to sell the policy to busy colleagues who didn't particularly want to learn."[35] Goodland's efforts eased tremendously when Barber Conable (1986–1991) replaced Alden Clausen (1981–1986) as president of the World Bank, because Conable prioritized environmental safeguards and supported Goodland's policies. One of Goodland's eight policies specifically targeted hydropower dams, which were the World Bank's most controversial type of project and were then absorbing the biggest portion of the Bank's energy lending.[36]

Goodland's commitment to implement environmental safeguard policies for large dams came from his experiences. During the 1980s, he had participated in the evaluation process of India's controversial Narmada River dams, and in 1989, he personally witnessed people displaced by Indonesia's Kedung Ombo Dam climb onto their thatched roofs to avoid being swept away by the floodwaters.[37] Even more formative than his experiences in Asia were his earlier experiences in Latin America, where he carried out environmental impact studies for a series of controversial large dams such as Sobradinho, Itaipu, and Tucuruí.

Goodland's Sobradinho environmental impact assessment reaffirmed the World Bank and IDB's apprehensions about population displacement, but otherwise praised the project. Goodland wrote that resettlement would be a huge challenge and should therefore be awarded all the necessary resources. The rest of his report pointed out other potential

35 WBOHP, "Transcript of Interview with Robert Goodland," 4.

36 World Bank, "Operational Directive 4.00 Environmental Policy for Dam and Reservoir Projects," April 28, 1989. The full text of this policy can be found in Annex B1-B4 of World Bank, "Introduction to Environmental and Social Assessment Requirements and Procedures for World Bank-Financed Projects," January 1997, WBGA.

37 WBOHP, "Transcript of Interview with Robert Goodland," 6, 34.

environmental impacts of the dam, but he was optimistic that the government could address them, and he concluded that overall Sobradinho would be a boon to the region.[38]

Goodland's Itaipu environmental reconnaissance report had been similarly glowing. He noted that the dam would have some detrimental social and environmental impacts but believed that programs could be implemented to compensate for these losses. Increased fishing production through pisciculture could compensate for agricultural losses, and forest and biological reserves surrounding the reservoir would make up for the biodiversity losses. He lamented the flooding of Guaíra Falls, but believed that the reservoir's scenic beauty could compensate for its disappearance. Moreover, like Dias Leite, he believed that Iguaçu Falls softened the opportunity cost of inundating Guaíra Falls because Iguaçu was easier to access and more photogenic, as it was just one large fall rather than a sprawling series of falls. Goodland concluded: "considering the project's enormous benefit in terms of its energetic potential, it is probable that the total sum of its negative environmental consequences will come to be relatively meaningless. The inexistence of a large Indigenous population, the current scarcity of silt, fish, and water use along the course of the river, combine to make this project an ecologically attractive proposal."[39] Goodland decided that Itaipu's benefits far outweighed its environmental consequences; he reached the same conclusion about Sobradinho a few months later.

In 1977, the Brazilian government followed the World Bank and IDB's lead and hired Goodland to carry out the environmental impact assessment of Tucuruí, which would flood large swaths of Amazonian rainforest. Goodland's Tucuruí report was more critical than his Itaipu and Sobradinho assessments, but as with these earlier reports Goodland provided a favorable assessment of the dam and approved it with only minor changes. His major criticisms concerned timing. He noted that relocation of Indigenous communities would take time to organize and implement and that there would be grave consequences if the process was rushed. He also pointed out that large tracts of the future reservoir area should be deforested before being flooded and that this process necessitated patience. Assuming that Eletronorte could properly relocate Indigenous groups, deforest the reservoir area, and step up malaria control measures,

[38] Goodland, *Sobradinho Hydroelectric Project Environmental Impact Reconnaissance*, 20, 92.

[39] Comissão Mista Técnica, "Estudo do Rio Paraná: Relatório especial no. 4."

Goodland believed that Tucuruí could provide immense benefit to the region. In addition to its power output, he believed the dam's engineering features would make it a tourist attraction similar to the Grand Coulee and Hoover Dams in the United States.[40]

It is likely that Goodland was holding back in his assessment of these Brazilian dams. The Brazilian government had already committed to building large dams (see Chapter 2) and was not going to redo its plans based on anything that Goodland said and told him so. Dr. Dario Gomes, who was the director of Eletronorte when it chose to build Tucuruí, confessed – with remarkable candor – that "on that occasion I said to him: 'look Goodland I want a report that says we can build the dam. I want to know the problems that we will have, but I want to know how to build the dam and to ignore what might happen in relation to environmental problems. I am not interested in a report that says it is going to flood the rainforest and we cannot build the dam'."[41] The protagonists in the Itaipu and Sobradinho cases were less forthcoming in admitting their efforts to influence the outcome of Goodland's studies, but it is almost certain that the same barriers shaped Goodland's experience at Itaipu and Sobradinho. National security concerns at Itaipu made it even more likely that the government told him it was going to build the dam regardless and that it wanted an environmental assessment that highlighted the problems it would have, rather than assess the environmental merits of Itaipu versus other options. Indeed, Goodland's report – the research for which was carried out prior to the April 1973 treaty – admits that he was operating under the assumption that the Brazilian and Paraguayan governments would choose the Itaipu High Dam option.[42]

The World Bank and IDB had deemed the Paulo Afonso IV scheme viable on economic grounds and so also had an interest in encouraging a positive assessment of Sobradinho's environmental impact. In 1973, the World Bank was only just beginning to implement environmental impact studies, and it was not yet a major factor in project assessment. In 1980, Goodland, then working directly for the World Bank, was banned from working on a controversial Brazilian highway project in the Brazilian Amazon – Polonoroeste – for saying that the scheme should not be funded

[40] Goodland, *Environmental Assessment of the Tucuruí Hydroelectric Project*, 11–55, 61, 81–98.

[41] "Entrevista do Dr. Dário Gomes por Paula Y. Stroh, 27 de fevereiro de 1992," 5, in "UHE Tucuruí estudo de caso: Entrevistas," September 1992, BELN.

[42] Comissão Mista Técnica, "Estudo do Rio Paraná: Relatório especial no. 4."

because of its impact on Indigenous communities. Upper management removed the project's chief architect for suggesting that river transport was more practical and hired two anthropologists to work on the project only to fire them immediately.[43] This was the context in which Goodland wrote his assessment of Sobradinho, and in 1973 he was not yet working directly for the World Bank, which meant he had even less power to take a stand against one of its projects and hope for meaningful change as a result.

Goodland's reasons for writing such cheerful assessments are unclear. Some of his optimism might be accounted for by the fact that he had yet to witness the tragedies of poorly executed dam projects and genuinely believed that if constructed properly the dams would be a boon to regional growth. Indeed, building large dams during the age of environmentalism was relatively new, and it is possible that Goodland sincerely believed that companies and governments would do more than pay lip service to environmental protection. It is also conceivable that he offered optimistic assessments of these projects for personal gain. He may have simply wanted the money and knew he had to produce a positive review to get it.

In Goodland's own telling, the Guatemalan government's murderous removal of Indigenous communities in the Chixoy Dam's reservoir area in the mid-1980s was a watershed moment in his feelings toward big dams.[44] The Guatemalan military dictatorship built the dam between 1976 and 1983 with World Bank and IDB funding. Between 1980 and 1982, government officers brutally raped and murdered over 400 Maya Achi as part of their efforts to clear the area for the reservoir to fill.[45] In his environmental impact assessment, Goodland had given the dam a glowing report, though he had not weighed the dam's social impact because the company building the dam told him that would be the work of another team. He felt betrayed when he discovered that no one ever carried out the social impact study and that the dam had led to the Río Negro Massacres, as the killings came to be called. He claims that this was the low point in his career and that it changed his attitude toward big dams.

[43] WBOHP, "Transcript of Interview with Robert Goodland," 9, 16. For more on Polonoroeste, see Rich, *Mortgaging the Earth*, 26–29, 120–127.

[44] WBOHP, "Transcript of Interview with Robert Goodland," 3.

[45] See "Guatemala's Chixoy Dam: Where Development and Terror Intersect," *The Guardian*, December 10, 2012.

In hindsight, Goodland came to see that most of the dams he had assessed positively were in fact social and environmental disasters. In 2006, he reflected on his experience with Brazil's power sector: "all the people who were kicked out of their homes by reservoirs became poorer, so hydro was a big poverty-creation machine ... poor people forcibly displaced by reservoirs didn't get compensated. Often, they had to squat without a home in plastic tents for five years before they could get resettled adequately."[46] This critical perspective came long after he published his reports, and it was his experience with these dams that pushed him to implement more comprehensive regulations within the World Bank and to become a critic of mega dams.

Timing also limited how effective Goodland's reports could be. In each case, Goodland was contracted fairly late in the planning process. In each case, the projects had already been approved and the government had already decided on the dam site. At Tucuruí, construction had started before Goodland began his work. Moreover, he was given very little time to carry out his studies. Goodland spent a couple of months surveying Sobradinho and Itaipu and only had one month to carry out the work for his Tucuruí report. Each of the reports contains a disclaimer that because of their hurried nature, the studies cannot be considered comprehensive impact statements but rather overviews of the principal issues at each dam that the government should use as a base for more detailed studies.[47]

Despite such shortcomings, both the international lending agencies and the Brazilian dictatorship got what they wanted out of the reports. They could cite them as evidence that they were taking the appropriate steps for dealing with the social and environmental impacts of their dams without having to redo their plans or scrap the projects altogether.[48] But because the reports were commissioned after the dam sites had already been approved, were done hastily, and, most importantly, were subject to censorship, they had little practical impact on the projects that they were studying.

The reports, however, were not all for naught. One result was a flood of environmental information about the dams. The reports covered everything from social issues associated with families being expelled from dam

[46] WBOHP, "Transcript of Interview with Robert Goodland," 6.

[47] Comissão Mista Técnica, "Estudo do Rio Paraná: Relatório especial no. 4;" Goodland, *Sobradinho Hydroelectric Project Environmental Impact Reconnaissance*; Goodland, *Environmental Assessment of the Tucuruí Hydroelectric Project.*

[48] See João Paulo Aguiar, "Depoimento concedido no âmbito do projeto meio ambiente e o setor de energia elétrica (Recife-PE)," 2003, 11, BCME.

sites to epidemic disease and expected changes in fisheries. They provided a wealth of information that the Brazilian government selectively used to create a series of environmental programs designed to address some of the negative impacts associated with its dams and to further showcase its burgeoning environmental conscience.

SELF-DEFENSE AND ENVIRONMENTALISM

What unfolded in the aftermath of Goodland's environmental impact studies was not a series of robust and wide-reaching environmental protection programs. The environmentalism of Eletrobras was selective and prioritized measures that both extended the lifespan of its reservoirs and extravagantly showcased its environmental consciousness.

The military regime's first priority was to protect its dams against adverse ecological impacts. Itaipu's environmental planning is exemplary in this regard. In 1975, Itaipu released its basic plan for environmental conservation, which was ostensibly a wide-reaching plan to mitigate the dam's ecological impacts. The report's opening paragraphs, however, revealed that its true intentions were more discriminatory. Notably, the first item on the report's list of priorities was to identify and study the "environmental problems that directly influence the functioning of the power plant, including those provoked by the dam itself."[49] The report identified other pernicious impacts, such as the depletion of fish stocks as a result of changing water chemistry, but it was far from a comprehensive environmental protection plan, and, like Goodland's report, it did not mention the Indigenous communities that Itaipu would displace. Its first priority was to identify environmental impacts that might harm the dam itself and to undertake measures to mitigate this damage.

Itaipu's forest reserves are a great example. The company's basic environmental plan for the dam outlined a program to reforest a belt surrounding the reservoir for the purpose of protecting the reservoir from siltation.[50] Creating a forest reserve around the perimeter of the reservoir would withhold the land from farmers whose work would engender erosion, and tree roots would help stabilize soils. Eletrobras engineers were influenced by the Tennessee Valley Authority in the United States, where studies showed that reforesting the reservoir margins reduced

[49] Itaipu Binacional, "Hidrelétrica de Itaipu plano básico para conservação do meio-ambiente," 1975, 1, BIB-FO.
[50] Ibid., 15–17.

sedimentation by as much as 90 percent and provided homes for aquatic and terrestrial fauna.[51] Operação Gralha Azul, as the program came to be called, reforested roughly 630 sq. km around the perimeter of the Brazilian side of Itaipu's reservoir in a belt whose average width is 200 meters.[52] As discussed further in the following section, the Paraguayan side retained most of its forest cover, so reforestation was unnecessary there.

That Itaipu planned its reforestation program primarily to preserve the longevity of its reservoir has not stopped it from marketing the measure as evidence of a strong environmental commitment.[53] The forested belt did bring ecological benefits, such as providing habitat for animals. But it also had negative consequences, such as excluding Indigenous communities from land that they had traditionally used for subsistence (see Chapter 4). Forests created for self-defense may have had some positive unintentional impacts, but they were programs designed to increase the lifespan of the reservoir, and such programs alone do not constitute robust environmental protection.[54]

Eletronorte had the same priorities at Tucuruí. The first comprehensive environmental protection plan for Tucuruí highlighted a series of detrimental impacts that might result from flooding tropical forests,[55] and as construction proceeded, Eletronorte began a surfeit of further studies based on Goodland's recommendations.[56] These studies covered a wide

[51] Dr. Cirilo Eduardo de Maira Machado, "Efeitos do Meio Ambiente sobre Barragens e Reservatórios," *Revista Brasileira de Energia Elétrica* 32 (October–December 1975): 19–43, 20.

[52] Nilson Monteiro, *Itaipu, A Luz* (Curitiba: Itaipu Binacional, Assessoria de Comunicação Social, 1999), 112. For more on Operação Gralha Azul, see Centro de Pesquisas Florestais Setor de Ciências Agrárias na Universidade Federal do Paraná, "Inventário florestal da região influência da Represa de Itaipu – Subprojeto reflorestamento, 1978; Itaipu, "Conservacionismo: Projeto Gralha Azul, de reflorestamento ciliar, Projeto Mymba Kuera, de salvamento da fauna," Comitê Brasileiro de Grandes Barragens (CBGB), 1980, 1, BIB-FO.

[53] Monteiro, *Itaipu, A Luz*, 112.

[54] For more on Itaipu's environmental programs, see Nathalia Capellini and Carlos Gomez Florentin, "Hydroelectric Dams and the Rise of Environmentalism under Dictatorship in Brazil and Paraguay (1950–1990): The Case of Itaipu," in Stephen Brain and Viktor Pál, eds., *Environmentalism under Authoritarian Regimes: Myth, Propaganda, Reality* (New York: Routledge, 2018), 51–74.

[55] Eletronorte, "Hidrelétrica de Tucuruí, diretrizes para os estudos dos problemas ambientais," 1979, 5–11, AELN.

[56] For an annotated bibliography of Eletronorte's many Tucuruí environmental studies, see Eletronorte, "UHE Tucuruí estudos e atividades de proteção ambiental realizados," Document 0426, AELN. Many of these studies were carried out as partnerships with

range of environmental issues, but, similar to Itaipu, the programs and amendments that Eletronorte prioritized were those that would preserve the power plant.

Water quality studies are exemplary. Eletronorte engineers learned early on that flooding tropical rainforest would engender a series of reactions that would change the chemistry of the water in the reservoir. This process unleashed a host of consequences, such as the release of hydrogen sulfide, methane, and other gases, and the acidification of reservoir water that could corrode turbines.[57] Of these consequences, the one that concerned Eletronorte the most was acidic water because it threatened turbines and thus Tucuruí's hydroelectric output.

In February 1983, Eletronorte's acidification nightmare became a reality at Curuá-Una, a large dam in the Amazon that Eletronorte had finished six years earlier.[58] Engineers surveying the dam found that acidic water had severely corroded the turbines and formed large holes in them.[59] The damage was so grave that Eletronorte had to shut down the plant completely while they made costly repairs. Curuá-Una was small enough that the stakes were low, but engineers were determined to avoid repeating the catastrophe at Tucuruí.

Luckily for Eletronorte, there was an easy, albeit expensive, fix: installing stainless steel turbines that were more resistant to corrosion. Forestalling corrosion was likely also the motivation for felling 100 sq. km of forest near the power plant. Goodland's report had recommended clearing at least 85 percent of the future reservoir area to reduce gas emissions, but Eletronorte only agreed to fell 30 percent. It hired an inexperienced company that was only able to cut down 5 percent of the area to be inundated before going bankrupt.[60] Eletronorte was apathetic

other research institutions. Two of the most prolific sets of studies were those done with INPA and those done with the Engevix and Themag companies. Copies of the INPA studies can be found in its library in Manaus, and most of the Engevix-Themag studies can be found in Eletronorte's archive in Brasília. Eletronorte also created partnerships with government agencies to address specific issues, such as infectious disease.

57 Philip M. Fearnside, "Environmental Impacts of Brazil's Tucuruí Dam: Unlearned Lessons for Hydroelectric Development in Amazonia," *Environmental Management* 27, no. 3 (2001): 377–396, 380–385.

58 Though Curuá-Una Dam meets the technical criteria to be considered a large dam, it is not among the cohort of powerful and high-impact dams identified in the introduction. Its installed capacity is 40 megawatts, and its reservoir size is 72 sq. km.

59 Eletronorte, "Relatório de viagem UHE Curuá-Una, 23/02/83 DE-002/83," Document 02519, AELN. This episode is discussed in Fearnside, "Brazil's Balbina Dam," 408.

60 Fearnside, "Environmental Impacts of Brazil's Tucuruí Dam," 381.

toward clearing the reservoir to reduce gas emissions but acted with greater urgency when it came to deteriorating water quality in a stagnant bay near the dam, known as the Caraipé branch, which had especially sluggish water that threatened to produce some of the most acidic water in the reservoir. Eletronorte itself felled the 100 sq. km of forest nearest the dam and completed the task swiftly.[61]

Eletronorte had studied all the environmental consequences associated with deteriorating water quality as a result of flooding tropical forests, but acted with urgency and modified project designs only when issues arose that jeopardized infrastructure. As with the forested belt surrounding Itaipu's reservoir, Eletronorte's studies and mitigation measures brought some benefit to the region's environment, but the company's priorities greatly limited the effectiveness of its environmentalism.

The military government was able to market some of these measures, such as reforestation, as effective environmentalism, but it realized that these defensive measures alone were not enough to satisfy domestic and international critics. For this, it would need to dramatically showcase its commitment to environmental protection. The second part of the Brazilian government's pharaonic environmentalism was ostentation.

OSTENTATIOUS ENVIRONMENTALISM

There is no protection measure that better epitomizes the Brazilian military government's environmental strategy for its large dams than animal rescue missions. Between 1973 and 1989, Eletrobras and its subsidiaries carried out a series of animal rescue missions that they used as evidence of their burgeoning environmental consciousness. An Itaipu Binacional preparatory report for its animal rescue missions (discussed below) concluded that the upcoming operation would, along with reforestation efforts, "substantiate the ecological concern evidenced by the Binational Entity."[62] The operation's final report similarly concluded that the rescue missions were evidence that Itaipu was helping care for the environment.[63] Itaipu defended this position publicly; one Itaipu ecologist interviewed for a report in a prominent magazine claimed that the rescue

[61] Eletronorte, "UHE Tucuruí: Estudo de caso relatório preliminar consolidado," July 1992, 86, AELN.

[62] Itaipu, "Conservacionismo," 1.

[63] Itaipu, "Operação de resgate faunístico margem esquerda, relatório final," undated, BIB-FO.

missions would "shut up all those saying Itaipu is not caring for the environment."[64]

Animal rescue missions continue to feature prominently in the official histories, museum tours, and gift shops of these electricity companies.[65] As of October 2018, visitors at Itaipu could buy wood ornaments, coffee mugs, and other merchandise decorated with submerged treetops full of animals that the gift shop calls "trees of life." Such marketing is an attempt to rewrite the story of the reservoir's filling as something environmentally constructive.

But such an agreeable narrative is false. Rescue workers were well intentioned, and the missions probably provided a net benefit to the fauna around the reservoirs. However, it is a mistake to consider these episodes as effective environmentalism because they had little practical impact on minimizing the harm that reservoirs inflicted on animals. Rescue missions also did nothing to compensate for the immense losses of flora that reservoirs submerged, most of which was uncatalogued and potentially contained valuable endemic species, such as plants that provide genetic material for developing pharmaceutical drugs. In the mid-1990s, about 30–50 percent of pharmaceutical drugs were derived from plant material (and about 60 percent of anticancer agents), yet less than 1 percent of tropical plants had been screened for potential pharmaceutical use.[66]

The idea for animal rescue operations spread from Southeast Africa to South America in the 1960s. In 1958, the aluminum giant Alcoa signed an agreement with the Surinamese government to build a big dam called Afobaka to power nearby aluminum smelters. Construction finished in 1964, forming the massive Brokopondo Reservoir, which flooded roughly 1,500 sq. km of rainforest. It was the first dam to flood such a large area of tropical rainforest, and neighboring governments that were contemplating building big dams in similar settings sent research teams to study

[64] Hélio Teixeira and Pedro Franco, "Cerimônia de adeus," *Veja*, September, 22 1982, 74–75.

[65] Monteiro, *Itaipu, A Luz*, 106–118; Memória da Eletricidade, *O meio ambiente e o setor de energia elétrica brasileiro* (Rio de Janeiro: Memória da Eletricidade, 2009), 92–93, 211–215.

[66] See Robert Mendelsohn and Michael J. Balick, "The Value of Undiscovered Pharmaceuticals in Tropical Forests," *Economic Botany* 49, no. 2 (1995): 223–228; Geoffrey A. Cordell, "Biodiversity and Drug Discovery – A Symbiotic Relationship," *Phytochemistry* 55 (2000): 463–480, 465; and Emily Roberson, *Nature's Pharmacy, Our Treasure Chest: Why We Must Conserve Our Natural Heritage* (Tucson: Center for Biological Diversity, 2008), 1.

its environmental impacts.[67] Over the following decades, Brazilian scientists studied the environmental consequences of Afobaka in an effort to mitigate, and perhaps avoid altogether, similar problems at dams in the Brazilian Amazon.[68] In addition to observing the dam's deleterious environmental impacts, Brazilian engineers also encountered an encouraging method by which the Surinamese government was generating positive publicity around the reservoir's environmental footprint: animal rescue operations.[69]

Afobaka Dam was the site of the first large animal rescue mission in the Western Hemisphere. Operation Gwamba, as the Surinamese mission was called, differed from its Southeast African predecessor in that the operation was not spearheaded by local rangers, but rather the International Society for the Protection of Animals (ISPA). The flora and fauna in the flooded area were also different. But in most other respects Operation Gwamba resembled Kariba's Operation Noah. For almost two years, North American John Walsh led a team of local recruits that captured more than 10,000 animals stranded on islands and treetops.[70] The event generated tremendous publicity and international sympathy, benefits that the Brazilian military government hoped to gain from carrying out such operations at its dams.

Brazil began flirting with the idea of rescue missions in April 1973, when biologists from the Herpetology Division of the Butantan Institute took advantage of the rising floodwaters of the Ilha Solteira Dam to capture snakes.[71] This is the first reference to a collective effort to remove animals from rising floodwaters of reservoirs in Brazil, but it was a campaign to gather specimens, not a rescue mission. Brazil's first formal animal rescue operations took place the following year, at the Promissão Dam, where the Companhia Energética de São Paulo (CESP), the Fundação Parque Zoológico de São Paulo, and the Instituto Butantan

[67] See Goodland, *Environmental Assessment of the Tucuruí Hydroelectric Project*, 81–84.

[68] For example, see Eletronorte, "UHE Balbina viagem do chefe do projeto ao Suriname de 21-10-77 a 02-11-77, BAL-10B-0105-RE," Document 1885, AELN.

[69] Hélio Emerson Belluomini et al., "Wildlife Rescue, Capture of Snakes and the Establishment of Anti-Ophitic Stations in Flooded Areas Destined for Brazilian Hydroelectric Power Plants," *Memoria Institucional Butantan* 40/41 (1976): 129–154, 152. These Butantan Institute publications are digitized and available on its website.

[70] See John Walsh and Robert Gannon, *Time Is Short and the Water Rises* (New York: Tower Publications, 1967).

[71] Alphonse Richard Hoge et al., "Nota prévia, lista das espécies de serpentes coletadas na região da Usina Hidroelétrica de Ilha Solteira-Brasil," *Memoria Institucional Butantan* 38 (1974): 167–178.

coordinated an effort to save animals from small islands the reservoir created.[72] Between 1975 and 1980, Brazilian hydroelectric companies teamed up with interested scientific institutions in at least eight more animal rescue missions at newly completed dams (Figure 3.1).[73]

These early rescue missions were carried out on small reservoirs, and they did not generate much media attention and were not widely publicized by the electricity companies themselves. Two of the dams – Coaracy Nunes and Salto Osório – flooded no more than 55 sq. km. The largest of these reservoirs was Água Vermelha, which covered 650 sq. km, less than half the size of Itaipu's surface area and less than one-sixth of Sobradinho's reservoir. In at least three cases, farming and logging had diminished existing fauna so much that rescuers admitted there was not much to save when the floodwaters came. In other cases, the topography was such that reservoirs rose without leaving islands that could trap animals.[74]

These rescue operations may not have been as big a spectacle as Operation Gwamba, but that did not stop their organizers from boasting. Rescuers saved 3,263 animals at Água Vermelha, and zoologists noted that the effort was "once again" dubbed Operation Noah's Ark, suggesting that other early rescue missions also used the biblical reference to describe their work. Perhaps the most important result of these early operations was the experience and training that the military government gained, which helped it stage larger rescue missions at its big showcase projects. For example, four Itaipu technicians participated in the planning and execution of the Água Vermelha rescue and used that experience to plan one of the most ostentatious rescue operations yet.[75]

In October and November 1982, Itaipu Binacional carried out a large animal rescue mission named Operation Mymba-Kuera, which means "rescue animals" in Guarani. The idea to rescue fauna as the reservoir filled was in the 1975 Basic Environmental Plan, and preparation began in 1978. That year, Itaipu diverted the Paraná River and it had rescue crews standing by in case the flooding associated with this procedure stranded animals. It did not, and there was no need to execute the first

[72] Belluomini et al., "Wildlife Rescue," 129.

[73] See Belluomini and Mario Paulo Autuori, "Methodology Applied in the Elaboration of Faunal Salvage in the Region of 'Água Vermelha', Hydroelectric Power Plant, Centrais Energéticas de São Paulo-CESP," *Memoria Institucional Butantan* 46 (1982): 119–138.

[74] Belluomini et al., "Wildlife Rescue," 137–153.

[75] Belluomini and Autuori, "Methodology Applied in the Elaboration of Faunal Salvage," 126.

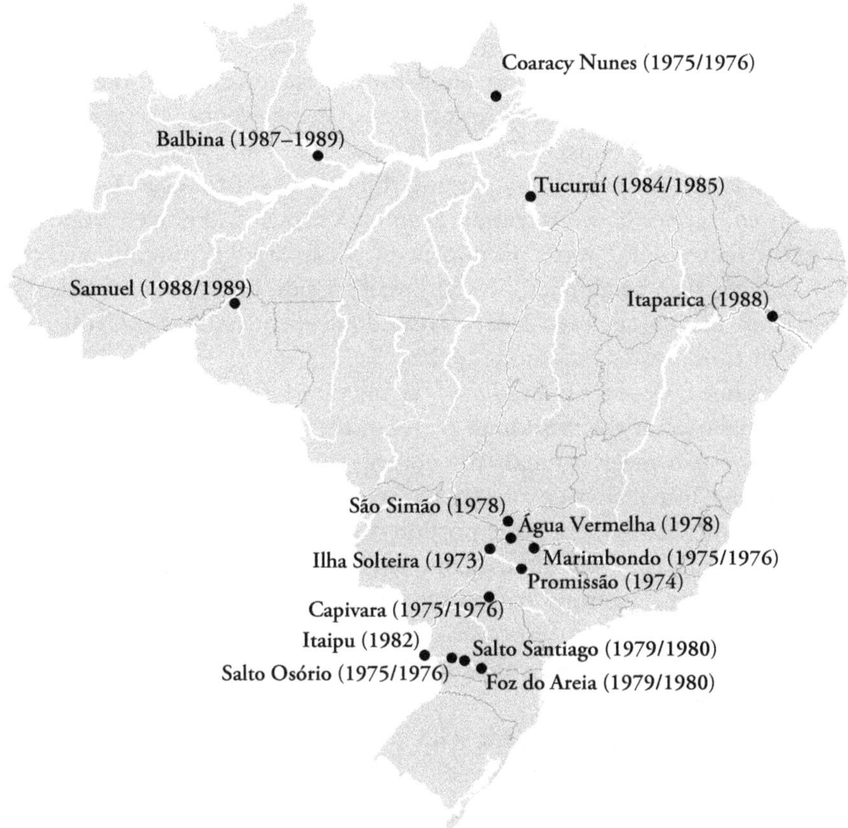

FIGURE 3.1. Animal rescue and specimen gathering campaigns, 1973–1989. Map by Geoffrey Wallace (G. Wallace Cartography & GIS).

phase of Mymba-Kuera. However, later that year, Itaipu did begin capturing animals in the future reservoir area near the dam site in anticipation of its filling, ramping up efforts starting in September 1982.[76]

The main operation began in earnest on October 13, 1982, when Itaipu closed the floodgates, and the floodwaters began rising. As Guaíra residents sat on the shore watching Guaíra Falls disappear, dozens of motorboats were speeding around the newly formed reservoir doing their best to help the animals stranded on treetops. In total,

[76] Itaipu, "Operação de resgate faunístico margem esquerda, relatório final," 2–5, and Itaipu, *Memoria de 1982*, 67, BIB-AS.

Brazilian and Paraguayan teams rescued nearly 34,000 animals from the floodwaters,[77] mostly snakes and monkeys.[78]

The work was grueling and dangerous. Rescue teams worked from sunrise until sunset capturing animals and spent evenings triaging and releasing them into refuges. Though the Brazilian and Paraguayan teams did not have to contend with charging rhinos and warthogs as Fothergill and his colleagues had, Operation Mymba-Kuera workers were exposed to many dangers, the most prominent of which were venomous snakes. This danger was heightened on the Brazilian side because grain fields – absent on the Paraguayan side – created excellent habitat for rodents, which in turn attract venomous snakes.[79]

That same difference in land use – farms versus forests – also created a great imbalance in the workload between the two teams. The Brazilian team had much fewer animals to capture, since farmers had deforested most of the Brazilian side of the reservoir prior to dam construction, displacing or destroying a large portion of animal habitat.[80] Paraguayan forests, in contrast, were almost completely intact, so rescue teams there had many more treetops to search and stranded animals to rescue. Paraguayan teams were responsible for 24,753 of the 34,000 animals captured, and they worked until late November, a month longer than the Brazilian teams.[81]

The operation was a big hit with the press. Over thirty-three news agencies covered the Brazilian side alone, some of which were international news agencies such as *Time, BBC News*, and the *Miami Herald*. A photographer for the United Nations took shots for a series on international wildlife protection. Reporters rode along with rescue teams where they experienced firsthand the arduous conditions under

[77] Itaipu, *Memoria de 1982*, 67. This figure does not include several thousand animals that Brazilian and Paraguayan teams captured in the future reservoir area before it began to fill.

[78] Roughly 60 percent of the animals captured on the Brazilian side were snakes. Paraguayan documents do not have the same data for comparison, but reports from excursions suggest that there too snakes were the most common animal captured. In addition to snakes and monkeys, rescuers also captured many lizards, hedgehogs, possums, coatis, and wild pigs. See "Palestra de zoólogo Hélio Martins Fontes Júnior, tema: fauna terrestre," in "1° seminário do ECOMUSEU de Itaipu sobre meio ambiente para professores de ciência e biologia," 1988, BIB-FO; "La difícil tarea de rescatar animales," *ABC Color*, October 22, 1982; "Rescataran 6 mil animales de aguas del río Paraná," *ABC Color*, October 27, 1982.

[79] Belluomini and Autuori, "Methodology Applied in the Elaboration of Faunal Salvage."

[80] Itaipu, "Projeto de resgate faunístico," undated, BIB-FO.

[81] Itaipu, *Memoria de 1982*, 67.

which rescuers labored to save animals.[82] These accounts earned much sympathy for the workers and Itaipu Binacional, which had coordinated the rescue efforts. Itaipu now had an environmental program to match the grandness of the dam itself. It was a South American Operation Noah, an environmental protection movement of biblical proportions. And it was comparatively cheap. Though Goodland had warned Eletronorte that rescue missions would be costly (he made no comment on rescue missions in his assessment of Itaipu), the final cost for the entire operation was roughly US$2 million, which was 0.0001 percent of the project's US$ 20 billion final cost.[83]

The spectacle was repeated on an even larger scale at Tucuruí. Between September 1984 and March 1985, more than 80 teams of workers rescued upwards of 280,000 animals from the treetops of the flooded Amazonian rainforest, which was nearly eightfold the number of animals captured at Itaipu.[84] Tucuruí's reservoir flooded a massive 2,850 sq. km of rainforest, which meant that there was much more wildlife there than at the grain fields that Itaipu flooded. An undated pre-operation promotional pamphlet boasted: "although similar operations have been carried out during the construction of other hydroelectric dams in Brazil and in other countries, nothing can compare to the conditions at Tucuruí, erected in a region hardly touched by civilization and therefore still occupied by flora and fauna in its near total abundance."[85]

82 Itaipu, "Operação de resgate faunístico margem esquerda, relatório final," 20. For more details, see Maria Emília Medeiros de Souza, "Mymba-Kuera," Undergraduate Thesis, Faculdade de Ciências Sociais Aplicadas de Foz de Iguaçu, 1987, 33, 41.

83 Final cost calculated by author. Itaipu's official documents do not list the total cost of the rescue operations, but newspapers reported that the Brazilian operation cost US$620,000 and that the Paraguayan operation cost 1.69 times as much because it required more resources. These figures suggest that the Paraguayan rescue mission cost a total of US$1,048,000 through the end of October, when the Brazilian rescue mission ended, or about US$21,000 per day during the fifty-one days between the September 11 start date and the end of October. Applying that daily rate to the twenty-four days in November the Paraguayan operation continued yields an additional US$504,000. See "Rescate de la fauna en el lado brasileño se capturaron 10,000 piezas," *ABC Color*, November 2, 1982; "El Rescate de la fauna es más importante en la margen paraguaya," *ABC Color*, October 22, 1982. For Goodland's remarks, see *Environmental Assessment of the Tucuruí Hydroelectric Project*, 1977, 76–77.

84 Of these, about 101,000 were reptiles, 100,000 were mammals, 67,000 were arachnids, and 4,000 were birds (numbers rounded). See Bento Melo Mascarenhas and Giuseppe Puorto, "Nonvolant mammals rescued at the Tucuruí Dam in the Brazilian Amazon," *Primate Conservation* 9 (1988): 91–93, and Cleber J.R. Alho, "Maneje com cuidado: frágil," *Ciência Hoje* 8, no. 46 (September 1988): 40–47.

85 Eletronorte, "Operação Curupira," undated, BELN.

Rescuers at Tucuruí captured a diverse cast of animals. Tortoises, sloths, iguanas, and monkeys were the most frequently caught animals excluding spiders. Rescue workers also found many armadillos, hedgehogs, and rodents.[86] Snakes abounded, but were not as prolific as at Itaipu. Nevertheless, like their counterparts at the binational dam, rescue teams worked under difficult conditions and had to contend with venomous and hostile animals. Working conditions were made even worse by the proliferation of mosquitoes as the reservoir slowly formed, which is discussed further in Chapter 5.[87]

Eletronorte used the display to market itself as an environmental defender. It named the effort Operation Curupira, which is the Tupi name for a mythical creature with backward feet that protects the animals of the rainforest by banging its club against trees to warn of coming dangers, such as storms and hunters.[88] Like Operation Noah, the title was attractive because of its mythological symbolism. A 1985 Eletronorte report concluded that Operation Curupira and Tucuruí's other environmental programs constituted "an ecological defense effort never before seen in the implementation of hydroelectric dams in any part of the country at any time."[89] It was the Brazilian military government's largest animal rescue mission and the height of its energy sector's ostentatious environmentalism.

These animal rescue missions produced more propaganda than practical benefit. The hydroelectric companies admitted that most animals fled as floodwaters rose and that their job was mostly to rescue tree-climbing species such as snakes, monkeys, and sloths.[90] Among the snakes and lizards captured at Itaipu, rescuers even noted that some could swim and had a good chance of saving themselves.[91] The dramatic loss of forest cover on the Brazilian side of Itaipu's reservoir had further diminished the number of animals there that needed saving.[92]

86 Alho, "Maneje com cuidado."

87 Eletronorte, "Reservatório da Usina Hidrelétrica de Tucuruí Operação Curupira relatório de viagem, TUC-10-26743-RE," September 18–October 5, 1984, Document 03492, AELN.

88 Eletronorte, "Operação Curupira."

89 Eletronorte, "Estudos realizados na área do reservatório da UHE Tucuruí, CIENT-KIT No. 03," March, 1985, Folder "Tucuruí," APF.

90 Eletronorte, "Operação Curupira"; "El Rescate de la fauna es más importante en la margen paraguaya," *ABC Color*, October 22, 1982.

91 See Itaipu, "Operação de resgate faunístico margem esquerda, relatório final," 13.

92 Itaipu, "Projeto de resgate faunístico."

Among the species that could not swim to safety, more animals were lost than rescued. One biologist at the Belém Primate Center claimed that Tucuruí's floodwaters killed most of the 100,000 primates estimated to have inhabited the reservoir area.[93] The true loss of fauna to reservoir floodwaters is unknowable because most animals were not cataloged, and it is possible that extinctions of endemic fauna occurred. Rescuers captured hundreds of thousands of animals that might have otherwise died, but most animals either escaped to uncertain fates or drowned.

Those animals that were saved often had a hard time staying alive in their new habitats. Rescuers released animals into areas that were often refuges in name only. In such tracts, human activity and hunting put a strain on animal populations. In 1988, concerned Eletronorte biologists wrote an internal report highlighting the dire state of Operation Curupira's four refuge areas. In just three years, farmers and loggers had reduced the four protected territories from 85,125 to 11,366 ha. The report noted that the first two release areas were "irreparably lost" because of human settlement and logging and that the fourth was down to 1,235 ha, just one-tenth of its original 14,950 ha.[94]

Even in areas that were true refuges, the survival chances of relocated animals were slim because these areas already had large animal populations to begin with, so the new animals had to compete with established animals for food and shelter. At Tucuruí, that pressure was exacerbated by the fact that most of the refuge areas were difficult to access and had only one or two entry points that did not penetrate the entire area, meaning that most of the animals were released into the same areas.[95]

[93] W. Robin Kingston, "Establishment of Primates Rescued from the Tucuruí Dam Flooding in the Brazilian National Primate Centre," *Primate Eye* 28 (1986): 18–20, 20, and Rogério Gribel, "Os mamíferos silvestres e as grandes barragens na Amazônia," in Efrem Jorge Gondim Ferreira et al., eds., *Bases científicas para estratégias de preservação e desenvolvimento da Amazônia: Fatos e perspectivas, Vol.* 2 (Manaus: INPA, 1993), 125–140. On primates and rescue missions broadly, see Amy Harrison-Levine et al., "Dams: Implications of Widespread Anthropic Flooding for Primate Populations," in Katarzyna Nowak et al., eds., *Primates in Flooded Habitats: Ecology and Conservation* (New York: Cambridge University Press, 2019), 285–292.

[94] Eletronorte, "UHE Tucuruí, plano de utilização do reservatório, plano de manejo das áreas de soltura, implementação efetiva das áreas de soltura A3 e A4, TUC-10-26483-RE," September 30, 1988, Document 00177, AELN.

[95] See Eletronorte, "Reservatório da Usina Hidrelétrica de Tucuruí Operação Curupira relatório de viagem," and Gribel, "Os mamíferos silvestres e as grandes barragens."

In 1988, ecologist Raimundo Henriques estimated that the reservoir's ecological refuges were hosting anteater populations that were more than three times their baseline.[96] Moreover, many relocated animals suffered tremendous trauma as a result of being captured, which weakened them and made them more vulnerable to predators.[97] Rescue missions simply did not constitute effective environmental protection.

Goodland had in fact advised against them. He had not mentioned animal rescue efforts at all in his Itaipu assessment and had explicitly discouraged them at Tucuruí. He wrote: "while photogenic and superficially laudable, expensive rescue operations for the release of animals elsewhere, such as Operation Gwamba ... are of dubious efficacy in this case."[98] He argued that displaced animals would have a hard time surviving in their new habitat and that many would die anyway as a result of wounds and stress from capture.

But Goodland's recommendations, like the rescue missions, were as much about publicity as they were about environmental protection, and the military regime took his advice selectively. The fact was that rescue missions were – as Goodland himself noted – photogenic and applaudable, and that was exactly what the military government needed to show the world that it was doing its best to ameliorate the environmental impact of its large dams without having to change their designs. Moreover, in contrast to displaced human communities, nonhuman animals – unknowledgeable about what was transpiring and unable to communicate concerns – mounted no resistance to the dictatorship's plans, which made them an ideal target for the spotlight. Animal rescue missions were pharaonic environmentalism at its finest.

Similar priorities shaped Sobradinho's social and environmental planning, though in some ways the semiarid environment the dam flooded made it exceptional. The Companhia Hidrelétrica do Rio São Francisco (CHESF) had also prioritized self-defense, and two of the principal concerns in the dam's earliest environmental studies were water shortages and sedimentation, which the company monitored closely. However, the company also paid close attention early on to the dam's social and

[96] Raimundo Paulo Barros Henriques, "Salvamento ou massacre?," *Ciência Hoje* 8, no. 46 (1988): 64–66.

[97] Waldinar Ranulpho e Maurício Azêdo, "Itaipu, o grande ecocídio," *Jornal do Commercio*, December 13, 1982.

[98] Goodland, *Environmental Assessment of the Tucuruí Hydroelectric Project*, 76–77.

environmental impacts, and the reservoir did not host an animal rescue campaign.[99]

Neither Eletrobras nor CHESF has explained why it chose not to carry out an animal rescue mission at Sobradinho, but the answer is likely that the semiarid Northeast was simply not a marketable stage for such ostentatious environmentalism. The desert was home to comparatively few animals to begin with, and, as Goodland noted in his report, human occupation along the riverbanks had all but destroyed most of the region's terrestrial fauna.[100] A decade later, in 1988, CHESF did carry out a rescue mission at the nearby Itaparica Dam, called Operation Saci, and the fact that this operation was not well publicized and captured little media attention compared to the rescue mission happening at Balbina at the same time (see Chapter 6) suggests that the Northeast was indeed a poor stage for showpiece rescue missions.[101] CHESF's motive for carrying out the program at Itaparica despite such unfavorable conditions was pressure from the World Bank, which had heightened in the mid-1980s, as discussed in the following section. Such pressures were milder when Sobradinho's reservoir filled.

These characteristics made Sobradinho somewhat unusual in terms of social and environmental planning, though not exempt from pharaonic environmentalism. Similar to Itaipu and Tucuruí, the military regime's priorities were first to protect its dams and second to find projects that would dramatically showcase the government's concern with mitigating social and environmental impacts, in this case through large resettlement projects (see Chapter 4) rather than animal rescue missions.

Eletrobras and its subsidiaries' pharaonic environmentalism did not constitute effective environmental protection. It was last-minute lip

99 For more on early attention to resettlement, see "Eng. Eunápio Peltier de Queiroz to Mário Penna Bhering, Assunto: Reservatório de Sobradinho," June 14, 1972, CHESF, AJPA, Recife; CHESF, "Reservatório de Sobradinho: Plano integrado de relocações das sedes dos municípios atingidos pelo reservatório, HE190-R1–1073," October 1973 and CHESF "Resettlement Plan for the Population Reached by Sobradinho's Reservoir, HE290-R01-0375," March 1975, Series "Modules," Box 2, Module C, Bookcase 3, Shelf 2, ACHESF, Recife.

100 Goodland, *Sobradinho Hydroelectric Project Environmental Impact Reconnaissance.*

101 Operation Saci captured 52,000 animals, 80 percent of which were lizards or other reptiles. See CHESF, "Operação Saci: Salvamento dos animais capturados em Itaparica," 1987, Box 4, Mod. B-E007, P05, ACHESF; Ricardo Braga and Raimundo Lima, "Bichos e plantas no doce mar do sertão," *Ciência Hoje 10* no. 56 (1989): 54–57, and World Bank, "Approaches to the Environment in Brazil: A Review of Selected Projects, Vol. IV," 64–65.

service to ecological stewardship that prioritized the longevity of dam projects and publicity over mitigating the impact of the reservoirs on the land and people. In the late 1980s, the international community came to the same conclusion.

SCRUTINY AND CONDEMNATION

The military government only succeeded in resolving one of the goals of its pharaonic environmentalism: protecting dam infrastructure and extending the longevity of its reservoirs. Most of the Brazilian military regime's large dams are not boondoggles.[102] The twentieth century is replete with stories of socially and ecologically destructive dams that failed to bring the benefits their boosters had promised. In some cases, economics and politics played a leading role. In the American West, the Bureau of Reclamation built a host of large irrigation and multipurpose dams between the 1910s and 1960s to provide land to small farmers. In most cases, the final cost of the dams was so expensive that the rates of irrigation water had to be raised to prices that only giant companies growing cash crops could afford.[103]

In other cases, the environment was responsible for the failure. High evaporation rates in the semiarid American West increased the salt content of the water in the Bureau of Reclamation's irrigation canals. Salinization decreased soil fertility and made some plots unfarmable. The irrigated cropland around Egypt's Aswan High Dam also suffers from salinization, and farmers have relied on chemical fertilizers to maintain productivity.[104]

Even more menacing is siltation, which affects hydroelectric, irrigation, and storage dams alike by reducing their storage capacity. China's rivers are some of the world's most heavily silted. Between 1957 and 1962, Chinese engineers pressed ahead with the dubious Sanmenxia Dam on the Yellow River despite warnings from critics that the dam would rapidly fill with sediment. The dam's skeptics were vindicated. Sanmenxia silted so quickly that over the subsequent decade the Chinese government had to rebuild it to flush sediments. The dam now has so many drainage gates

[102] The most glaring exception is Balbina Dam, the subject of Chapter 6.

[103] Donald Worster, *Rivers of Empire: Water, Aridity, and the Growth of the American West* (New York: Oxford University Press, 1985); Reisner, *Cadillac Desert.*

[104] Hesham Abd-El Monsef et al., "Impacts of the Aswan High Dam after 50 years," *Water Resources Management* 29 (2015): 1873–1885.

that its hydroelectric potential has been reduced tremendously.[105] Another region that suffers from high sedimentation rates is the Caribbean, where steep mountains and fierce storms exacerbate erosion.[106]

The Brazilian military regime's dams – with the exception of Balbina – are highly productive power plants whose infrastructure has not yet suffered heavily from self-imposed environmental impacts.[107] To be sure, sedimentation plagues Brazilian reservoirs, and, in at least one case, Itaipu, it has had localized impacts on fishermen (see Chapter 5). But none of Brazil's reservoirs have silted as rapidly as Chinese and Caribbean reservoirs. Brazilian environmentalist José Lutzenberger once estimated that Itaipu's reservoir would fill with silt in thirty years.[108] However, no rapid siltation problems have thus far come to light, and in 1993, Eletrobras estimated that the reservoir had at least one hundred years of useful life remaining.[109] Some of this prolonged lifespan is undoubtedly due to Itaipu's reforestation program, though it is impossible to say just how much credit the program deserves. Amazonian dams have contended with acidic water, but stainless-steel turbines appear to have thus far held

105 Shang Wei, "A Lamentation for the Yellow River: The Three Gate Gorge Dam (Sanmenxia)," in Dai Qing, ed., *The River Dragon Has Come! The Three Gorges Dam and the Fate of China's Yangtze River and Its People* (Armonk: M.E. Sharpe, 1998), 143–159, and Xiangli Ding, "'The Yellow River Comes from Our Hands': Silt, Hydroelectricity, and the Sanmenxia Dam, 1929–1973," *Environment & History* 27, no. 4 (November 2021): 665–694.

106 See M. Frenette et al., "Cas Historique de Sédimentation du Barrage Péligre, Haïti," *Canadian Journal of Civil Engineering* 9 (1982): 206–223, and Johnson, "'Thirsty Sugar Lands'," 337–365.

107 This claim appears true at the time of writing but might not be true in the future and is qualified by data limitations. The most recent sedimentation studies available for Itaipu and Sobradinho are from the 1990s, and no sedimentation surveys were available in Eletronorte's archive. Eletrobras and its subsidiaries have not publicized information that suggests that the dams' turbines or foundations have been degraded. Such information generally comes to light through independent or government-funded researchers, or through the companies themselves when the problem becomes acute, as was the case with siltation at Chinese and Caribbean dams.

108 Shaw William Miller, *An Environmental History of Latin America* (Cambridge: Cambridge University Press, 2007), 165. In 2001, Fearnside issued a similarly grim prognosis for Eletronorte's Tucuruí Dam, suggesting that agricultural practices in the surrounding watershed would engender erosion that might reduce the reservoir's usable lifespan to forty years. There is as yet no evidence from Eletronorte that confirms or refutes his estimation, though the reservoir is nearing forty years and complete siltation does not appear imminent. See Fearnside, "Environmental Impacts of Brazil's Tucuruí Dam," 385.

109 Newton de Oliveira Carvalho, "Avaliação do assoreamento do Reservatório da UHE Itaipu, PR," 1993, BIB-FO.

up against corrosion. Though the Brazilian mining sector's tailings dams have recently been failing with horrific consequences, available evidence suggests that the energy sector's dams are structurally sound.

As a result, the Brazilian military government can boast of having produced some of the grandest and most productive power plants the world has seen. When completed, Itaipu was the largest hydroelectric dam in the world – only China's Three Gorges Dam has outstripped it – with the capacity to produce 14,000 MW of energy, enough to power Portugal or New York City. As of this writing, the dam alone provides around 10 percent of the energy consumed in Brazil and around 90 percent of Paraguay's. It has produced more energy over its lifetime than any other power plant in the world.[110] Tucuruí is also near the top of the list of the world's top ten most productive hydroelectric plants, and the São Francisco River dams' combined capacity is roughly 10,400 MW, which, if considered a single dam, would make it the fourth largest hydroelectric plant in the world. The immense benefits that Brazilians have reaped from this energy had been possible largely thanks to the successful management of nature in defense of power plant infrastructure.

The military regime can also boast of having helped create one of the least carbon-intensive electrical grids in the industrial world. There are no accessible statistics for the dictatorship's dams' total carbon savings, but there is enough information to make very rough estimates. Itaipu has, as of April 2023, produced 2.7 billion MWh, and using a conservative estimate for average hydropower emissions suggests that the dam has saved somewhere between 1.4 and 2.4 billion tons of CO_2-equivalent over its lifetime compared to producing the same amount of electricity using fossil fuels.[111] These savings equal Brazil's entire CO_2 emissions for

110 www.itaipu.gov.br/en/press-office/faq (last accessed April 2023).

111 Author's calculations. The Intergovernmental Panel on Climate Change (IPCC) reports that hydropower produces between 4 and 14 gCO_2-e/kWh. The International Hydropower Association (IHA) puts the figure at 18.5 gCO_2-e/kWh (excluding emissions caused by construction activity). I have used the IHA's more conservative figure. Converting Itaipu's total production into kWh, multiplying it by 18.5 gCO_2-e/kWh, converting that number to tons, and then adding Itaipu's construction emissions (11 million tons CO_2-e), puts Itaipu's total lifetime emissions at about 66 million tons CO_2-e. For fossil fuels, I have used natural gas and coal as bookends for the range of emissions. The IPCC puts average natural gas emissions at 490 g CO_2-e/kWh. Doing the same math (without adding construction emissions) puts the total natural gas emissions for producing the same amount of electricity as Itaipu at 1.404 billion tons of CO_2-e. The difference, rounded, is 1.4 billion tons CO_2-e. The IPCC puts average coal emissions at 820 g CO_2-e/kWh. Doing the same math (without adding construction emissions) puts the total coal emissions for producing the same amount of electricity as

the 1980s, or, at rates from the 2010s, about four or five years' worth of emissions.[112]

These data can help us get a rough estimate for the entire sector. Extrapolating Itaipu's data to the entire cohort of 25 big dams identified in the Introduction is fraught with complications. For starters, these dams began producing power at different times, so their emission timelines are different. Moreover, not all dams produce power at the same rate, and the output of dams in the semiarid Northeast is often well below their installed capacity. Dams in the Amazon Rainforest generally emit at least as much greenhouse gas as coal-fired power plants (see Chapter 6), so they cannot be included.

Despite these complications, the calculation can produce a helpful ballpark estimate for dictatorship's dams' greenhouse gas emissions savings. The military government installed roughly 36,000 MW, excluding the greenhouse-gas-emitting Amazonian dams. The ratio between Itaipu's installed capacity and that of the dictatorship's big dams outside the Amazon is 2.57, and multiplying Itaipu's 1.4 to 2.4 billion tons of CO_2-e lifetime savings by this factor yields lifetime carbon savings of between 3.6 and 6.2 billion tons of CO_2-e for the dictatorship's big dams compared to producing the same amount of electricity using fossil fuels. The middle point in that range is about 5 billion tons.

Some perspective helps highlight the magnitude of these savings. During the 2010s, the entire Brazilian energy sector – which includes its fossil fuels – emitted about 456 million tons of CO_2 on average per year.[113] In 2015 and 2016, the entire country's greenhouse gas emissions for both years combined were 4.37 billion tons of CO_2-e.[114] That means by electrifying using hydropower instead of coal, the Brazilian

Itaipu at 2.132 billion tons of CO_2-e. The difference, rounded, is 2.4 billion tons of CO_2-e. Itaipu's production figures come from www.itaipu.gov.br/en/press-office/faq (last accessed April 2023). Both the IPCC and IHA figures for both hydropower and coal are listed in IHA, "Study Shows Hydropower's Greenhouse Gas Footprint," in IHA, *2018 Hydropower Status Report* (London: IHA, 2018), available at www.hydropower.org/publications/2018-hydropower-status-report (last accessed April 2023).

112 Greenhouse Gas Emission and Removal Estimating System [Sistema de Estimativas de Emissões e Remoções de Gases de Efeito Estufa, SEEG] database, http://plataforma.seeg.eco.br/total_emission (last accessed April 2023).

113 Author's calculations based on data from International Energy Association (IEA) database on Brazil's energy sector, www.iea.org/countries/brazil (last accessed April 2023).

114 http://plataforma.seeg.eco.br/total_emission (last accessed April 2023). Most of Brazil's greenhouse gas emissions during these years came from deforestation, ranching, and agriculture.

government has spared the Earth's atmosphere more than the equivalent carbon emissions of an entire decade's (2010–2019) worth of its energy production, or a full two years' worth of its entire country's recent greenhouse gas emissions. Such gains are an impressive benefit of the dictatorship's non-Amazonian hydropower dams.

But, at the time, carbon emissions did not factor into the ecological assessment of dams. When reservoir floodwaters settled in the 1970s and 1980s, fighting climate change was not yet a priority of the burgeoning environmental movement. Scientists had long been warning about climate change, but the phenomenon did not become a broader public concern until the late 1980s and 1990s. Eletrobras and its subsidiaries did not build dams with carbon emissions in mind and did not advertise emissions savings as a benefit of their projects. In turn, critics did not judge the environmental impacts of dams based on their carbon footprints, but rather a host of other social and environmental consequences. [115]

Eletrobras had a harder time addressing the second goal of its environmental program: convincing the world that it could mitigate the social and environmental impacts of its big dams, which are the subject of Chapters 4 and 5, respectively. The problem was that both goals were simply not compatible given the chosen dam sites and the conditions under which the dams were constructed. The dictatorship's political priorities demanded building giant dams fast and without public discussion, and the results were a series of reservoirs that had huge environmental footprints (see Chapter 2). The military government committed to these dams before environmental impact studies were undertaken and refused to alter designs in light of these reports' findings. Pharaonic environmentalism was the government's failed attempt to win over environmentalists while stubbornly remaining committed to its chosen dam sites.

Its failure had serious consequences. In the international arena, Eletrobras' foundering on environmental protection discouraged multinational banks from lending it money. In the late 1980s, the World Bank cancelled a US$350 million power sector loan – the second of the

[115] Fearnside might be an exception, but even his early writings about Balbina mentioned its impact on climate only in passing. It was not until 1995 that he began writing extensively about greenhouse gas emissions at reservoirs. See Fearnside, "Brazil's Balbina Dam," 409; Philip Fearnside, "Hydroelectric Dams in the Brazilian Amazon as Sources of 'Greenhouse Gases'," *Environmental Conservation* 22, no. 1 (Spring 1995): 7–19.

decade – for environmental and economic reasons.[116] The Brazilian government had approached the World Bank in January 1984, requesting help with its financial difficulties. Brazil had borrowed heavily over the preceding decade and was struggling to pay its debts because the soaring price of oil after the 1970s energy crises and rising interest rates sent Brazil into an economic recession. From 1984 to 1986, Brazil and the World Bank hammered out an agreement for a loan that was ostensibly for the power sector, but was designed primarily to help with general debt-crisis economic restructuring.[117] In 1986, the World Bank approved a US$500 million loan to the Brazilian power sector (Loan 2720-BR).[118] The Bank, however, was sensitive to the criticism that was then mounting about the social and environmental impacts of the large dams built by the Brazilian government, which had returned to civilian rule in 1985. The World Bank required that the loan be disbursed in two segments, subject to the government's successful resolution of a controversial resettlement program at Itaparica and its production of an environmental master plan outlining how it would address the environmental impacts of future dams. The Brazilian government's progress on Itaparica and its master plan were both lackluster (see Chapter 4), but good enough to trigger the release of the loan's second tranche in December 1986.[119]

But the World Bank's lenience stopped there. The mounting evidence of the government's dismal environmental record, combined with deteriorating economic conditions, convinced the Bank to cancel the second loan altogether. The World Bank complained that "the electric sector … pays only rhetorical attention at higher managerial levels to sociological and environmental variables, as reflected in some of the projects presently under construction … ." The pharaonic environmentalism of the military regime and the civilian administration that replaced it had failed to satisfy

[116] Goodland, "Annex 2: Brazil's Power Sector Loan and Nuclear Energy," in WBOHP, "Transcript of Interview with Robert Goodland," 54–55; World Bank, "Project Performance Audit Report, Brazil Power Sector Loan (Loan 2720-BR)," 23–28, 61.

[117] WBOHP, "Transcript of Interview with Robert Goodland," 6.

[118] World Bank, "Report and Recommendation of the President of the International Bank for Reconstruction and Development to the Executive Directors on a Proposed Loan in an Amount Equivalent to US$500.0 Million to the Federative Republic of Brazil for an Electric Power Sector Loan, Report No. P-4335-BR," May 29, 1986, WBGA.

[119] World Bank, "Project Performance Audit Report, Brazil Power Sector Loan (Loan 2720-BR)," 24–26.

the World Bank, which understood that more sound protection measures would involve altering dam designs or shelving some of the most controversial plans altogether. The World Bank report continued, "the electric sector lacks mechanisms for introducing energy and site-selection alternatives, based upon sociological and environmental consideration, in the feasibility and design stages of projects [and] lacks institutional means for increasing local, regional, and national participation in the electrical planning and decision-making process."[120] The Brazilian government had failed to convince the leader of the international lending community that it could satisfactorily ameliorate the social and environmental impacts of its large dams. In response, the multinational development bank began curtailing lending, stalling the second US$350 power sector loan on environmental grounds for a few years before formally canceling it when macroeconomic conditions worsened in May 1989.[121] In 1990, the World Bank issued its last loan to the Brazilian power sector. It was a US$232 million loan for the Itaparica Dam's resettlement program, a final attempt to help Brazil's state-owned electricity company and its subsidiaries improve their environmental reputation.[122]

At the national level too, the military government's failed ostentatious environmentalism had consequences that hurt the dam-building industry. Popular environmentalism had been growing in Brazil since the early 1970s, and the environmental lobby gained the power to influence legislation after 1985, when the military regime ceded power to a civilian administration (see Chapter 1). The damage that unfolded as a result of the military regime's large reservoirs convinced the environmental lobby that no such dams should be built in the future. In 1986, the Brazilian government began requiring that any new dams undergo a rigorous environmental impact assessment process. Environmental protection was further enshrined in the 1988 constitution, which gave special attention to Indigenous rights and stated that mining and hydropower produc-

[120] World Bank, "Staff Appraisal Report Brazil Electric Power Sector Loan" May 27, 1986, 53–54, WBGA.

[121] World Bank, "Project Performance Audit Report, Brazil Power Sector Loan (Loan 2720-BR),"48–50.

[122] The original loan was approved the year prior and totaled US$132 million. The World Bank approved an additional US$100 million in 1990. See World Bank, "Management Response to Request for Inspection Panel Review of Itaparica Resettlement and Irrigation Project Loan 2883-BR," 1997, 1, WBGA.

tion on Indigenous land can only take place with congressional authorization after hearing from affected communities.[123]

The combined result of disappearing funding and stringent environmental legislation was a dramatic ebb in the Brazilian government's dam-building ambitions. It had neither the resources nor the public approval to continue building its modern temples. The military regime's environmentalism may have helped ensure that its dams were not structural boondoggles, but it was unable to convince its critics, at home or abroad, that big dams were environmentally sound.

CONCLUSION

During the 1970s–1980s, the rise of international environmentalism pushed the Brazilian military regime to engage in a brand of environmentalism designed to convince its critics that it could mitigate the social and environmental impacts of its big dams without having to choose different sites, adjust designs, or delay construction. I have labeled the program that the government implemented "pharaonic environmentalism" because it constituted a form of environmentalism that prioritized the spectacle of caring for the environment more than actual social or environmental care and showed a weakness for grand gestures. The military government's ostentatious environmentalism had two priorities: protecting dam infrastructure from adverse ecological impacts and programs that would dramatically showcase its environmental consciousness.

Such environmentalism was not entirely useless, but it had limited effectiveness compared to choosing alternative dam sites or modifying project designs to minimize environmental impacts. Waiting until environmental impact studies were completed before selecting dam sites and opening plans to public scrutiny might have produced a more meaningful brand of environmentalism. But doing so would have compromised the military regime's political goals (see Chapter 2). What transpired was a watered-down version of environmental protection that was ultimately ineffective at dealing with the social and environmental consequences that the government's big dams unleashed. The military regime's failure to protect the land and people affected by its reservoirs discredited its dam-building program in the eyes of international and domestic critics. This

[123] República Federativa do Brasil, *Constituição da República Federativa do Brasil de 1988* (Brasília: Congresso Nacional, 1988), Título VIII, Da Ordem Social, Capítulo VIII, Dos Índios, Art. 231, §3°. Accessed through ACD.

disrepute led to the evaporation of funding from multinational banks and the establishment in Brazil of forceful legal measures for environmental protection that curtailed subsequent dam construction.

By the late 1980s, the reservoirs of Sobradinho, Itaipu, and Tucuruí had filled, and two particularly dubious dams, Itaparica and Balbina, were nearing completion. Pharaonic environmentalism was not enough to mitigate the impact of these dams, and in the following years, they became tragic stories of social and environmental catastrophe. It is to the disasters that unfolded during and after the reservoirs filled that the following chapters turn.

4

Negotiating with Floodwaters

Impacts on Displaced Communities, 1970s–1990s

INTRODUCTION

Third time is the charm, so they say. The idiom meant little to the Parakanã, who would have preferred that the military government got their resettlement right the first two times. The Parakanã are an Indigenous community that inhabited a large region near the Tocantins River in the state of Pará, in the Brazilian Amazon. Throughout the 1970s, the military regime had sent the National Indian Foundation [Fundação Nacional do Índio, FUNAI] to relocate various Parakanã settlements to make way for agricultural colonization projects, and by the end of the decade, a second threat menaced the group: the Tucuruí Dam. Its reservoir's floodwaters, which were due in 1984, would inundate 875 sq. km of their land, which included two settlements with a combined population of around 200 people.

Eletronorte had done a poor job of preparing the group to relocate. It created a program to assist the group in 1978, but cancelled it the next year. In 1981, the Parakanã decided to relocate by their own accord to an area that Eletronorte had set aside on the margins of the future reservoir. The following year, Eletronorte used helicopters to move the rest of the group. But when the reservoir filled, mosquitoes proliferated and made the land uninhabitable (see Chapter 5). The government, unable to control the mosquito swarms, had to relocate the Parakanã yet again, this time to a site further from the reservoir. Eletronorte dragged its feet, and it was only after the group staged protests at the company's headquarters in Brasília between 1991 and 1993 that it agreed to send financial support and to build infrastructure.[1]

The relocation catastrophe that had befallen the Parakanã was not an isolated incident. Tucuruí's reservoir had also flooded the land of about 32,000 non-Indigenous farmers, and the dam's subsidiary infrastructure (such as transmission lines) had brought similarly devastating consequences to a handful of other Indigenous groups.[2]

Big dams in the Northeast and Southeast caused disasters of even greater magnitude, displacing hundreds of thousands more people and flooding the land of Indigenous groups. Finding suitable land for resettlement was even more difficult in those places; the semiarid climate in the Northeast made much of the land uninhabitable, and most of the land in the Southeast was already occupied by non-Indigenous Brazilians. These factors exacerbated the troubles for displaced families, who had received derisory indemnity and gained concessions from the government only after sustained protest.

This chapter is about the fate of the families whose land the dictatorship's big dams flooded. It covers the twenty-year period from the late 1970s to the late 1990s, the first half of which the military government's reservoirs filled and expelled farmers and Indigenous communities, and the second half of which these families adjusted to their new homes. Some families remained on land near the reservoir, while others started their lives anew in distant commercial farming communities or cities. Some benefited from these new living conditions, but most watched their standard of living deteriorate after being displaced.

1 Antônio Carlos Magalhães, "Os Parakanã: Quando o rumo da estrada e o curso de águas perpassam a vida de um povo," master's thesis, Universidade de São Paulo, 1982; Comissão Pro-Índio de São Paulo, "Os povos indígenas e a hidrelétrica de Tucuruí," July 12, 1989, Folder BR.PA.00.5d (Energético Tucuruí), ACIMI; Henri Acselrad and Maria das Graças da Silva, "Conflito social e mudança ambiental na barragem de Tucuruí," in Henri Acselrad, ed., *Conflitos Ambientais no Brasil* (Rio de Janeiro: Relume Dumará, 2004), 175–193. For more on the Parakanã, see Magalhães, "Os Parakanã: Espaços da socialização e suas articulações simbólicas," (PhD Diss., Universidade de São Paulo, 1994).

2 See Sônia Barbosa Magalhães, "Lamento e dor: Uma análise sócio-antropológica do deslocamento compulsório provocada pela construção de barragens," (PhD Diss., Universidade Federal do Pará, 2007); Maria Gracinda Carvalho Teixeira, *Energy Policy in Latin America: Social and Environmental Dimensions of Hydropower in Amazônia* (Aldershot: Ashgate, 1996), 192–227, 235–252; Philip M. Fearnside, "Social Impacts of Brazil's Tucuruí Dam," *Environmental Management* 24, no. 4 (1999): 483–495; Acselrad and Silva, "Conflito social e mudança ambiental," and the website "Memória dos Atingidos de Tucuruí," https://memoriasocialtucurui.org/ (last accessed April 2023).

This chapter has two arguments. First, it argues that the Brazilian military government largely ignored the social costs of its big dams because it felt pressure to build them quickly and cheaply and because it believed that its pharaonic environmentalism would satisfy its critics. Such indifference to the fates of displaced people stemmed in part from the fact that most were poor farmers and Indigenous people, marginalized groups with little political clout. Yet, as this book argues, the military dictatorship greatly underestimated the organizational power of the then social justice-oriented Catholic Church, and as stories of catastrophe surfaced and spread, tremendous domestic and international criticism poured in that pressured the government to mitigate the social costs of its dams.

This chapter also argues that the reservoirs had an unintended consequence that dovetailed with the dictatorship's dream of national integration: the forced integration of ousted populations into the rapidly industrializing and urbanizing national economy. Prior to being expelled by reservoirs, many communities made their livings through self-sustaining subsistence economies in the floodplains and forests that lined riverbanks. The government did not provide substitutes for these resources, and once the reservoirs flooded them, these groups were forced to engage in commercial farming or move to cities. The government provided few other resources to help displaced communities transition, and such indifference made this subsidiary social engineering project disastrous.

LOSING THE FLOODPLAIN

The human costs of Sobradinho weighed heavily on the minds of its architects during the planning stages. On June 14, 1972, Sobradinho's director, Eunápio Peltier de Queiroz, wrote the president of Eletrobras to alert him to the social impacts of the dam. Queiroz warned that the reservoir would displace tens of thousands of people, most of whom were landless farmers – meaning they did not hold a legal title for their land – who made their living from the fertile soils of the São Francisco River's floodplain. Land of such high quality was rare in the semiarid Northeast, and Sobradinho's reservoir would inundate large stretches of it. The scarcity of good land limited the options that Sobradinho's directors had for relocation. There was fertile land near the city of Petrolina in an area called Project Massangano, but the company building the dam, the São Francisco River Hydroelectric Company [Companhia Hidrelétrica do

Rio São Francisco, CHESF], had reserved it for an irrigation project for wealthy landowners.[3]

In addition to the issue of finding suitable land, Queiroz worried that displaced farmers would have a hard time adjusting to rainfed or irrigated agriculture, not to mention footing the bill for expensive irrigation projects. Most were accustomed to the seasonal rhythms of the river that brought water and sediments, along with a freshet of fish.

Everything considered, resettling these farmers seemed to be a herculean task that Queiroz believed warranted support from other agencies. He closed his letter with a plea for help, writing that inaction would "permit disordered outmigration and consequent marginalization, exposing them [displaced farmers] to beggary, delinquency, trafficking, and prostitution, which is obviously unacceptable."[4] Queiroz worried that if resettlement efforts were not given proper attention and resources, Sobradinho risked further impoverishing those it would displace.

CHESF had a legal obligation to compensate those displaced by its dam. Article 161 of the military regime's 1967 Constitution stated that anyone displaced by government projects was entitled to fair compensation. But the text was vague and did not elaborate on what criteria would be used to define fair compensation: "the government may expropriate rural property upon payment of fair compensation, determined according to the criteria established by the law, in special public debt securities ..."[5] Article 3 suggested that indemnity would take the form of cash payments, except in rare cases, but even that was not explicit. Such uncertainty had the potential to fuel rancorous disputes, which Sobradinho's architects feared would arise should the issue not be treated with proper care and attention.

The World Bank was similarly apprehensive about Sobradinho's social impact. The reservoir had a big footprint, and as the environmental movement gathered momentum, Bank staff were growing increasingly

[3] Bishop Dom José Rodrigues, "Depoimento na Câmara dos Deputados, em Brasília no dia 29 de junho, 1978," in *Caminhar Juntos* no. 22, July 1978, BDJR. *Caminhar Juntos* was a report that Rodrigues published monthly about events in the region. See also World Bank, "Approaches to the Environment in Brazil: A Review of Selected Projects, Vol. IV: The Middle and Lower San Francisco Valley," April 30, 1992, 71, WBGA.

[4] "Eunápio Peltier de Queiroz to Mário Penna Bhering, RE: Reservatório de Sobradinho," June 14, 1972, Folder "CHESF, Reservatório de Sobradinho, Reassentamento de Populações, Dados e Informações," AJPA.

[5] República Federativa do Brasil, *Constituição do Brasil de 1967* (Brasília: Congresso Nacional, 1967), Título III, Da Ordem Econômica e Social, Art. 157, §1°. Accessed through ACD.

anxious about the social and environmental impacts of their projects. Robert Goodland's report (see Chapter 3) only heightened their concerns. He had concluded: "the relocation of the displaced population is more complicated than construction of the dam; the relocation process must therefore start at once and must be accorded all attention and resources necessary, commensurate with its importance."[6] The stakes were high, so much so that in 1974, World Bank president Robert McNamara personally reviewed resettlement plans and visited Northeastern Brazil to inspect the reservoir site and resettlement conditions.[7]

The situation McNamara found upon landing in the Northeast was grounds for concern. The number of those who would be displaced was much higher than Queiroz had estimated. There were roughly 12,000 families, about 72,000 people, living in the area that Sobradinho's reservoir would flood, which spanned four municipalities: Casa Nova, Sento Sé, Remanso, and Pilão Acordo. Most lived in the floodplain and practiced subsistence agriculture, which they supplemented with trade in local markets. The majority were illiterate, and few were connected to regional or national economies or social networks.[8] They were unprepared to move and had few resources to make a living in barren lands outside of the floodplain.[9]

CHESF's resettlement plans were lackluster in the face of this impending social crisis. It had produced detailed studies about resettlement, but these studies greatly underestimated the number of people who would be displaced, and their recommendations did not translate into

6 Goodland, *Sobradinho Hydroelectric Project Environmental Impact Reconnaissance*, 1973, 30, General Collection, US-LOC.

7 World Bank, "Approaches to the Environment in Brazil, Vol. IV," 31.

8 "Queiroz to Bhering, RE: Reservatório de Sobradinho."

9 For more on those displaced at Sobradinho, see Lygia Sigaud, *Efeitos sociais de grandes projetos hidrelétricos: As barragens de Sobradinho e Machadinho* (Rio de Janeiro: Museu Nacional-UFRJ, 1986); Ana Maria Lima Daou, "Políticas de estado e organização social camponesa: A barragem de Sobradinho," master's thesis, Museu Nacional and UFRJ, 1988; Ana Maria Lima Daou, "Efeitos sociais da Barragem de Sobradinho: A experiência dos camponeses do Brejo de Dentro," *Revista Geográfica* 120 (1994): 27–40; Ruben Alfredo de Siqueira, "Do que as águas não cobriram: Um estudo sobre o movimento dos camponeses atingidos pela barragem de Sobradinho," master's thesis, Universidade Federal de Paraíba, 1992; Edcarlos Mendes da Silva, "Desterritorialização sob as águas de Sobradinho: Ganhos e desenganos," master's thesis, Universidade Federal da Bahia, 2010; Ely Souza Estrela, *Sobradinho: a retirada de um povo* (Salvador: EDUNEB, 2010); Ana Luiza Martins Costa, *Uma retirada insólita* (Rio de Janeiro: IPPUR/UFRJ, 2013); and Adzamara Amaral, *Sento Sé: memória de uma cidade submersa* (São Paulo: Lisbon International Press, 2020).

effective resettlement measures.[10] CHESF offered displaced families three options. First, they could move anywhere they wanted, and it would pay them indemnity for their property, travel expenses, and a small one-time payment to help them get started in their new homes. Second, they could move into the regional backlands and receive the same benefits, plus additional financial help to get their new farms started. Lastly, they could move to a new farming community called Serra de Ramalho in the municipality of Bom Jesus de Lapa, located along the São Francisco River about 700 kilometers upstream from Sobradinho. There, CHESF and the National Institute of Colonization and Agrarian Reform [Instituto Nacional de Colonização e Reforma Agrária, INCRA] were building infrastructure for a small town and demarcating land.[11]

Though CHESF hoped that this site would absorb much of the displaced population, it was highly unpopular because of its distance from the reservoir area. Indeed, CHESF's own surveys found that most people preferred to stay near the margins of the newly formed reservoir because the region had immense value in terms of social networks and nonmaterial connections to the land.[12] Therefore, the majority of families that CHESF was dealing with consisted of poor farmers who planned to relocate to barren land on the edge of the reservoir, and its plan to accommodate them was to build basic infrastructure in new towns and indemnify families.

CHESF was in too much of a hurry to properly assess property values. It began by sending legal representatives to farms to negotiate indemnity with individual families, many of whom did not possess the title to their land. CHESF's legal team almost always undervalued properties, and when families refused to accept their offer, construction workers would drive bulldozers to farms to intimidate families.[13] CHESF also threatened

[10] See CHESF, "Reservatório de Sobradinho: Plano integrado de relocações das sedes dos municípios atingidos pelo reservatório, HE190-R1–1073," October 1973; CHESF, "Resettlement Plan for the Population Reached by Sobradinho's Reservoir, HE290-R01–0375," March 1975, Series "Modules," Box 2, Module C, Bookcase 3, Shelf 2, ACHESF; Convênio CHESF-ANCARBA, "Reservatório de Sobradinho projeto alternativas de produção, estudos complementares," 1975; CHESF, "Plano de desenvolvimento para a região do Lago de Sobradinho," undated, Series "Geração," Box T1212, ACHESF.

[11] Rodrigues, "Pronunciamento na Comissão Parlamentar de Inquérito destinada a investigar as causas e consequências das cheias do Rio São Francisco," in *Caminhar Juntos* no. 54, April 1981, BDJR. This speech can also be found at the ASF and ACPT.

[12] CHESF, "Resettlement Plan for the Population."

[13] The original Portuguese is "tratores," translated here as bulldozer.

that soon the floodwaters would come and that it would be in the farmers' interests to accept its offer and get something rather than have to leave when the floodwaters came with nothing. Such pressure led some families to capitulate immediately. However, others protested, sitting in front of bulldozers until they received a just indemnity.[14]

Many families that resisted ended up persuading CHESF to raise the amount it offered for their land, but the price was often still well short of what farmers thought was fair. Farmers who resisted too much sometimes lost everything. José Gonçalves de Freitas, a farmer from Pilão Acordo, refused CHESF's offer for one of his two pieces of land because he thought the land was worth much more than what they claimed. Instead of paying him its original price, CHESF decided to pay him nothing.[15]

The local government was of little help. In 1976, President Geisel had declared the four municipalities surrounding the reservoir to be "areas of national security." The designation meant that the generals appointed mayors, who were submissive to CHESF and did little to support the families its reservoir was displacing.[16]

That these farmers were able to fight CHESF at all was largely thanks to the help of a key ally: the Catholic Church. Between 1958 and 1963, Pope John XXIII had reoriented the Catholic Church internationally to be more engaged in social justice issues. This reorientation was well received in Latin America, where socioeconomic inequality was severe and social movements were emerging to curb it. The Catholic Church in Brazil had supported the March 1964 coup that ushered in the military regime, but the dictatorship's policies had exacerbated inequality so much that within a few years of the coup, Brazilian bishops had turned against the military regime and – along with the Catholic Church elsewhere in Latin America – were more committed than ever to fighting socioeconomic inequality. In 1968, the Second Episcopal Conference of Latin America met in Medellín, Colombia, and officially committed to fighting for social justice. It was the birth of Latin America's liberation theology movement.[17]

[14] Rodrigues, "Documento dos Três Bispos," *Caminhar Juntos* no. 10, April 1977, 6–7, BDJR.

[15] Comissão Pastoral da Terra (CPT), *Ilhas de resistência: Os lavadores no Vale do Rio São Francisco* (Goiânia: CPT, 1979), 18–20, ACPT.

[16] Rodrigues, "Flashes da pastoral na diocese de Juazeiro, a segunda ferramenta: a política," *Caminhar Juntos* no. 102, April 1985, 3, BDJR.

[17] Margarete Pereira da Silva, "O bispo de Juazeiro e a ditadura militar," in Grimaldo Carneiro Zachariadhes, ed., *Ditadura militar na Bahia: Novos olhares, novos objetivos, novos horizontes* (Salvador: EDUFBA, 2009), 241–249. See also, Scott Mainwaring, *The*

This reformist church was at the forefront of helping those displaced by dams in Brazil. There were two ecclesiastical organizations in particular that were the most heavily involved in the social conflicts generated by the dictatorship's dams. The first was the Indigenist Missionary Council [Conselho Indigenista Missionário, CIMI], which the National Conference of Brazilian Bishops [Conferência Nacional dos Bispos do Brasil, CNBB] founded in 1972 in Brasília. The organization fought exclusively on behalf of Indigenous communities and often acted as the intermediary between the federal government and Indigenous communities when the government agency charged with this task, FUNAI, proved inept. Though the organization was headquartered in Brasília, it created regional branches in almost all states that became involved in local conflicts concerning Indigenous communities. The second important ecclesiastical organization was the Pastoral Land Commission [Comissão Pastoral da Terra, CPT], which the CNBB founded in 1975 in the mid-western city of Goiânia. Like CIMI, the organization's headquarters remained in Goiânia, but within a few years the organization had created regional branches in nearly all states that became involved in the social conflicts of their regions. In addition to documenting and publicizing state violence against rural farmers, CPT leaders organized impacted people into communities that could act collectively to stand a better chance of winning concessions from the government.

Ecclesiastical organizations like CIMI and CPT were able to organize without repression from the government both because they were connected to the Catholic Church and because during the late 1970s the military government was gradually loosening its authoritarian grip over the country through its abertura policies (see Chapter 1). Ecclesiastical groups had an easier time confronting the dictatorship to begin with, and abertura allowed them to be even bolder in their denunciations. Within just a few years, both CIMI and CPT had become tremendously influential. Even religious leaders involved in social justice advocacy who were not working directly under the supervision of either group coordinated with them. Whether working independently or with CIMI or the CPT, bishops across Brazil were becoming heavily involved in helping farmers and Indigenous groups displaced by the military regime's large reservoirs fight for fair indemnity.

Catholic Church and Politics in Brazil, 1916–1985 (Stanford: Stanford University Press, 1986).

Nowhere more so than at Sobradinho. In 1975, Bishop Dom José Rodrigues took over the diocese in Juazeiro, the urban center nearest Sobradinho, and was immediately thrown into the conflict between CHESF and the farmers that Sobradinho would displace. Rodrigues had been involved with the Catholic Church since the 1940s – he was ordained in 1950 – but for most of his career he was more interested in its internal workings than in the socioeconomic issues of Brazilian society. But what he witnessed at Sobradinho radicalized him. He remembered, "I was nominated to the Bishop of Juazeiro, without knowing the situation there. Upon arriving I heard of a dam called Sobradinho and started visiting the municipalities … I heard the clamor and dramatic appeals of the impoverished and I could only be on the side of them. So, in the name of Christ and the bishop, I became increasingly involved."[18] Upon witnessing the unfolding social crisis at Sobradinho, Rodrigues stepped into the front lines of Brazilian liberation theology.

Rodrigues led a movement from the local diocese that helped those displaced fight for just indemnity. It consisted of two principal parts. The first was legal protection. The diocese hired lawyers who could represent farmers against CHESF in court, and the presence of these lawyers during negotiations helped pressure CHESF into fairly compensating farmers without having to go to court in the first place. Second, the diocese helped displaced families organize collectively and published their struggle widely, the latter of which attracted allies and brought negative press to CHESF.[19]

Rodrigues' campaign was tremendously influential in getting CHESF to step up its efforts to justly compensate those displaced. When CHESF began closing Sobradinho's floodgates in February 1977 to fill the reservoir, the fate of many displaced people remained unresolved, and many were expelled without indemnity or a place to go.[20] Between September that year and the following February – as Sobradinho's reservoir continued to fill – Rodrigues wrote a series of letters to CHESF's directors, alerting them to the horrible living conditions that displaced families were enduring. In addition to his own writing, he forwarded letters that community members themselves had written about their dire

[18] Silva, "O bispo de Juazeiro e a ditadura militar," 256.

[19] Rodrigues, "Pronunciamento na Comissão Parlamentar de Inquérito."

[20] Sobradinho's reservoir filled in two phases. The first, a partial filling, occurred between February and December 1977, and the second, the complete filling, occurred between January and July 1978.

situations.[21] He also published stories of the hardships suffered by displaced families in print and recounted them on local radio programs. His first letters to CHESF went unanswered, so he decided to write directly to the president of Eletrobras. In March 1978, the president responded, saying that the federal government understood the gravity of the situation and Eletrobras would launch a program called Project Sobradinho to help displaced families. The project consisted of allocating additional money for infrastructure improvements in the new towns that CHESF built and paying stipends to displaced families that were struggling.[22]

The project came much too late to avoid disaster. Many were still in their homes and unprepared to move when the floodwaters came. When the water rose, it stranded herds of livestock that families were trying to move to higher ground. On April 3, 1978, people of the displaced community of Marcos penned a desperate letter to CHESF, saying that more than one hundred head of cattle were stuck in the water and would die soon if not helped immediately. They exclaimed, "these cattle are our life, and if they die, it will greatly damage our livelihood. Until now, we've not received any help, though two times representatives from our community have spoken with you ... we ask respectfully that you help us, because the situation is very critical."[23] While Eletrobras was busy planning rescue missions to save snakes and monkeys from treetops at Itaipu's reservoir (see Chapter 3), it was looking the other way as herds of livestock were swept away by Sobradinho's floodwaters. Ironically, having boats of rescue workers on hand might have been more effective at Sobradinho than at the other sites because rising floodwaters swept away herds of animals that were vital to the livelihood of displaced communities.

The drowning herds at Marcos were not an isolated incident. Altogether, thousands of animals were carried away by the floodwaters of Sobradinho, and the diocese estimated that 80 percent of the region's cattle were lost to floods or to the thirst, hunger, and disease that plagued animals after leaving the floodplain.[24]

Even those who escaped the rising waters with their herds intact left their homeland for uncertain fates and struggled to survive in the barren lands on the edge of the reservoir. Most families ended up camping under

[21] Rodrigues "Drama das populações transferidas da beira do rio para as caatingas," *Caminhar Juntos*, no. 18, February 1978, BDJR.

[22] Rodrigues, "Carta alvissareira da Eletrobrás," *Caminhar Juntos*, no. 20, April 1978, BDJR.

[23] Rodrigues, "Pronunciamento na Comissão Parlamentar de Inquérito."

[24] CPT, *Ilhas de resistência*, 18–20.

plum trees or tarps in shanty towns on the edges of urban areas or trying to farm the infertile soils along the reservoir. The meager financial indemnities were not enough to help – starting a new farm on marginal desert land was expensive, and few had the necessary resources to do so successfully.

Infrastructure in new towns and farming communities was also lacking. In many cases, there was no potable water and residents became sick after drinking putrid water from the reservoir, which contained, among other unpleasantries, the decaying carcasses of thousands of drowned cattle.[25] In general, CHESF did a better job of relocating the city centers of Casa Nova, Sento Sé, Remanso, and Pilão Acordo, but these urban areas also suffered tremendously from delays in infrastructure.[26]

Some tried their luck at Serra de Ramalho, but living conditions there proved equally catastrophic and most returned to the reservoir's margins with familiar stories of inadequate land and failing infrastructure. Though CHESF had originally hoped the project would absorb upwards of 4,000 families, only 1,013 families, or about 8.65 percent of those displaced, resettled there.[27] Resettlement had become the nightmare that Queiroz had dreaded.

Heavy rainfall in the watershed upstream made a bad situation worse. In 1979, the rainy season (which generally peaks in February through May) was unusually wet. The torrents came just after Sobradinho's reservoir had filled and was at maximum capacity. President Geisel was scheduled to inaugurate Sobradinho on March 10, 1979, and engineers were keeping the reservoir full so that the President would see South America's largest reservoir in all its grandeur. The combination of heavy rains and a full reservoir was disastrous, despite CHESF's attempts to euphemize the calamity as "controlled floods." On January 24, 1979, CHESF began releasing water at nearly three times the normal rate; over the course of the next two months, that number shot up to four and then six times the normal rate.

[25] Rodrigues, "Sobrevivência das populações transferidas," *Caminhar Juntos*, no. 20, April 1978, 3, BDJR; CPT, *Ilhas de Resistência*, 24–25; Rodrigues, "A situação dos pobres agricultores, pecuaristas e pescadores nas bordas do lago de Sobradinho e poluição no Vale do São Francisco," *Caminhar Juntos*, no. 96, October 1984, BDJR.

[26] Rodrigues, "Pronunciamento na Comissão Parlamentar de Inquérito."

[27] João Paulo Aguiar, "Reservatório de Sobradinho, síntese do reassentamento de populações," June 2001, Folder "CHESF, Reservatório de Sobradinho, Reassentamento de Populações, Dados e Informações," AJPA.

The floods lasted three months and inundated the downstream city of Juazeiro, destroying 800 homes and displacing 4,689 people. The floods also destroyed relocated communities on the edge of Sobradinho's reservoir. Many were displaced for a second or third time and once again lost their harvests. The following rainy season was almost nearly as wet, and between January and March 1980, a second set of floods brought more hardship to farmers on the edges of Sobradinho's reservoir. The irony that the inauguration of a dam designed to control the São Francisco's floods was immediately followed by some of the worst floods in the river's history was not lost on Sobradinho's critics. Since then, CHESF has managed Sobradinho's reservoir in a manner that has thus far prevented such flooding, but that is a small condolence to those who suffered the consequences of these floods.[28]

In response to the bad publicity surrounding the deteriorating living conditions of displaced families, CHESF initiated another wave of relief efforts in 1982. On September 10, CHESF and the Bahian government created a land titling program that would provide squatting farmers with capital in the form of land ownership. The program's results were impressive. Within three years, CHESF and the Bahian State Government had issued 5,806 land titles to families, which roughly translates to land for about 34,800 people.[29]

But as with Project Sobradinho, it was too little too late. Most farmers had become so impoverished as a result of the traumatic move that upon receiving the title for their land they immediately sold it to wealthy landowners and moved to urban centers or began working as laborers on nearby farms.[30] In 1992, the World Bank released a report that reflected on the Sobradinho experience and concluded that most displaced families had suffered a net loss as a result of the project: "although official reports tend to gloss over the subject, the livelihoods of many displacees from Sobradinho and the lower valley had, in effect, been ruined,

28 Rodrigues, "Pronunciamento na Comissão Parlamentar de Inquérito"; José Francisco da Silva, Presidente Confederação Nacional dos Trabalhadores na Agricultura, "Depoimento sobre as cheias," April 2, 1981, Folder BA-0502 Sobradinho, ACPT.

29 "CHESF to Instituto de Terras da Bahia," October 18, 1985, Folder "CHESF, Reservatório de Sobradinho, Reassentamento de Populações, Dados e Informações," AJPA.

30 Governo de Bahia et al., "Barragem de Sobradinho palestra na V Curso Internacional de Planejamento e Desenvolvimento Rural Integrado," Salvador, December 1985, 57, Folder "Sobradinho," BDJR; World Bank, "Approaches to the Environment in Brazil, Vol. IV," 74.

especially those of smallholders and sharecroppers who received little or no compensation or were ineligible to join the new projects."[31] CHESF's relief efforts were much too late to mitigate the hardships of displacement, and the result was that many dispossessed families became increasingly impoverished.

To be sure, not everyone suffered from the dam. João Paulo Aguiar, Sobradinho's director during construction, is the foremost proponent of the narrative that Sobradinho vastly improved the lives of most people that it displaced. He used to remark that he would quit his job without any severance pay if 90 percent of the following generation were not better off than they were before. He recognized that there was incredible turmoil and hardship in the short term and admitted frankly that building a dam entails inevitable aggression against the people that it displaces. But Aguiar argued that in the long term, Sobradinho greatly improved the lives of the majority of the people living in the region because it improved their standard of living and granted them access to the amenities of industrial life.[32] Indeed, many benefited. The World Bank's 1992 report noted that urban centers "have become quite prosperous and provided better living standards for a large proportion of their expanding populations."[33]

But the argument that Sobradinho entailed a net benefit for displaced people overlooks the fact that its benefits largely went to those who had the interest and resources to successfully adapt to urban living. Most dispossessed farmers wanted to continue to live off the land, and many suffered impoverishment. Of the roughly 12,000 families displaced, about 32 percent moved to new cities. The rest moved to rural communities or settled in more isolated areas of the semiarid desert that surrounded the reservoir.[34]

Sobradinho fundamentally changed the way the region's inhabitants related to their environment. The dam electrified cities and industries in

[31] World Bank, "Approaches to the Environment in Brazil, Vol. IV," 91.

[32] Memoria da Eletricidade, "João Paulo Aguiar, depoimento concedido no âmbito do projeto meio ambiente e o setor de energia elétrica (Recife-PE)," 2003, 12–13, BCME.

[33] World Bank, "Approaches to the Environment in Brazil, Vol. IV," 75.

[34] In total, roughly 70 percent of those displaced stayed near the reservoir and just 9 percent went to the Serra do Ramalho agricultural project. The rest accepted cash payments and moved elsewhere. See CHESF, "Quadro, destino das famílias atingidas pelo reservatório de Sobradinho e reassentadas pela CHESF," Folder "CHESF, Reservatório de Sobradinho, Reassentamento de Populações, Dados e Informações," AJPA; World Bank "Approaches to the Environment in Brazil, Vol. IV," 70–76.

distant lands and also imposed urbanization and industrialization on the region. It destroyed the only natural asset that allowed people to sustain themselves directly from the land: the floodplain. The seasonal pulses of the river brought water and fertile sediments, along with an abundance of fish, to smallholders and landless farmers. Sobradinho replaced these natural rhythms with the rhythms of industrialization, which kept the richest soils inundated and diverted irrigation water to mechanized farms that could afford it. The military government was in such a hurry to get their dam built that they allocated few resources to helping poor farmers adapt to these alien forms of relating to the environment.

CHURCH AND STATE IN THE SOUTH

On a cool afternoon in June 1980, a group of farmers sat playing cards at a bar in Porto Mendes, a town alongside the Paraná River that the Itaipu Dam would flood within a few years. They hardly glanced up from their game when they heard the sound of pickup trucks pulling up outside, but they dropped their cards in astonishment when they saw who got out of the car. It was Bishop Dom José Brandão de Castro, the regional coordinator for the CPT's Northeastern branch in Sergipe. Bishops were a rare sight in rural communities in Western Paraná, and this was the first time that most of these farmers had seen a bishop outside of an occasional baptism.

Even more astounding was the motivation for his visit and the speeches he began delivering. Pastor Gernote Kirinus, who had founded the Western Paraná branch of the CPT, had invited Brandão de Castro to the region to raise awareness about the horrors that might befall farming communities after Itaipu's reservoir filled. Brandão de Castro had witnessed the calamities that Sobradinho had brought to rural farmers in Bahia, and Kirinus thought that sharing his experience with farmers in Western Paraná might encourage them to organize to forestall similar tragedies at Itaipu. During his visit, Brandão de Castro traveled along the Paraná River, giving speeches in dancehalls and bars to attentive farmers. He denounced the ills of the capitalist system that was producing big dams and exclaimed that the government would not come to the aid of displaced farmers because it was the government that was orchestrating these projects. He shared the depressing experience of Sobradinho, where the military regime's bulldozers and trucks came to level the homes of farmers and remove them from the area. He argued that since the government would not help them it was up to farming communities to organize and demand fair compensation for having their homes flooded by Itaipu's reservoir.[35]

The ecological and socioeconomic conditions of farming communities in Western Paraná were different from those in the Northeast, but the threat both communities faced was nearly identical. Unlike the basin of the São Francisco, the watershed of the Paraná River enjoys abundant rainfall throughout most of the year, so farmers living in the region were not so dependent on the floodplain for their livelihood. Land along the riverbank was some of the region's best, but there was a greater abundance of fecund soils outside of the floodplain than in the Northeast. The timeline of Brazilian occupation was also different. Brazilians from the eastern parts of the state only began settling in the area around Itaipu in the 1960s.[36] Though many of these farmers were similarly poor and landless, a greater number had capital and access to the machinery needed to clear land and mechanize harvests than those around Sobradinho. Thus, most farmers in Western Paraná did not have the same attachments to the land and its natural rhythms as their Northeastern counterparts, both because the environment was more forgiving and because they had only recently settled there and had greater access to capital. But such differences did not make Brandão de Castro's horror stories any less vivid, and his inflammatory speeches struck a chord with Itaipu's farmers.

Brandão de Castro's visit was among the sparks that ignited an influential protest movement against Itaipu, but it was the work of the Catholic Church in the preceding years that had built up large amounts of tinder that were ready to be lit.[37] In 1973 and 1974, Itaipu carried out a census of the lands that would be flooded by the dam and began the expropriation and indemnification process for 8,257 families on the Brazilian side of the reservoir, which was about 40,000 people, 80 percent of whom were rural. It started with the area that became the construction site,

[35] Pastor Werner Fuchs, Personal Interview, August 23, 2018, Curitiba. Pastor Fuchs was the coordinator of the CPT in Western Paraná after Kirinus. Bishop Dom José Brandão de Castro's visit is also mentioned in Gerd Kohlhepp, *Itaipú Basic Geopolitical and Energy Situation: Socio-economic and Ecological Consequences of the Itaipú Dam on the Rio Paraná* (Braunschweig: F. Vieweg, 1987), 59; Juvêncio Mazzarollo, *A taipa da injustiça: Esbanjamento econômico, drama social e holocausto ecológico em Itaipu, 2ª edição* (Curitiba: CPT Paraná, 2003), 77; and Blanc, *Before the Flood*, 2019, 79–80.

[36] See Ruy Christovam Wachowicz, *História do Paraná, 10ª edição* (Ponta Grossa: Editora UEPG, 2010).

[37] Excellent overviews of the protest movement can be found in Mazzarollo, *A taipa da injustiça, 2ª edição*; Guiomar Inez Germani, *Os expropriados de terra e água: O conflito de Itaipu, 2ª edição* (Canoas: Editora da ULBRA, 2003), and Blanc, *Before the Flood.*

which went smoothly because Itaipu needed the land quickly and paid high prices willingly.[38] In 1976, the company began indemnifying those in the future reservoir area and, starting in 1978, accelerated the process.

Itaipu's attitude toward families in the area the reservoir would flood was much different than its attitude toward those at the construction site. With the dam's construction proceeding apace, it felt little pressure to willingly pay high sums to farmers who had to leave the area anyway when the reservoir filled. Itaipu's agents offered families a nonnegotiable sum for their land and its improvements, which were often well under market value. One displaced woman remembered: "I had four hectares and then some. We were indemnified, or rather, robbed. They came with their proposals, and we did not accept them. So, they told us that we would have to negotiate with the courts, or with the floodwaters ..."[39]

In these early years of the expropriation process, few families felt compelled to protest because they were afraid that criticizing Itaipu might earn them a trip to prison under the repressive dictatorship. The fear was not unfounded. Local journalist Juvêncio Mazzarollo spent nearly two years in jail for publicizing both Itaipu's misdoings and the military regime's attempts to silence him through intimidation.[40]

But there was one institution that could stand up to Itaipu without fear of retribution: the Catholic Church. It was church against state, once again, this time in the South. CPT founded its Paraná branch in 1976 and its Western Paraná branch in 1977, just in time to come to the aid of the farmers that Itaipu was expelling from the riverbanks of the Paraná. The CPT organized assemblies for farmers and, in August 1978, published an influential report, written by one of its journalists, Antônio Carlos Moura, titled *The Mausoleum of the Pharaoh [O mausoléu do faraó]* that highlighted the derisory indemnities that Itaipu was issuing.[41] The national press spread the document widely, which raised awareness about the plight of the displaced farmers across the country. Thus, prior to June 1980, when Bishop Brandão de Castro visited the region, there

[38] Mazzarollo, *A taipa da injustiça, 2ª edição*, 48–50. [39] Ibid., 90.

[40] See Blanc, "The Last Political Prisoner."

[41] CPT, *O mausoléu do faraó: A Usina de Itaipu contra os lavadores do Paraná* (Goiânia: CPT, 1978), ACPT. The following year, in 1979, Carlos Moura wrote another report for the CPT about the experience of displaced farmers at Sobradinho called the *Ilhas de resistência [Islands of Resistance]*, which is cited in the preceding section. Sobradinho's reservoir had filled between January and July 1978 – around the time Moura was writing his Itaipu report – and his experiences at Sobradinho greatly influenced his writing about Itaipu.

were rumblings of protest in Western Paraná, but there was no unified movement.

On July 14, 1980, the farmers' movement, newly christened as the Justice and Land Movement [Movimento Justiça e Terra, MJT], organized its first major protest: a campout (lasting between July 14 and 29) outside Itaipu's office in Santa Helena, a town along the Paraná that Itaipu's reservoir would soon flood.[42] The encampment became national and international news, and it pushed Itaipu to raise the minimum price for land and to readjust the price every 120 days to better match market prices.[43] For a short period, the victory satisfied farmers, but within months, the market price shot up, and Itaipu's payments were again insufficient to resettle on neighboring lands.

The farmers staged a second occupation. On March 17, 1981, 800 farmers caravanned to Foz de Iguaçu, where they planned to occupy the dam's construction site. The military stopped them at the juncture where the main highway met Itaipu's access road, and they decided to set up camp there (the protest lasted 54 days). By early May, Itaipu was willing to negotiate with farmers to once again raise the price they were offering. The solution was hardly satisfying, but it was enough to appease protesting families, who by this time were tired and exhausted.[44]

The aftermath of the Foz de Iguaçu protest was similar to that of the Santa Helena occupation. For a short period, farmers were satisfied, but land speculation raised prices again in a matter of months. This time, there were only a few hundred farmers who had not been indemnified. In March 1982, a group of farmers threatened to stage a protest, and Itaipu Binacional, weary of more bad press as it was preparing to inaugurate the dam, quickly raised prices.

Though Itaipu and the state government had refused to set aside land in Paraná for displaced farmers, after March 1981, they did request that INCRA put aside some land for these farmers in agricultural projects already underway. One of these projects was in the municipality of Arapoti in Northeastern Paraná, and the others were in the Amazonian states of Acre and Rondônia. Arapoti provided land for about 350 families, or about 1,750 people, which was about 4 percent of those displaced, and the land proved to be a poor substitute for the highly fertile soils farther

42 CPT Paraná, "Terra e justiça exigem os agricultores," *Poeira* 3, no. 12 (July–August: 1980), 6–12, BCEFURIA. *Poeira* is a monthly report that the CPT of Paraná began publishing in 1978 to highlight issues affecting the state's rural communities.

43 CPT Paraná, "Terra e justiça exigem os agricultores," 6–12.

44 Mazzarollo, *A taipa da injustiça, 2ª edição*, 98–120.

west. Those in Amazônia did not fare much better. The living conditions were so rough that many chose to return to Western Paraná, often bringing with them the malaria parasites that they contracted in the Amazon (see Chapter 5).

Itaipu's reservoir also displaced a much smaller group of non-Indigenous farmers on the Paraguayan side of the reservoir, many of whom were recent immigrants from Brazil. Political repression in Paraguay was so severe that these farmers did not organize in the same manner as their Brazilian counterparts and thus left almost no paper trail.[45]

LOSING THE FORESTS

The group that Itaipu most greatly impoverished was the Avá Guaraní, a subgroup of the Guarani Indigenous community that inhabited the banks of the Paraná River. Violent European incursions diminished their numbers over the preceding centuries, but there was still a significant Indigenous presence along the Paraná River when construction started on Itaipu.[46] In 1975, Itaipu carried out a survey and found that there were around 401 Indigenous families (about 2,108 people) in the future reservoir area.[47] These numbers understated the true population, which Indigenous communities themselves, working with ecclesiastical NGOs, later estimated to be around 680 families (about 3,600 people).[48] Nearly all of these Avá Guaraní families lived in communities on the forested Paraguayan side of the reservoir, except one group that lived on about 15 sq. km (1,500 hectares) near the Ocoí tributary on the Brazilian side. The economy of the Avá Guaraní was based primarily on hunting, gathering, and fishing, and communities had a spiritual attachment to the land that sustained them.[49] Given their connection with the land, they were reluctant to move and only did so when the floodwaters were imminent.

[45] See Ibid., 153–159.

[46] See Mariblanca Barón, *Los Ava Guaraní Paranaenses: Un etnocidio sistemático*, Series Biblioteca Paraguaya de Antropología, Vol. 107 (Asunción: Centro de Estudios Antropológicos de la Universidad Católica, 2017).

[47] Bartomeu Melià y Ramón César Bejarano, "Características de los indígenas actuales," in Itaipu, "Investigaciones históricas, socioculturales y arqueológicas del área de Itaipú," December 1975, BIB-AS.

[48] Barón, *Los Avá Guaraní Paranaenses*, 65.

[49] For more on the traditional economy of the Avá Guaraní, see Bartomeu Melià, *Itaipú: Deforestación y desarraigo de la nación Guaraní* (Asunción: El Lector, 2018); Maria Lucia Brant de Carvalho, "Das terras dos Índios a Índios sem terras, o estado e

By late 1973, both Paraguay and Brazil had laws stating that any Indigenous group displaced by a dam was entitled to be compensated with land of equal or greater quality. In 1968, Paraguay passed Ley No. 63/68, which ratified the International Labor Organization's (ILO) 1957 Indigenous and Tribal Populations Convention. Article 12 stated that governments had the right to remove Indigenous communities from their traditional land in the interest of national development but that, "when in such cases removal of these populations is necessary as an exceptional measure, they shall be provided with lands of quality at least equal to that of the lands previously occupied by them, suitable to provide for their present needs and future development."[50] On December 19, 1973, the Brazilian military government promulgated Lei No. 6.001/73, known as the Indian Statute, which outlined the various rules that governed the relationship between the country's Indigenous communities and the state. Article 20 proclaimed that the Brazilian government had the right to acquire Indigenous territory; however, paragraph three clarified that displaced communities would have a right to "land that was equal to the area they were removed from, including in terms of its ecological conditions."[51]

Despite having instated these laws, both military regimes ignored their legal obligations to Indigenous communities when it came to the Avá Guaraní. On the Paraguayan side, Itaipu director Enzo Debernardi argued that the ILO's convention did not apply to Itaipu, and he refused to offer anything more than minimal cash payments as indemnity.[52]

It took great effort to persuade him and his colleagues otherwise. As the floodwaters approached, Paraguayan NGOs stepped up the pressure on the government to set aside land for the Indigenous communities

Os Guarani do Oco'y: Violência, silêncio e luta," (PhD Diss., Universidade de São Paulo, 2013).

[50] ILO, *C107 – Indigenous and Tribal Populations Convention, 1957 (No. 107)*. www.ilo.org/dyn/normlex/en/f?p=NORMLEXPUB:12100:0::NO::P12100_INSTRUMENT_ID,P12100_LANG_CODE:312252,en (last accessed April 2023).

[51] República do Brasil, Lei No. 6.001/73, articulo 20, parágrafo 3. www.jusbrasil.com.br/topicos/11724011/artigo-20-da-lei-n-6001-de-19-de-dezembro-de-1973 (last accessed April 2023).

[52] Equipe Nacional de Misiones de la Conferencia Episcopal Paraguaya (ENM) "Avá Chiripá afectados por Itaipú esperan resoluciones," *Dialogo indígena Misionario (DIM)* 3, no. 7, May 1982, ACONAPI. The ENM is a Paraguayan ecclesiastical NGO founded in 1969 whose mission is to provide support to Indigenous groups. In the 1980s, it began publishing a monthly report called DIM that highlighted the issues facing the country's Indigenous communities. In 1994, ENM changed its name to Coordinación Nacional de Pastoral Indígena (CONAPI).

that would be displaced by Itaipu, a campaign that dovetailed with nationwide protests to strengthen the rights of Indigenous groups.[53] The effort produced results: On December 18, 1981, the Paraguay military government promulgated law number 904/81, the Statute of Indigenous Communities.[54] Article 15 reaffirmed what the ILO convention required governments to do in cases when Indigenous land needed to be flooded for national development: "they will be given suitable land of at least equal quality to land previously inhabited and will be conveniently indemnified for the value of land improvements and for the harm and damages suffered as a consequence of displacement."[55] With a second law instituted in defense of Indigenous communities, Itaipu, and the government agencies it coordinated with, now had a harder time brushing off the protests of the Avá Guaraní. Itaipu continued to complain that the law obligated the government, not necessarily Itaipu, to acquire land for the Avá Guaraní, but by January 1982, it had agreed to purchase land to resettle them.

The Brazilian military government was equally apathetic about upholding its legal obligations to the Avá Guarani. When Itaipu accelerated its indemnification process in 1978, it sent an anthropologist to Ocoí who concluded there were no Indigenous people living there, a result that pleased Itaipu because this meant that it would only have to compensate those living there with financial indemnity.[56]

Yet, subsequent reports soon confirmed that Itaipu's reservoir would in fact flood Indigenous land on the Brazilian side.[57] The Avá Guarani

53 "Reunión de Equipo Nacional de Misiones de la CEP sobre la situación de comunidades Avá-Chiripá afectadas por los trabajos de Itaipú Binacional, March 5–6, 1982," Folder "Itaipu-FUNAI No' Ovusu," APIES. The Pastoral Indígena de Espirito Santo (PIES), led by Mariblanca Barón, is a group of nuns who have been working with the Avá Guaraní since before Itaipu was built and are committed to helping them regain the land owed to them in compensation for what they lost to Itaipu's floodwaters. They have a small library in Nueva Esperanza, Paraguay, that contains troves of information about the history and culture of the Avá Guaraní.

54 For more on the history of Paraguayan legislation concerning Indigenous communities, see Mariblanca Barón, "Legislación paraguaya respecto a tierra y al territorio indígena," *DIM* 23, no. 66, February 2010, ACONAPI.

55 Republica de Paraguay, *Ley No. 904/81, Estatuto de las comunidades Indígenas*, Articulo 15. www.cultura.gov.py/marcolegal/ley-90481-estatuto-de-las-comunidades-indigenas/ (last accessed April 2023).

56 "Avá Guarani to Barber Conable," September 18, 1986, Folder BR.PR.GN5d Guarani Itaipu, ACIMI.

57 See, for example, Edgard de Assis Carvalho, "Avá Guarani do Ocoí-Jacutinga," 1981, FSI, CEPEDAL.

later estimated that there were about 100 families, or 500 people, living at Ocoí when construction began at Itaipu, and that pressures from INCRA and Brazilian farmers had cut the population to around 22 families, or 110 people, by 1979.[58] That year, the remaining families at Ocoí demanded that Itaipu grant them land as compensation for displacement, launching a fight for just indemnity that paralleled – but was independent from – the MJT.[59]

Itaipu's original offer was derisory: 20 hectares and an additional 80 hectares in the forests surrounding the reservoir that they could use to hunt, but not to farm or build on.[60] Severely restricted access to land that totaled a mere 6.6 percent of their original territory was a far cry from the legal requirement that the government provide land of equal quantity and quality. The Avá Guarani complained, but Itaipu only raised its offer by 21 hectares. On March 22, 1982, the Avá Guarani wrote FUNAI, saying that they knew the law obligated the Brazilian government to indemnify them for their land and that 121 hectares was vastly insufficient to compensate them for the 1,500 hectares they would lose. The group concluded: "we are not leaving because we want to, but because of a government project. This is the same government that made the law guaranteeing rights to Indians and created FUNAI to comply with this law. Why, then, is FUNAI not obeying the law?"[61] Yet Itaipu remained intransigent, raising its offer by only small amounts. Those at Ocoí grew desperate because the floodwaters were coming in October. On May 12, 1982, they concluded they had no choice but to accept Itaipu's latest offer of about 250 hectares of land on the edge of the reservoir.

Once the reservoir filled, the Avá Guaraní on both sides of the reservoir found themselves living in precarious conditions. They were living on land that was inadequate to support them. Itaipu had taken about 2,350 sq. km from them – 1,350 sq. km for the reservoir and 1,000 sq. km for the protected areas that surrounded it – but it had put aside only

58 See "Avá Guarani to Barber Conable."

59 MJT offered occasional support; see MJT, "Documento de reivindicações aprovado na assembleia de agricultores em Itacorá, dia 16 de março de 1981," Folder BR.PR.GN5d, ACIMI. For more on the overlap between these two movements, see Blanc, *Before the Flood* and Elaine Pereira Rocha, *Canal de desvio: Um estudo da experiência de agricultores e índios no confronto com a Itaipu Binacional* (São Paulo: E-Manuscrito, 2021).

60 "Avá Guarani de Barra do Ocoí to Coronel Paulo Moreira Leal, Presidente de FUNAI, and General Costa Cavalcanti, Presidente da Itaipu Binacional," February 5, 1982, Folder BR.PR.GN5d, ACIMI.

61 "Avá Guarani de Barra do Ocoí to Coronel Paulo Moreira Leal," March 22, 1982, Folder BR.PR.GN5d, ACIMI.

56 sq. km, spread across four sites in Paraguay and one in Brazil.[62] This was not very much land for 688 Indigenous families, or 3,440 people, whose economy and culture were intimately tied to land that was more than forty times greater than their current holdings. Worse still, the new land was already occupied by other Guaraní. Overpopulation of these small areas put tremendous pressure on the land and led to material impoverishment and infighting.[63] Though groups on both sides suffered tremendously, it was the group on the Brazilian side that was the first to loudly denounce Itaipu's wrongdoing and fight for just compensation.

The Avá Guarani at Ocoí vociferously protested their abysmal living conditions. For more than four years after the reservoir filled in October 1982, the Avá Guarani of Ocoí wrote letters to Itaipu that went unanswered.[64] Having gotten only silence from Itaipu and the federal government, the Avá Guarani decided to take their complaints to international allies. They understood that the ambitions of the Brazilian electricity sector gave them leverage that could help them regain their land: international condemnation. During the mid-1980s, the Brazilian government was negotiating with the World Bank to secure loans for new hydropower projects, and environmental safeguards weighed heavily in the discussions. The Avá Guarani knew that the Bank was ready to withhold funds from the Brazilian power sector if it could not satisfactorily mitigate the adverse impacts of its big dams (see Chapter 3).

On September 18, 1986, the Avá Guarani of Ocoí wrote World Bank president Barber Conable. Their letter alerted him to their precarious living conditions, as well as the fact that shoreline erosion was shrinking their landholdings (they estimated they had lost about 40 hectares already). The situation was dismal, and the group asked Conable to pressure Itaipu to return the land it owed them. They concluded: "we want Itaipu to indemnify the rest of the land they stole from us. We had

62 On the Paraguayan side: Kirito Pyahu, 10 sq. km, bought by the Conferencia Episcopal de Paraguay with resources from German churches; Yukyry, 20 sq. km, bought by the Asociación Indigenista Paraguaya and the Instituto de Bien Estar Rural; Itabo, 13 sq. km, and Limoy, 10 sq. km, both bought by Itaipu. On the Brazilian side: Santa Rosa do Ocoí, about 2.5 sq. km, the majority of which INCRA paid for. See Barón, *Los Avá Guaraní Paranaenses*, 84–85, 103.

63 "Avá Chiripa afectados por Itaipú, ¿dónde están?" *DIM* 3, no. 8, December 1982, ACONAPI, and Enrique Gaska, "Del canto al llanto: Situación de las comunidades indígenas afectadas por la Represa de Itaipú," *Acción* no. 157, September 1995, ACONAPI.

64 See "Avá Guarani to Itaipu Binacional," November 25, 1985, and "Avá Guarani to Itaipu Binacional," March 4, 1986, Folder BR.PR.GN5d, ACIMI.

1,500 hectares of land, with forests, fish, and game, not 210 hectares with poison [pesticide pollution] and malaria."[65] It was a desperate appeal to a potential international ally. The World Bank had not funded Itaipu, but it had loaned, and was continuing to loan, a lot of money to the power sector and was then working to clean up its environmental reputation.

The World Bank responded positively. On November 14, 1986, its international relations department wrote João Centurião, the chief of the Avá Guarani of Ocoí, saying that despite not having financed the Itaipu Dam, the Bank would do its best to help them resolve their land dispute with Itaipu.[66] Within a few months, the World Bank wrote Eletrobras,[67] and, unsatisfied with its response, it sent Goodland and anthropologist Shelton Davis to inspect the situation. Goodland had carried out the environmental impact study that Itaipu used to build its environmental mitigation program, though this report had, devastatingly, not mentioned Indigenous communities (see Chapter 3).[68] On February 19, 1987, Goodland and Davis visited Ocoí and produced a report that confirmed the awful living conditions in the community. The situation was so obviously grim and had such a straightforward solution that Davis was actually optimistic that the problem could be resolved without protracted legal battles.[69]

But Itaipu continued to withhold land. The World Bank's intervention pushed Itaipu to meet with the Avá Guarani, but it refused to grant them more land, instead offering access to amenities such as hospitals and electricity as compensation.[70] When the Avá Guarani refused, Itaipu began offering land, but it was either barren or too far from Western Paraná.[71] Such indemnification could not compensate for forests, and the Avá Guarani wrote the Bank twice more to clarify their position and to solicit further support: "World Bank executives, we want forested land

65 "Avá Guarani to Conable."

66 "Banco Mundial to Sr. Cacique João Centurião," November 14, 1986, Folder BR.PR.GN5d, ACIMI.

67 "Banco Mundial to Mario Penna Bhering, Presidente de Eletrobras," January 28, 1987, Folder BR.PR.GN5d, ACIMI.

68 See Comissão Mista Técnica, "Estudo do Rio Paraná: Relatório especial no. 4."

69 "Shelton Davis to Júlio Geiger," March 14, 1987, Folder BR.PR.GN5d, ACIMI.

70 "Ata de reunião realizada no escritório da diretoria jurídico em Foz de Iguaçu," March 27, 1987, and "Avá Guarani to Banco Mundial," May 4, 1987, Folder BR.PR.GN5d, ACIMI.

71 CIMI, "Ocoí, Luta pela terra–10 anos," September 8, 1992, Folder BR.PR.GN.4b Luta Pela Terra, ACIMI.

because ... cleared land cannot support us. In the forest there are materials to build houses. Itaipu offers us concrete houses but we do not want to live that way ... you have said that you will fight alongside us to help us achieve our goals and what we need is land."[72] There were certainly parts of the industrial world that interested the Avá Guarani, but they did not want to assimilate into urban and industrial Brazil and could only maintain their traditional economy with forested land. They had a legal right to such land and were growing tired and frustrated of Itaipu and the Brazilian governments' efforts to circumvent their constitutional obligation to compensate them with land of equal quantity and quality.

But even the World Bank could not help the Avá Guarani. In December 1986, the Bank had released the second tranche of the 2720-BR power sector loan, which diminished the leverage that it had over the Brazilian government.[73] The government's failure to satisfactorily compensate the Avá Guarani of Ocoí for the land it flooded was certainly one of the factors, among many others, that encouraged the World Bank to delay and ultimately cancel the second power sector loan in the late 1980s (see Chapter 3).

Itaipu's continued intransigence was eroding the Avá Guarani's material well-being and culture. In 1992, CIMI surveyed the Indigenous community at Ocoí to assess living conditions ten years after the floodwaters had displaced the group. Exhaustion and despair abounded. For a decade, the Avá Guarani had been living on land that was 16 percent of their former territory and was completely without forests, which had provided all the resources the group needed to survive. Childbirth was expanding families, and community members felt increasingly compelled to work as laborers on neighboring farms to make a living. They were desperate for land: "we cannot keep waiting, if we wait any longer, we will die because the land is diminishing as families are growing. Disease and hunger are also on the rise and it is becoming increasingly necessary to work for farmers, and all this together is destroying the way of being that is distinctly Guarani."[74] Itaipu had impoverished the Indigenous group so much that it was losing its culture and sense of being.

The impoverishment of the Avá Guarani of Ocoí highlights the racism that characterized the Brazilian military government's pharaonic

[72] "Avá Guarani to Banco Mundial," July 17, 1987, Folder BR.PR.GN5d, ACIMI.

[73] World Bank, "Project Performance Audit Report, Brazil Power Sector Loan (Loan 2720-BR)" iv.

[74] CIMI, "Ocoí, Luta pela terra–10 anos."

environmentalism. Itaipu and the federal government's standard response to the Avá Guarani's demands was that forested land no longer existed in the area.[75] The Avá Guarani agreed that such land was scarce. Indeed, industrial farms had so thoroughly deforested the region that few large stretches of forested land remained. But there was still plenty of land for the Avá Guarani in Western Paraná; the problem was that the government had set aside this land for other purposes.

Consider Itaipu's forest reserves, which were created in part to house animals rescued from rising floodwaters (see Chapter 3). Itaipu had set aside nearly six times as much land for these animals as for the Avá Guarani, and the total costs of Operation Mymba-Kuera and the net worth of the reserved land were eight times what Itaipu spent on indemnifying and buying land for the Avá Guarani.[76] In 1982, indigenist Wilmar D'Angelis wrote an article denouncing these disparities. He concluded: "the numbers are clear, indisputable. Itaipu is condemning Indigenous people to death, ripping them from the resources necessary for their survival. This has an internationally known name: genocide ... will the men and women of Brazil continue to permit ... this absurd inversion of values that saves snakes and rats and condemns indigenous people to death?"[77] Operation Mymba-Kuera grabbed international headlines, but for those impacted, the rescue mission did a terrible job of masking the dam's more insidious impacts.

There was one other area where there was a lot of intact forest: the Iguaçu National Park. The Brazilian and Argentinian governments had created their respective Iguaçu National Parks in 1930s,[78] and, in Brazil, the park took on additional significance once the government decided to build Itaipu because it softened the opportunity cost of flooding Guaíra Falls. The military regime promoted domestic and international tourism at Iguaçu and invested heavily in the park's infrastructure in an effort to distract attention from the disappearing Guaíra Falls (covered in Chapters 2 and 5). In 1981, the government expanded the park's acreage for a second time, and in 1986, UNESCO declared it a World Heritage Site.[79] Thus, the rise of Iguaçu Falls as an international tourist destination

75 "Avá Guarani to Banco Mundial."

76 Johnson, "Temples of Modern Pharaohs," 238.

77 Wilmar D'Angelis, "Itaipu, os índios e os bichos," *Luta Indígena* 18, December 1982, 6. *Luta Indígena* was a periodical that CIMI Regional-Sul published, and copies can be consulted at the BCEFURIA.

78 See Freitas, *Nationalizing Nature*, 2021.

79 https://whc.unesco.org/en/list/355/ (last accessed April 2023).

was in many ways part of the military regime's program for blunting the environmental impact of Itaipu.

It was the most forested area in Western Paraná and perfect land for the Avá Guarani. In the early 1990s, after a government commission failed to find suitable land for the group, the Avá Guarani created their own commission that identified the Iguaçu National Park as the best site for resettlement because it had forests and was close to their homeland. The Iguaçu National Park has an area of 1,700 sq. km, so the group's original request of 15 sq. km would have taken only 0.8 percent of the park's area. Indeed, their impact would be so minimal that the group decided to raise their request to 50 sq. km, about 3 percent of the National Park. Their request gained a lot of support from other Indigenous groups throughout the country. Between April 25 and 30, 1992, more than 200 leaders from different Indigenous groups at the Meeting of Brazil's Indigenous People and Organizations [Encontro dos Povos e de Organizações Indígenas do Brasil] in Brasília signed a document in support.[80]

But the Brazilian government refused. It seemed that setting aside land for an Indigenous group that was uninterested in assimilating into industrial society was not a flashy enough means of showing off social and ecological conscience. It also risked subverting long-standing racial inequalities and the racist ideas created to justify them, which kept Indigenous people oppressed and enriched wealthier and whiter Brazilians. That is, justly indemnifying the Avá Guarani would set a precedent that might give leverage to other Indigenous groups demanding compensation for land that the government (or private companies and individuals) stole from them for resource extraction or other purposes.

The rise of international environmentalism had demanded action on the part of Itaipu, and institutional racism in both societies encouraged it to prioritize saving animals and expanding national parks over the Avá Guarani. Ten years after Itaipu's floodwaters expelled them from their homes, the Avá Guarani of Ocoí continued to live impoverished on barren land on the edge of the reservoir while tens of thousands of tourists visited Itaipu and the Iguaçu National Park annually.

80 "Avá Guarani pede providências imediatas aos órgãos competentes do governo para demarcação de suas terras no Parque Nacional de Iguaçu, Paraná," April 28, 1992, Folder BR.PR.GN.4b Luta Pela Terra, ACIMI.

SOBRADINHO NEVER AGAIN

Back in the Northeast, the government was getting ready to inundate more of the floodplain. In 1986, CHESF was finishing construction on its Itaparica Dam, which is about 300 kilometers downstream from Sobradinho on the São Francisco River. Though the Brazilian government ultimately chose to build its big storage dam at Sobradinho, it held on to the idea of building a shorter dam at Itaparica for power generation (see Chapter 2). The 1973 energy crisis gave the Itaparica project a sense of urgency, and in 1975, CHESF approved the dam and began building the construction site.[81] Financial difficulties intermittently halted the project. Construction on the dam itself began in 1979, and in January 1988, its reservoir began filling.[82]

The dam's environmental footprint was similar to Sobradinho's. The dam's 834 sq. km reservoir would inundate the land of about 7,500 families, or 45,000 people, living on a fecund floodplain measuring 400 sq. km. There was a greater concentration of wealthy farmers in the area that Itaparica would flood, but wealth inequality was sharp, which meant that most of the population living there was landless.[83]

Farmers whose land Itaparica would flood had been watching the project unfold with tremendous apprehension. At the same time that CHESF announced its decision to build the dam, stories began pouring in from upstream municipalities about the catastrophes at Sobradinho. When CHESF began approaching farmers in 1976 to indemnify them for their land, they feared that they would suffer the same fate as their upstream neighbors.

But the farmers at Itaparica were determined to stop the horrors of Sobradinho from repeating. They launched a protest movement with the slogan "Sobradinho Never Again," using a phrase ("nunca mais") that was then gaining resonance throughout the Southern Cone in reaction to dictatorship, serving as the title of at least two influential reports on authoritarian violence.[84] At the height of protest in 1986, the movement

[81] CHESF, *Relatório de Diretoria 1976*, 21, ACHESF.

[82] CHESF, *Relatório de Diretoria 1979*, 21, ACHESF.

[83] At that time, 1.7 percent of landowners owned 72 percent of the total farmland. See World Bank, "Approaches to the Environment in Brazil, Vol. IV," 94.

[84] For more on these protests, see Centro de Defesa dos Direitos Humanos do Submédio São Francisco (CDDHSSF), *Itaparica: A dor de um povo gerando energia* (Petrolândia: CDDHSSF, 1992) (consulted at BDJR); Maria Lia Corrêa de Araújo et al., eds., *Sonhos submersos ou desenvolvimento? Impactos sociais da Barragem de Itaparica* (Recife:

proclaimed: "We do not want more promises. We want concrete measures that give us a sense of security that our land rights and survival will be respected. We are not going to permit that Itaparica repeat what happened at Sobradinho."[85] Stories of resistance had made it downriver along with the stories of socioeconomic calamity, and the farmers at Itaparica set out to avoid the tragedies that befell those displaced at Sobradinho.

Itaparica farmers created a local worker's union to lead the fight. In 1979, local bishops and NGOs helped farmers organize the Lower-Middle São Francisco Worker's Syndicate [Polo Sindical do Submédio São Francisco], which was a conglomerate of more than thirteen local unions. The syndicate held its inaugural protests when construction started.[86]

The movement's early years had drawn a lot of attention to the plight of the farmers, but did little to alter CHESF's plan, which was to move displaced farmers to plots of irrigable land that it had chosen without consulting them. The farmers protested that these plots were too small and had poor soil that could not be irrigated as the company claimed. In 1984, the farmers wrote an open letter to CHESF in which they asked to meet to negotiate better terms for indemnity and resettlement.[87]

Still nothing changed. The civilian administration that replaced the military regime in 1985 did nothing to help the soon-to-be displaced farmers. On October 8, 1986, the farmers penned another letter, repeating their claims and complaining that CHESF had not yet started

Massangana, 2000); Parry Scott, *Negociações e resistências persistentes: Agricultores e a barragem de Itaparica num contexto de descaso planejado* (Recife: EDUFPB, 2009); and Luciano Silva de Menezes, "A Barragem de Itaparica e os atingidos de Petrolândia-PE," master's thesis, Universidade do Estado da Bahia, Juazeiro, 2017. On the phrase "nunca mais," see Comisión Nacional sobre la Desaparición de Personas, *Nunca más: Informe de la Comisión Nacional sobre la Desaparición de Personas* (Buenos Aires: EUDEBA, 1984), and Paulo Evaristo Arns, *Brasil: Nunca mais* (São Paulo: Arquidiocese de São Paulo, 1985).

85 "Manifesto dos trabalhadores rurais atingidos pela Barragem de Itaparica, a onze meses da inundação," December 1, 1986, Folder 19-Barragem de Itaparica, ACPT.

86 See Confederação Nacional dos Trabalhadores na Agricultura (CONTAG), "Posição do movimento sindical de trabalhadores rurais a respeito das reivindicações dos trabalhadores rurais da Barragem de Itaparica," November 30, 1979; "Boletim dos trabalhadores rurais atingidas pela Barragem de Itaparica," January 1980, Folder 19-Barragem de Itaparica, ACPT.

87 Polo Sindical do Submédio São Francisco, "Diretrizes básicas para reassentamento dos trabalhadores rurais atingidos pela Barragem de Itaparica, Pernambuco, Bahia," May 10, 1984, Folder 19-Barragem de Itaparica, ACPT.

the relocation process even though the dam was nearing completion.[88] But these outcries had little impact; CHESF continued to drag its feet and refused meetings.

The workers and their families, emboldened by the end of military rule, staged a dramatic response. On December 1, 1986, more than one thousand men, women, and children caravanned to Itaparica to occupy the construction site.[89] When they arrived at the entrance to Itaparica, they were met by police armed with machine guns. The situation was tense, but the women in the group were undaunted. They picked up stones to defend themselves and continued marching toward the soldiers who soon received orders to stand down to the approaching multitude.[90] Having made it past the guards, the group set up camp at the worksite, impeding machinery and halting construction. They remained camped at the construction site in a state of unease until December 6, when the movement's leaders met with CHESF and emerged victorious.[91]

CHESF agreed to nearly all of the movement's demands. It would start resettlement immediately and outline clear timetables for building infrastructure on new plots. The company also agreed to increase irrigated plot acreage and pay families monthly stipends until their first harvests.[92]

The occupation succeeded in part because of World Bank pressure. Over the previous years, Itaparica farmers had drawn so much attention to their plight that the World Bank specifically included the successful resolution to their conflict in its requirements for the disbursement of the second half of its Brazilian power sector loan (see Chapter 3). The

88 "Denúncia de Itaparica, a 12 meses da inundação," October 8, 1986, Folder 19-Barragem de Itaparica, ACPT.

89 "Manifesto dos trabalhadores rurais atingidos pela Barragem de Itaparica."

90 Edvalda Pereira Torres Lins Aroucha, "Ecologia das mulheres do São Francisco," in Juracy Marques, ed., *Ecologias do São Francisco* (Self-Published, 2006), 358–365. I consulted this rare book at BFUNDAJ.

91 See "Mil trabalhadores ficam alojados em Itaparica," *Jornal do Commercio*, December 3, 1986; "Trabalhadores invadem barragem de Itaparica," *Tribuna da Bahia*, December 4, 1986; "Trabalhadores paralisam as obras na área de Itaparica," *Diário de Pernambuco*, December 4, 1986; "Itaparica: Povo não aceita promessa para sair," *Diário de Pernambuco*, December 5, 1986; "Itaparica: Permanece impasse com trabalhadores," *Diário de Pernambuco*, December 6, 1986; "Polo Sindical do Submédio São Francisco aos colaboradores, entidades sindicais, movimentos populares, movimentos de igreja, bispos, partidos públicos, parlamentares etc.," December 11, 1986, Folder 19-Barragem de Itaparica, ACPT.

92 See "Acordo entre a CHESF e trabalhadores rurais atingidos pelo reservatório da Usina Hidrelétrica de Itaparica, com a presença do Ministério das Minas e Energia," December 6, 1986, Folder 19-Barragem de Itaparica, ACPT.

December 1986 occupation came on the heels of a World Bank visit to assess whether or not CHESF had satisfactorily followed through on its commitment to resolve the complaints of displaced families.[93] CHESF found itself in a bind. At that time, the Northeast was facing energy shortages and rationing, and halting construction on Itaparica in the midst of such a crisis was unpopular. But violently repressing displaced families would have generated equally bad press and jeopardized access to international funding.

The decision to concede to the farmers enabled CHESF to resume construction and maintain access to funding. The World Bank released the second tranche of its sector loan on December 22, 1986, just weeks after the company signed its agreement with the displaced farmers.[94] Together, displaced families and their allies had pushed CHESF to make unprecedented concessions that resulted in a much smoother resettlement process at Itaparica.

The dam's floodgates closed in February 1988, and by April, relocation was complete. The resettlement process was carried out without dramatic stories of crisis and tragedy. Of the total number of displaced rural people, 76.7 percent chose to move to irrigation projects, 11.8 percent (most of whom were wealthy farmers) opted for cash compensation and free transport elsewhere, 8.1 percent chose to move into cities, and 3.1 percent moved to farming collectives (the remaining 0.3 percent moved to fish farms).[95] Thus, the vast majority of farmers relocated to plots of irrigable farmland, a huge improvement over the horror stories that had followed the completion of Sobradinho's reservoir. Moreover, the 1986 agreement mandated that the government had to issue land titles to farmers regardless of whether or not they had owned land previously, so the Itaparica resettlement project had resulted in an unprecedented land reform that redistributed land from wealthy landowners to smallholders.

But resettling displaced people was only half the challenge. CHESF still had to build irrigation systems on the displaced farmers' new land, and construction was proceeding slowly. By 1992, four years after Itaparica's reservoir had filled, not a single irrigation system had been built, and as late as 1999, eleven years after the reservoir had filled, CHESF had still not finished, having completed only 69 percent of the irrigation

93 World Bank, "Approaches to the Environment in Brazil, Vol. IV," 90.
94 World Bank, "Project Performance Audit Report, Loan 2720-BR," i, 24, 57–58.
95 World Bank, "Approaches to the Environment in Brazil, Vol. IV," 91–92.

projects.[96] In its evaluation report, the World Bank concluded that the resettlement project had been a costly misadventure due to the delays and high costs of irrigation.[97]

Monthly stipends paid by CHESF kept the displaced families from slipping into poverty. In 1986, CHESF had agreed to pay a monthly stipend to farmers until their irrigation systems were functioning. Over the subsequent years, these temporary payments became indefinite and were the salvation of the Itaparica farmers, whose fates otherwise would have been the same as those at Sobradinho.

Yet, living off monthly stipends was not the solution that displaced farmers wanted. They remained without work and suffered from idleness. They had become so dependent on the payments that they were anxious about transitioning to irrigated agriculture when the time came, all the more so because high output on their new land would require costly fertilizers. Most had no experience with irrigation and were ten years out of farming practice.

The result was that Itaparica resettlement, despite the high hopes of its architects, entailed the same forced integration into the rapidly urbanizing and industrializing economy as those at other dams. Itaparica's displaced were much better compensated than their counterparts at other dams, but losing the floodplain meant losing their way of interacting with the land.

LOSING THE ARCHIPELAGO

Forced urbanization with scarce resources proved especially ruinous for the Tuxá, an Indigenous group whose material and cultural well-being were rooted in land that Itaparica flooded. The group – whose current population is about 2,000 people – had lived in the São Francisco River basin since before Portuguese colonization.[98] When Portuguese cattle

[96] See CDDHSSF, *Itaparica*, 24, and World Bank, "Operations Evaluation Department ICR Review of Itaparica Resettlement and Irrigation Project L2883-BR and L2883–1-BR," May 19, 1999, WBGA.

[97] World Bank, "Recent Experience with Involuntary Resettlement, Brazil-Itaparica, Report No. 17544," June 2, 1998, 33–35, WBGA.

[98] For more on the Tuxá and the impact of the Itaparica Dam, see Ricardo Dantas Borges Salomão, "Etnicidade, processos de territorialização e ritual entre os índios Tuxá de Rodelas," master's thesis, Universidade Federal Fluminense, 2006; Juracy Marques, "Cultura material e etnicidade dos povos indígenas do São Francisco afetados por barragens: Um estudo de caso dos Tuxá de Rodelas, Bahia, Brasil," (PhD Diss., Universidade Federal da Bahia, 2008); Felipe Sotto Maior Cruz, "'Quando a terra sair', os índios Tuxá de Rodelas e a Barragem de Itaparica: Memórias do desterro, memórias

ranchers and missionaries began moving inland along the São Francisco in the 1600s, the Tuxá, and other Indigenous groups in the region, retreated upriver. By the 1640s, the Tuxá, who were then composed of about 600 people, had settled on 123 sq. km along the shores of the São Francisco River near Itaparica Falls, in a village the Portuguese called Rodelas. The settlement was named after Francisco Rodelas, the leader of the Indigenous group, who, in that decade, helped the Portuguese fight off the Dutch. Between 1700 and 1703, the Portuguese crown officially recognized the land as Tuxá territory.[99]

For centuries, the Tuxá practiced floodplain agriculture along the margins of the São Francisco and on thirty or so fertile islands in the middle of the river, whose area totaled around 15 sq. km (Figure 4.1).[100] The Tuxá supplemented their agriculture with other activities such as fishing, hunting, gathering, raising livestock, producing ceramics, and extracting wood from the riparian forests. During this period, the group had much contact with Brazilian society, but remained economically and culturally independent.[101]

The Tuxá's landholdings diminished dramatically at the end of the nineteenth century and the beginning of the twentieth century. In 1889, the monarchy fell, and the new government abolished the Direitoria dos Índios, which had protected Indigenous land rights. Brazilian ranchers began to usurp large areas of the group's territory. Seu Armando, a religious leader from Rodelas, reflected on that era: "from there, they began taking, and taking, and taking ... and the Indians retreated from their land. They did not want to die; they did not have guns, and if they resisted, they knew they would die."[102] In some cases, the Tuxá fought

da resistência," master's thesis, Universidade de Brasília, 2017; Antônio Fernandes de Jesus Vieira, "Os índios Tuxá na rota do desenvolvimento: Violações de direitos," master's thesis, Universidade de Brasília, 2017; and Nelson Eduardo Bernal Dávalos, "Vulnerabilidade socioambiental e os impactos do reassentamento e dos eventos climáticos extremos sobre o povo indígena Tuxá de Rodelas-Bahia," (PhD Diss., Universidade de Brasília, 2021).

99 Ministério Público Federal Procuradoria da República Bahia (MPF-PRB), "Ação civil pública com pedido de liminar, contra a CHESF e a União Federal na defesa de direitos dos índios Tuxá, que perderam suas terras tradicionais quando do enchimento da Barragem de Itaparica, situada na fronteira dos Estados de Bahia e Pernambuco, em 1986," Salvador, June 30, 1999, 1, AUT.

100 Nássaro Antônio de Souza Nasser, "A economia Tuxá," master's thesis, Universidade Federal de Bahia, Salvador, 1975, 23, and Cruz, "'Quando a terra sair'," 42–46.

101 Sheila Brasileiro, "O processo de reassentamento do povo Tuxá," 2001, Folder BR.BA.TF.1b (Histórico), ACIMI.

102 Cruz, "'Quando a terra sair'," 53–55.

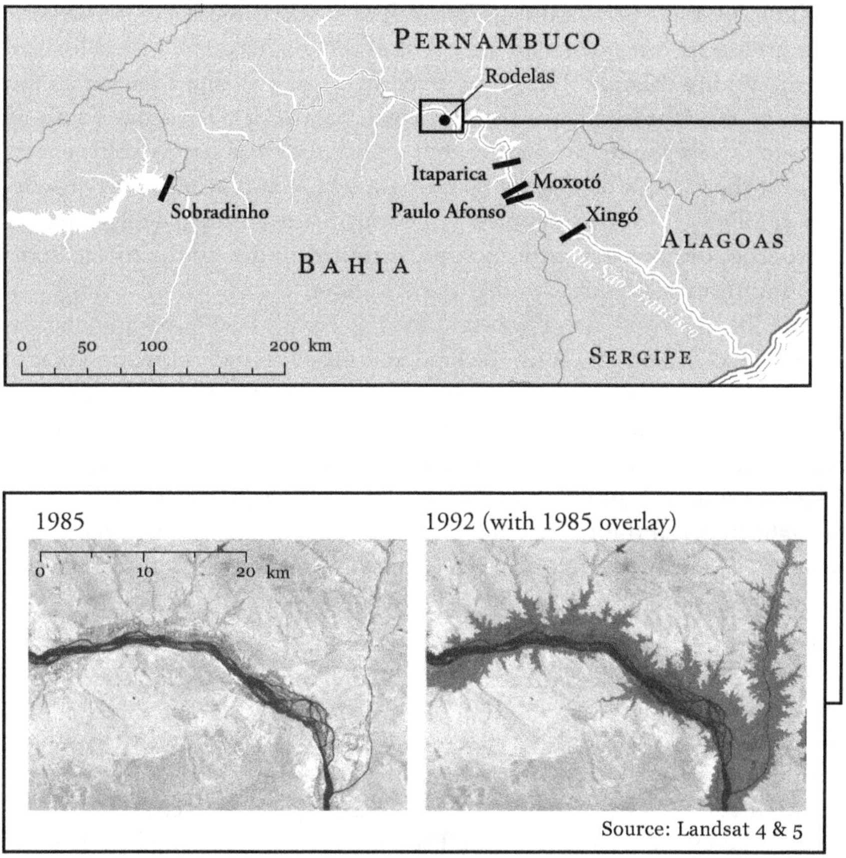

FIGURE 4.1. Tuxá land flooded by the Itaparica Dam's reservoir in 1988. Map by Geoffrey Wallace (G. Wallace Cartography & GIS).

back and retook land, only to face more waves of armed farmers who had the support of the Brazilian government.

In response to this violent land grabbing, between the 1930s and 1940s, the leadership of this group pressured the Indian Protection Service [Serviço de Proteção aos Índios, SPI] – a government agency created in 1910 – to help them defend their land. The Tuxá's protests were only partially successful: In 1945, the SPI established an office in Rodelas to help retake one important island, but it did not help them recover the rest of their land.[103]

[103] Ibid., 55–61 and Vieira, "Os índios Tuxá na rota do desenvolvimento," 49.

Thus, by the 1940s, land grabbers had reduced the Tuxá's ancestral archipelago to just one small island of about 50 hectares, called Ilha de Viúva [Widow Island]. The group lived in a small village adjacent to the river and used the island mostly to cultivate foodstuffs. The fecund soils of the island were highly productive, but 50 hectares was barely sufficient to support the group's demographic growth, which at that time was around 240 families, or 1,100 people.[104] Output soon became dependent on electricity-powered irrigation pumps, installed in the 1950s to facilitate planting further inland from the island's shore.[105]

The Ilha de Viúva was important for the Tuxá's livelihood and culture despite its deficient size. With the land and the collective cultivation that it facilitated, the Tuxá achieved economic autonomy and solidarity, two fundamental elements of their culture.[106] Moreover, the island was the place where the spirits of their ancestors lived and where they carried out religious rituals. Sustaining the entire group with just one small island brought material difficulties, but, for many Tuxá, the nonmaterial gains related to the way of living that the island sustained were more important. In the words of Tia Dôra, who lived in Rodelas at that time, in spite of the material difficulties, "we were happy and did not know it."[107]

Itaparica would inundate the last vestiges of the Tuxá's traditional land, an impact the dam's architects acknowledged from the beginning. Between January 26 and February 1, 1976, CHESF sent agents to meet with the Tuxá and carry out more than twenty interviews that it used to design a resettlement program. The Tuxá made it very clear that their principal concern was ecological: They needed to be resettled on land of similar quantity and quality as their original territory. The Tuxá self-identified as floodplain farmers and wanted to relocate somewhere near the river where they could farm. Additionally, they had grown accustomed to sending their children to Brazilian schools and placed great

104 Numbers rounded, see Fabiana Lima dos Santos, "Reconhecimento do território do povo Tuxá de Rodelas, Bahia," September 2, 2010, Justiça Federal Brasília, AUT.

105 CHESF, "Os Tuxá de Rodelas-Bahia, índios cujas terras serão inundadas pelas águas do reservatório de Itaparica, HE373-R21-0376," March 1976, 26–30, Document TI879, ACHESF; Salomão, "Etnicidade, processos de territorialização e ritual entre os índios Tuxá de Rodelas," 83; and Brasileiro, "O processo de reassentamento do povo Tuxá," 3–4.

106 On Tuxá economy and society during this period, see Nasser, "A economia Tuxá"; Elizabeth Cabral Nasser, "A sociedade Tuxá," master's thesis, Universidade Federal de Bahia, Salvador, 1975; and Orlando Sampaio Silva, *Tuxá: Índios do Nordeste* (São Paulo: Annablume, 1997).

107 Cruz, "'Quando a terra sair'," 125–130.

value on education, so they also wanted to relocate somewhere with access to Brazilian schools, which they had at Rodelas.

CHESF's report optimistically stated that both these goals were feasible and suggested that the two options that best conformed to the Tuxá's desires were to move them alongside Rodelas or to move them to an alternative area of their choosing in a similar ecological setting with access to education. The report concluded: "in whichever case, ecological characteristics of the new area should be considered, which should give the Indigenous group resources for development and progress and be in accordance with the where the Tuxá themselves choose to resettle."[108] In contrast to the situation at Itaipu, CHESF acknowledged that the Tuxá had rights to land of equal quality and quantity.

CHESF worked with FUNAI to hammer out an agreement. At that time, the Brazilian government's Indigenous policy was tutelary, and FUNAI signed accords in the name of Indigenous groups (the 1988 Constitution changed that policy).[109]

In 1986, CHESF and FUNAI negotiated an agreement – revised in the following year – that established the official terms for resettlement. During the preceding years, disagreements over indemnity and the resettlement location had caused political infighting that resulted in the splintering of the group into two main factions (and a third smaller one), shattering the group's social organization and solidarity. The accords solidified these divisions.[110] When it came time to move, 82 of the 214 Tuxá families displaced by Itaparica would move to Ibotirama, Bahia, an area far from Rodelas along the upper stretches of the São Francisco River. A smaller group would move to Inajá, Pernambuco, which was closer to Rodelas, but far from the river. The remaining families (the majority of those displaced) would move to a neighborhood on the outskirts of the new city of Rodelas and wait for the government to demarcate land and complete irrigation projects as compensation for the land the reservoir would flood.[111]

In 1988, with the dam completed, the Tuxá watched with heavy hearts as Itaparica's floodwaters covered the islands that had sustained them for hundreds of years. The remaining families relocated alongside the town of

108 CHESF, "Os Tuxá de Rodelas-Bahia, índios cujas terras serão inundadas," 44.

109 Cruz, "'Quando a terra sair'," 112.

110 FUNAI, "Relatório Tuxá," May 1984, 7–11, Document TFD02, AISA.

111 Santos, "Reconhecimento do território do povo Tuxá de Rodelas, Bahia," and Ely Souza Estrela, "Assentamentos indígenas no Médio São Francisco: O caso dos Tuxá e Pankarú," *Revista Travessia* 39, no. 1 (2001): 42–46.

Rodelas and waited eagerly for the Brazilian government to demarcate and deliver cultivable land to them.

But the Brazilian government never delivered on its promise. Over the subsequent decades, CHESF dragged its feet on demarcating and buying land for the Tuxá, and when it came under criticism, it pointed the finger at other federal agencies, such as FUNAI, that shared responsibility. However, the Tuxá never stopped pressuring CHESF to follow through on its agreement, and in 1991, they got the Federal Prosecution Service involved. On May 5, 1994, CHESF sat down with the Tuxá to renegotiate their agreement for the third time, though this agreement, like those that came before, did not produce the land that the Tuxá desperately needed to survive.

During these years, the group sustained themselves through monthly stipends from CHESF called Temporary Maintenance Stipends [Verbas de Manutenção Temporária, VMT], whose value was set to equal two and a half minimum wage salaries for each family. The company originally planned for the stipends to last just nine months, but, without land, the stipend became the principal source of income for the group for almost two decades. As time passed, the impact of the stipends diminished because CHESF did not extend payments to new families. New agreements in 2004 and 2006 (discussed further in Chapter 7) ended the VMTs, substituting them with a different payment system that was phased out in the following years.[112] The money from the VMTs helped the group sustain themselves during a difficult period, but it was not even close to sufficient compensation for the flooded land. The resulting desperation brought high rates of depression and alcoholism, among other ills.[113]

The Brazilian government's negligence toward its agreement to secure land for the Tuxá to replace the land Itaparica flooded constitutes a form of cultural genocide. Like the Avá Guarani, whose supporters had used the term to describe the group's suffering, the Tuxá have endured community disintegration and tremendous material and cultural impoverishment after Itaparica inundated their homeland. The Tuxá had made it clear to CHESF's agents in 1976 that farming was integral to their

112 Dávalos, "Vulnerabilidade socioambiental," 109.

113 MPF-PRB, "Ação civil pública," 3–4; Jorge Bruno Souza and Sheila dos Santos Brasileiro, "Diagnóstico socioeconômico das comunidades Tuxá de Rodelas, Ibotirama e Banzaê, na Bahia, e da comunidade Tuxá de Inajá em Pernambuco, Ministério Público Federal Parecer Pericial No. 041/2012," May 10, 2012, Justiça Federal Brasília, AUT.

identity. By not finding replacement land, the Brazilian government has been destroying what it means to be Tuxá.

The group is resilient and has fought to ensure that their culture survives the government's willful neglect, but cultural preservation gets more difficult with each year that passes without land. One of the older generation's biggest worries is that their children have grown up without learning how to farm, fish, and collect food and medicinal plants. Chief Bidu, after commenting on hunting and fishing before the dam, lamented that the current situation has been "calamitous for the Tuxá, because the children who have grown up since the dam no longer have the love for fishing, hunting, and for the land … today the change is so great that we live off the supermarket … because we do not have anything else to rely on."[114] The threat of losing cultural knowledge and traditions becomes ever greater the more time passes without land.

The Brazilian government has long recognized this fact. In 1999, the Public Prosecutor of the State of Bahia released a statement on the conflict, which they argued needed immediate resolution. The report noted: "there has been damage to culture and learning because new indigenous generations have been distanced from contact with land and are not familiarized with agricultural techniques, hunting, fishing, and other traditional activities."[115] A decade after Itaparica's reservoir filled, the government recognized that reneging on its promise to find compensatory land was causing devastating cultural loss.

The Tuxá are among the groups that have suffered the most from dam-induced displacement because subsequent governments have done so little to ameliorate their hardship despite the fact that all parties agree that the group deserves land. Their story is otherwise similar to the other communities displaced by the dictatorship's big reservoirs: The Tuxá were forced off their land so that Brazil could industrialize and urbanize and were forced to participate in this urban and industrial society against their will and without the resources needed to successfully adapt.

Displacement is standard for big reservoirs, and the number of those displaced by the Brazilian military regime's dams is somewhere near the middle of the range for such projects. The dictatorship's biggest dams dispossessed tens of thousands of people each, which is a lot compared to dams built in sparsely populated areas that displace a few hundred or thousand people. On the other end of the spectrum, Brazil's dam

114 Vieira, "Os índios Tuxá na rota do desenvolvimento," 72.
115 MPF-PRB, "Ação civil pública," 3.

displacement pales in comparison with reservoirs in India and China, two densely populated countries with lots of dams. India's Sardar Sarovar Dam on the Narmada River displaced between 200,000 and 500,000 people, and China's Three Gorges Dam on the Yangtze displaced more than 1.3 million people.[116]

The material hardship and trauma that dam displacement caused in Brazil parallel similar tragedies across the world. In most places, such calamities have been particularly acute for Indigenous communities because reservoirs often flood land with cultural and spiritual significance (in addition to useful resources) and because governments have consistently neglected Indigenous peoples. Between the 1920s and 1960s, the US government built dams throughout the American West that displaced Indigenous groups and inflicted tremendous material and psychological hardship on them.[117] The Gwembe Tonga, displaced by Zimbabwe and Zambia's Kariba Dam, suffered similar consequences.[118] Some amount of trauma afflicts most displaced people, though the degree of suffering often depends on what measures companies and governments take to ameliorate the damage.[119]

CONCLUSION

Between the late 1970s and the late 1980s, the Brazilian government finished most of its big reservoirs, which expelled hundreds of thousands of people from their homes and impoverished many of them. The government did little to improve the standard of living of displaced people over the subsequent decades.

The military regime was in a rush to finish its dams and designated few resources to helping dispossessed people, despite having produced

[116] For Sardar Sarovar, see Thayer Scudder, "India's Sardar Sarovar Project (SSP)" unpublished manuscript, 2003, available at http://docshare04.docshare.tips/files/13861/138617771.pdf (last accessed April 2023).

[117] Dina Gilio-Whitaker, *As Long as Grass Grows: The Indigenous Fight for Environmental Justice, from Colonization to Standing Rock* (Boston: Beacon Press, 2019), 60–62 and 179–180n22.

[118] See Elizabeth Colson, *The Social Consequences of Resettlement: The Impact of Kariba Resettlement upon the Gwembe Tonga* (Manchester: Manchester University Press, 1971), and the website of the Gwembe Tonga Research Project, https://anthropology.as.uky.edu/cligget/gwembe-tonga-research-project (last accessed April 2023).

[119] For a comparative perspective on dam displacement focusing on India and the United States, see Rob Nixon, *Slow Violence and the Environmentalism of the Poor* (Cambridge: Harvard University Press, 2011), 150–174.

voluminous material about the socioeconomic conditions of the populations its dams would inundate and issuing recommendations for how to successfully resettle them. Construction proceeded much faster than resettlement planning, and in almost all cases, the government failed to satisfactorily indemnify displaced people before reservoirs filled. To cut costs, dam-building companies offered derisory indemnity and only raised their offering price after displaced communities protested.

Individuals and organizations associated with the Catholic Church helped the displaced, organizing protest campaigns and broadcasting their hardships. The publicity they generated caught the attention of the World Bank, which was then loaning large sums of money to the Brazilian power sector. The dam-induced expulsions came at a moment when both the Catholic Church and the World Bank were becoming sympathetic to the plight of dispossessed people, which made them willing allies. Domestic and international pressure helped ensure that the resettlement process at one of the dictatorship's last dams, Itaparica, was not as tragic as its predecessors. Yet, this pressure was not enough to ensure that the government followed through on its promises to displaced communities in the years after the dams were finished. Such neglect caused particularly grave problems for affected Indigenous communities, who lost valued land to the reservoirs and were never justly compensated.

The Brazilian military regime built its big dams to further industrialize and urbanize population centers far from the dams themselves, yet their construction also brought about socioeconomic changes locally. Reservoir construction forced ill-prepared displaced communities to assimilate into the national economy. Though some adapted and prospered, many had neither the resources nor the willpower to do so. The military government's big dams also engendered significant changes in the nonhuman environment that came to have significant consequences for local communities. It is to these changes that the next chapter turns.

5

Environmental Transformations

Impacts on National Parks, Fish, and Malaria, 1970s–1990s

INTRODUCTION

The São Francisco River dazzled its nineteenth-century visitors. Beginning in the 1850s, a flood of geologists and engineers descended upon the river to assess ways to improve its potential as a commercial highway. Though chiefly concerned with reengineering the river, some could not help but be impressed with its scenic qualities. Nearly everyone who stopped at Paulo Afonso extolled the beauty of the falls, which were so big that engineers simply deemed that stretch of the river unnavigable. Though not as grand, Itaparica Falls also caught the attention of visitors. There, the river cut through a series of sunken channels, sculpting polished black cliffs and ridges that were exposed as the water receded during the dry season.[1]

To some engineers, such areas of astounding natural beauty were worth preserving. In 1880, North American geologist Orville Derby – a long-term resident of Brazil who later headed the country's first official geologic survey of its mineral resources – published a report on his trip along the river with a team of engineers. He wrote: "before I move onto other matters, I cannot refrain from drawing to the attention of the government the practice underway in the United States of preserving areas that encompass natural wonders, like Yellowstone and Yosemite, as

[1] Orville A. Derby, "Reconhecimento geológico do Vale de S. Francisco," 12, in *Relatório de W. Milnor Roberts, Engenheiro Chefe da Comissão Hidráulica sobre o Exame do Rio São Francisco desde o Mar até a Cachoeira de Pirapora (extensão de 2,122 quilômetros), feito em 1879–1880* (Rio de Janeiro: Typographia Nacional, 1880). Sampaio, *O Rio São Francisco e a Chapada Diamantina*, 62.

National Parks ... I suggest that the areas around Paulo Afonso and Itaparica could be conserved in the same manner."[2]

Though discounted at the time, his suggestion eventually found a receptive audience when the Paulo Afonso Dam was under construction in the 1940s (see Chapter 2). The Brazilian government recognized the benefits of protecting some choice areas. It heeded Derby's old recommendation, and on November 24, 1948, President Eurico Dutra created the Paulo Afonso National Park, a 170 sq. km area of land surrounding the falls.[3]

But the park never functioned as a conservation unit. As construction proceeded on the dam, Paulo Afonso and neighboring towns expanded into the park with little regard for its borders. Indeed, the dam itself was within the park's boundaries, and the Forest Service made no effort to restrain the activities of the São Francisco River Hydroelectric Company [Companhia Hidrelétrica do Rio São Francisco, CHESF], arguing that the dam made the area more appealing because it showcased symbiosis between nature and culture.[4]

Indeed, the dam was the area's primary tourist attraction. Brazilians from cities in the South and Northeast visited the dam and adjacent waterfall in busloads after the first dam was completed in the 1950s. It was CHESF, not the National Park, that absorbed these visitors. The park had no visitor center or tour guides, and just one small building to house its administrative staff. In contrast, CHESF built a visitor center and offered guided tours.[5] One of these tour guides, Antônio Souza, affirmed that the areas under the company's jurisdiction were the most appealing. He remembers taking the occasional guest to other areas of the park, but most of his tours centered on the dam and the adjacent falls.[6] Thus, during its first decade, Paulo Afonso National Park was little more than a signpost; a conservation unit in name only, poorly maintained, and falling apart.

In the 1960s, environmentalists recommended revitalizing the park. Zoologist Bertha Lutz suggested that the government downsize the park

2 Derby, "Reconhecimento Geológico do Vale de S. Francisco," 12.

3 Eurico Dutra, *Decreto No. 25.865, de 24 de novembro de 1948, cria o Parque Nacional de Paulo Afonso*, ACD.

4 Wanderbilt Duarte de Barros, *Parques Nacionais do Brasil* (Rio de Janeiro: Ministério de Agricultura, 1952), 75–76. I am grateful to my colleague Filipe Oliveira for sharing his digitized copy of this book with me.

5 Interview with Antônio Galdino, April 8, 2019, Paulo Afonso.

6 Antônio Souza began working for CHESF in 1948, first as a receptionist and then as a tour guide. As of April 2019, he was still working for the visitor's center (now run by the city) and giving tours of the dam. Interview with Antônio Souza, April 9, 2019, Paulo Afonso.

and redirect resources to build research and visitor centers near the falls, the principal natural feature of the area worth preserving.[7]

Rather than revitalize the park, the military government did away with it altogether. The military regime's vision for that stretch of the São Francisco included building dams that might flood areas of the park and upstream reservoirs that might withhold water from the falls (see Chapter 2). The military government saw little benefit in preserving the falls. On June 2, 1969, General Artur da Costa e Silva abolished the Paulo Afonso National Park.[8]

This chapter is about the environmental changes that unfolded after the military government finished building its big dams. It covers the period from the late 1970s to the late 1990s, the years in which the military government's big reservoirs filled and set in motion a series of massive ecological transformations. Some of these changes set the stage for disaster. In anticipation of losing the renowned Guaíra Falls, tourists descended in droves upon the Sete Quedas National Park in the years before Itaipu's reservoir flooded it. During this period, the government had stopped maintaining the park's infrastructure, and in 1981, one of its bridges collapsed, killing more than thirty people.

Once settled, the floodwaters of these reservoirs set in motion profound ecological changes that had consequences for the river's fish and insects, and the people who interacted with them. Reservoirs created nutrient booms that sustained large schools of fish and also created a class of insolvent displaced people who turned to fishing for income. Overfishing and busts abounded at reservoirs where economic opportunities were limited and fisheries were unregulated. Some reservoirs also created propitious habitat for the malaria-carrying mosquitoes, which menaced lakeside residents. Malaria surged at Itaipu, though a fragile disease ecology and the efforts of the health department limited the damage. Rival non-malaria-transmitting mosquitoes helped forestall an epidemic at Tucuruí, but not without exacting their own toll: For eight

[7] Bertha Lutz, "Proposta de transformação do Parque Nacional de Paulo Afonso em Monumento Natural, apresentada ao Conselho Florestal Federal, em 11 de Junho de 1962," Document BR.MN.BL.O.CFF.62 in the records of the Conselho Florestal Federal at the Museu Nacional, Rio de Janeiro. I am grateful to my colleague Filipe Oliveira for sharing his digitized copy of this document, the original version of which burned in the September 2018 fire, an immense tragedy that destroyed nearly all of the Museu Nacional's collection, including the records of the Forest Council.

[8] Artur da Costa e Silva, *Decreto-Lei No. 605, de 2 de junho de 1969, Extingue o Parque Nacional de Paulo Afonso e dá outras providências*, ACD.

years, dense swarms of these mosquitoes relentlessly plagued people relocated along the reservoir margins.

This chapter builds on the argument that the Brazilian military government largely ignored the environmental costs of its big dams because it was under pressure to build its dams quickly and cheaply and because it believed that its pharaonic environmentalism would satisfy its critics. The military regime's limited interventions did little to ameliorate the damage of these ecological transformations. To be sure, not all changes spelled disaster for local communities. Indeed, reservoir fisheries boomed in the decades following the formation of reservoirs, and malaria outbreaks on the edges of reservoirs were comparatively mild. Nevertheless, disregard for the environment led to a series of local catastrophes that drew attention to the impacts of big dams.

Moreover, the military regime's mitigation efforts did little to ensure the outcomes that were propitious, and its negligence was often responsible for the disasters. Fishing booms were the result of natural ecological changes, and malaria outbreaks at Itaipu were more easily curtailed because the National Health Department's agency for fighting malaria had been active in the area for decades and was well prepared to curb the scourge along the reservoir. Where favorable ecological changes and interdepartmental support were lacking, disregard for the nonhuman environment and displaced people had more damaging consequences. Neglect led to crumbling infrastructure at the Sete Quedas National Park, compelled thousands of displaced people to fish in a commercial system that encouraged overfishing and paid them little, and led to the resettlement of people at Tucuruí amid unbearably dense swarms of mosquitoes without the resources to relocate. Such results were so disastrous that public outrage began to build against big dams.

VISIT BEFORE THEY ARE GONE

The dictatorship's decision to build Itaipu entailed flooding Guaíra Falls, and the generals tried to soften the impact by highlighting that the neighboring Iguaçu Falls would be unaffected and could be promoted as a tourist site (see Chapters 2 and 4). That the government would spare Iguaçu Falls did little to console those who wanted to protect Guaíra Falls, which was an astonishing twenty times the volume of Iguaçu; indeed, it was the largest waterfall in the world by volume. It was a true natural wonder, a sprawling collection of hundreds of waterfalls that observers grouped into seven prominent clusters. The thundering falls supported a rich ecosystem, and the islands between the falls were popular hiking, swimming, and picnic spots.[9] In 1980, a research group from the Universidade Federal de Paraná in Curitiba released a report about Itaipu that mourned the loss of the waterfalls. They wrote: "Whoever has had the pleasure of visiting Sete Quedas has certainly experienced the sensation of feeling all their tensions released by the immanent forces of that entire complex of ecstatic beauty, as if they had received the baptism that renews and replenishes the living spirit of the mind."[10] For many visitors, Guaíra Falls exuded a unique magic for which there was no substitute.[11]

For most of its existence, Guaíra Falls was seldom visited because it was difficult to reach. The area surrounding the falls was relatively unpopulated until 1949, when the Brazilian government created a military post there to protect the border. Two years later, the Mate Laranjeiras Co., a large yerba mate company that had been the economic heart of the area for the preceding century, decided to give the urban center it owned to the State of Paraná, who created the municipality of Guaíra. The creation of the military post and the municipality encouraged migration, and people began moving en masse to build houses and hotels near the falls.[12] Despite this growth, tourism remained limited because Guaíra was far from Brazil's main population centers and principal highways. To get there, visitors had to travel down 300 kilometers of dusty roads that turned to deep mud after rains.[13] In contrast, Iguaçu Falls was a short distance from the town of Foz de Iguaçu, which was closer to the main roads that ran east toward coastal population centers.

In the 1950s and 1960s, the state and federal governments invested in infrastructure to make Guaíra Falls more accessible, but their efforts were

[9] Edineia Maria Gasparini et al., "Itaipu e as Sete Quedas: responsabilidade de poucos à espera do julgamento de muitos," Universidade Federal do Paraná, Setor de Ciências Sociais Aplicadas, Curso de Administração, 22, 1980, 44–50, unmarked folder, FEM, CEPEDAL; "Ernst Mann to *O Estado de Paraná*," August 1, 1978, Folder "Correspondências enviadas por Ernest Mann," FEM.

[10] Gasparini et al., "Itaipu e as Sete Quedas," 22.

[11] See "Para eles, as Sete Quedas são mais belas do que Iguaçu," *Revista Manchete*, May 30, 1970, and Mann, "Itaipu: obra tocada por 60 mil mãos, fortaleza de segurança," November 25, 1978, unmarked folder, FEM. See also, Daniella Feteira Soares, "Paisagem e Memória: Dos Saltos de Sete Quedas ao lago de Itaipu," master's thesis, Universidade Federal do Rio de Janeiro, 2001, and Ana Paula dos Santos, "Lago de memórias: A submersão das Sete Quedas," master's thesis, Universidade Estadual de Maringá, 2006.

[12] Mann, "Guaíra–A Cidade," Folder "Livro," FEM.

[13] "Guaíra exige mais atenção," *O Estado de São Paulo*, February 27, 1970.

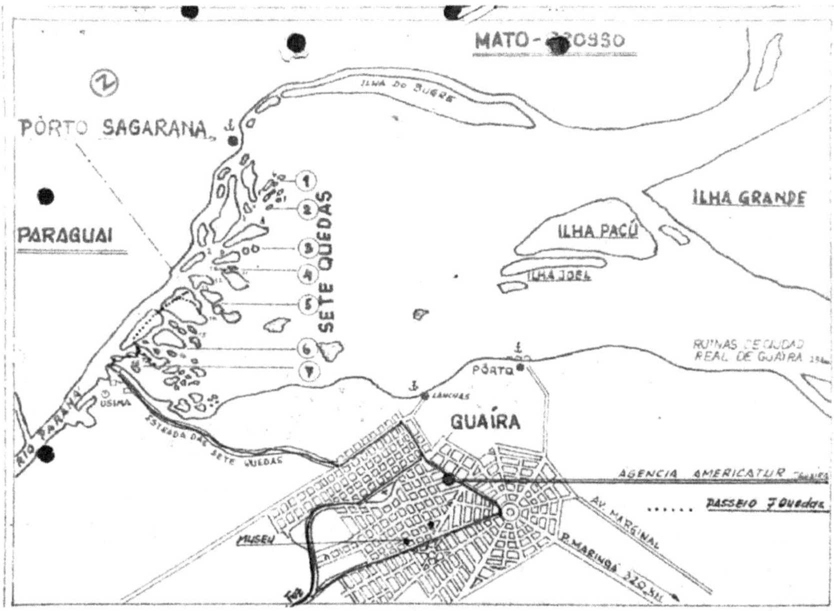

FIGURE 5.1. Sete Quedas National Park and Guaíra, undated, circa 1960s–1982. The faint dotted lines across the lower third of the falls mark the path that tourists could take to reach the fifth fall. Source: Folder 13, Fundo Ernst Mann, Núcleo de Pesquisa e Documentação sobre o Oeste do Paraná, Marechal Cândido Rondon, Brazil.

limited and did not do much to attract attention to the falls. On May 30, 1961, the Brazilian government took an assertive step when it created the Sete Quedas National Park [Parque Nacional de Sete Quedas, PNSQ] to protect the falls and encourage tourism (Figure 5.1).[14]

The idea to create a park at Guaíra Falls was an old one. In 1876, abolitionist André Rebouças had proposed a park that would encompass both Guaíra and Iguaçu Falls. He was greatly influenced by Yellowstone National Park, which the US government had created in 1872. His park never came to fruition, however, and the idea remained dormant until the late 1950s and early 1960s, when the Brazilian government began creating a host of national parks, including Sete Quedas.[15]

[14] Jânio Quadros, *Decreto No. 50.665, de 30 de maio de 1961, cria o Parque Nacional de Sete Quedas e dá outras providências*, ACD.

[15] See André Rebouças, *Excursão ao Salto do Guaíra, o parque nacional* (Rio de Janeiro, 1876); Pádua, *Um sopro de destruição*, 270–271; and Silva, "O Conselho Florestal Federal, 101–129.

The park was underfunded and understaffed. The Ministry of Agriculture's Forestry Service was in charge of managing the park until 1967, when the government created the Brazilian Institute of Forestry Development [Instituto Brasileiro de Desenvolvimento Florestal, IBDF], and the park came under its jurisdiction. On paper, the park also included land upstream of the falls, but the park had ill-defined boundaries, and staffing shortages meant there were few resources to enforce ecological preservation in the area beyond the falls. The park's director often complained to visitors about the degradation of local flora and fauna and, in the late 1970s, remarked that his position was comparable to a father of a stillborn child.[16]

From the beginning, local residents did more than the government to invest in infrastructure at the falls. PNSQ began at the southern end of the falls and consisted of a series of bridges that granted visitors access to the different clusters of falls (Figure 5.2). Local residents built most of the bridges, which in the early years extended only to the southernmost clusters. In 1964, the Rotary and Lions Clubs of Guaíra formed a commission to raise money to build a bridge to the fifth cluster, the largest and most beautiful of the falls. They completed the bridge in 1965; later, after floods destroyed this bridge, the government built a sturdier one.[17] The government invested little in the infrastructure within the park, and locals lamented the fact that there were neither informative plaques about the falls nor maps indicating points of interest. Few tourists actually made it to the fifth fall for lack of information on how to get there.[18]

The IBDF was also doing a poor job of promoting the National Park. Government publications invited international tourists to visit the beaches in Bahia and the Iguaçu Falls, but never mentioned Guaíra Falls.[19] Apparently, much of Brazil's own population thought that Iguaçu and Guaíra Falls were actually the same thing, and the press made no effort to try to distinguish between them.[20] PNSQ was ripe for tourism, but for more than a decade and a half after its creation, it remained difficult to access and mostly unknown.

The news that Guaíra Falls would be flooded further decreased investment from the government, which was unwilling to fund a park that

[16] "Restos do parque," *Veja*, October 27, 1976.
[17] Mann, "As 'Sete Quedas' do Rio Paraná," 4, Folder "Livro," FEM.
[18] Gasparini et al., "Itaipu e as Sete Quedas," 21–22.
[19] Mann, "O abandono das Sete Quedas," 1974, Folder "Artigos originais escritos por Ernest Mann," FEM.
[20] Gasparini et al., "Itaipu e as Sete Quedas," 41; "Mann to *O Estado de Paraná*."

FIGURE 5.2. Sete Quedas National Park photographs, September 1982. The top photo shows bridges across the sixth and seventh falls, which were the southernmost of the major clusters and the first that visitors encountered after entering the park. The bottom photo shows visitors at the fifth fall. Photographs by Jorge Somensi, published in Elio J. Brunetto and Jonas D. Muraski, eds., *Memórias de Sete Quedas* (Concórdia: Equiplan Serviços Gráficos Ltda., 1983).

would soon be underwater. Investing in the park would also draw attention to the environmental costs of Itaipu, which might incite protests that the military government was working hard to prevent.[21] Guaíra naturalist Ernst Mann (introduced at the beginning of Chapter 2) wrote the IBDF letters asking it to maintain the park's bridges, clear debris from pathways, put up signs to direct tourists, and install shelters for people to escape the rainstorms.[22] The government was slow to respond and ultimately did nothing to clean up the park.[23] Government neglect was so rife that residents had grown accustomed to organizing raffles to raise money to pay for improvements.[24]

Such disinvestment also made it difficult for tourists to reach the park. The government had started to build an airport and pave the highways that near Guaíra in the late 1960s, but abruptly halted construction on both projects.[25] One resident claimed that "Guaíra is nothing more than a postcard for the majority of the people from Paraná, who have never been. Getting there is a drama, a dangerous adventure, whose price few venture to pay ... abandoned, unprotected, without adequate infrastructure to promote itself ..."[26] On June 4, 1981, the government gave up altogether and terminated the PNSQ, though the falls remained open to visitors.[27]

Neglect was exacerbated by revenue misuse. In 1981, 194,000 tourists visited PNSQ, and the IBDF charged twenty cruzeiros to enter the park during the first half of the year and forty in the second half (after the park was abolished). The IBDF thus accumulated at least 12 million cruzeiros in entry fees alone and likely made even more money through the fees that vendors had to pay to operate in the park.[28] PNSQ was generating revenue, but none of it was used to maintain or improve the park's infrastructure.

The news that Guaíra Falls would be flooded also had the effect of drawing more attention to the PNSQ and encouraging more visitors. In Brazil's large cities, especially in Curitiba and Rio de Janeiro, word

[21] José Laurentino Gomes, "Guaíra não esquece seu maior acidente e lembra das lendas," *Jornal Ilha Grande*, January 30, 1982.

[22] Mann, "O abandono das Sete Quedas," 2, and Mann, "Os caminhos nas 7 Quedas," Folder "Artigos originais escritos por Ernest Mann," FEM.

[23] "Mann to *O Estado de Paraná*." [24] Gomes, "Guaíra não esquece."

[25] "Guaíra exige mais atenção."

[26] "Guaíra abandonada," *Jornal de Curitiba*, March 10, 1970.

[27] João Figueiredo, *Decreto No. 86.071, de 04 de junho de 1981, extingue o Parque Nacional de Sete Quedas, criado pelo Decreto No. 50665, de 30 de maio de 1961, e dá outras providências*, ACD.

[28] Gomes, "Guaíra não esquece"; "Gozar, sem maior responsabilidade," undated and unidentifiable newspaper, Folder "7QUE," FEM.

got out about the falls' impending disappearance.[29] Among other forms of publicity, cartoonists painted murals on the sides of restaurants, alerting people to the fact that Guaíra Falls would soon be underwater.[30] Advertisements promoting the park read "visit Sete Quedas before they disappear," and people poured into Guaíra in busloads to see one of the country's most impressive national wonders before it was gone.[31]

The combination of neglected infrastructure and massive waves of tourists proved disastrous. On the morning of January 17, 1981, the Roosevelt Bridge – the first large suspension bridge that tourists crossed upon entering the park – collapsed. Two Guaíra residents, Joel Onofre and João Mandi, bravely piloted a boat to the base of the fall where people were stranded, but many had already been swept away by the waters. About seventy people were on the bridge when it collapsed, and over half died as a result of the accident.[32]

Blame flew in all directions, but it was ultimately the fault of a government that had refused to maintain the bridges and monitor how many people were on them. Residents who had been complaining for years about the lack of attention paid to the falls had seen the disaster coming. One local paper wrote: "the fact is that the Sete Quedas National Park, recently extinct by presidential decree, has been abandoned for a long time, and Guaíra residents, in some sense, had predicted that a great accident would happen at the falls, sooner or later." Indeed, Mann had written to the IBDF just one week before the accident, asking it to maintain the bridges, but the agency never responded.[33] The article continued: "Some time ago ... the mayor assured that he would ask the IBDF to improve the safety of the bridges, but it always responds that any investment would be useless, since the park would be underwater in the month of October, after Itaipu's reservoir filled."[34]

Tourism waned in the weeks and months after the bridge collapsed, but picked back up thereafter. The swell of visitors grew to unprecedented

29 "Ecologistas protestam contra a destruição das Sete Quedas," *Jornal Ilha Grande*, July 18, 1981.

30 "Painéis pintados chamam para o 'Quarup' de Guaíra," undated and unidentifiable newspaper, Folder "7QUE," FEM.

31 Gasparini et al., "Itaipu e as Sete Quedas," 23; Hélio Teixeira and Pedro Franco, "Cerimônia de adeus," *Veja*, September 22, 1982.

32 "A tragédia das 7 Quedas," undated and unidentifiable newspaper, Folder "7QUE," FEM.

33 "Ernst Mann à *Folha de Londrina*," January 18, 1989, Folder "Correspondências enviadas por Ernest Mann," FEM.

34 "A tragédia das 7 Quedas."

numbers in the months before the falls were flooded. During the second to last weekend before the government closed the area to tourists, the city of Guaíra was packed to its limits. There were five-hour lines to cross the bridge into the park, and the city was running out of potable water. Guaíra expected up to 20,000 visitors the following weekend, and residents converted all of the city's schools, hospitals, and even the local daycare center into dormitories.[35] The military government had been neglecting the region to keep from drawing attention to the looming environmental disaster, but that did not keep tourists from flocking to visit Guaíra Falls before they were gone.

Though the military government was unable to prevent tourists from visiting, it succeeded in silencing any grassroots environmental movements that might have formed to protest Itaipu and save the falls. On June 6, 1968, the military government promulgated Decreto-Lei No. 5449, which made all areas within 150 kilometers of the country's international borders areas of national security and put them under the control of the military. That year, the military government extended indefinitely the tenure of Guaíra mayor Kurt Walter Hasper, who had been elected to the position in 1964 (he remained mayor until June 1985).[36] Hasper ensured that the local military prohibited all protests against Itaipu, and he made it known that officers would arrest any demonstrators.[37] Hasper himself may have mourned the loss of Guaíra Falls, but he agreed with his superiors that it was a necessary sacrifice to ease geopolitical tensions along the border (see Chapter 2).[38] Censorship was rife in Guaíra, and in 1973, Hasper shut down Mann's newspaper *Espelho de Guaíra* because he had criticized the government for not paving the highways to Guaíra.[39]

Protest started to build in the late 1970s as the military government's abertura program gradually loosened authoritarianism (see Chapter 1). Still, those in Guaíra were cautious. Earlier that year, Mann had written to *O Estado de Paraná* to clarify some doubts about Guaíra Falls and

35 "Guaíra, uma cidade já pronta para a despedida," unidentifiable newspaper, September 18, 1982, Folder "7QUE," FEM.

36 See Maria Helena Barbosa, "Memórias do regime militar em Guaíra," undergraduate thesis, Universidade Estadual do Oeste de Paraná, Marechal Cândido Rondon, 2003, and Mara Dhulle dos Santos Silva, "Guaíra: A cidade em tempos de ditadura civil-militar (1964–1985)," master's thesis, Universidade Estadual do Oeste de Paraná, Marechal Cândido Rondon, 2016.

37 Gasparini et al., "Itaipu e as Sete Quedas," 43–44.

38 Ibid., 44.

39 Lima, "Guaíra nos tempos de Ernst Mann," 36–38.

began by saying that he had never written before because he was scared that he would be censored. He wrote: "the winds appear to have changed and its now okay to say something, but the sky is still full of clouds and no one can predict what tomorrow will be."[40] In April 1979, Mann wrote to the military command in Guaíra to assure them that his calls for Itaipu to compensate the city for the loss of the falls would not incite demonstrations against the government or military.[41]

Nevertheless, abertura facilitated a small movement to save the falls. On March 15, 1979, the Guaíra Municipal Chamber held a meeting in which, for the first time, it openly criticized Itaipu for flooding the falls.[42] Momentum built over the next two years as concerned citizens from Brazil's largest cities rallied to save the falls. In September 1982, Rio de Janeiro-based lawyer Sulema Mendes, who was born and raised in Paraná, belatedly joined the chorus and tried to lodge a civil case that would require Itaipu to lower the dam's height.[43] Like Octávio Marcondes Ferraz and his supporters in the Senate (see Chapter 2), she argued that no other country would flood such a grand natural wonder, and she lamented the fact that dictatorship had impeded protests against the dam.[44]

Her argument held some weight. Though dams have flooded countless scenic landscapes, some of the world's most treasured waterscapes have been spared. Consider two of North America's most revered natural monuments: Niagara Falls and the Grand Canyon. In the early twentieth century, engineers proposed building a dam downstream from Niagara Falls that would flood it, but the proposal was discounted – both for its logistical issues and impact on the falls – in favor of smaller diversion dams that would preserve the falls while channeling water to downstream power houses. The idea for a downstream dam resurfaced in the postwar period, but engineers quickly dismissed it in order to preserve such an

40 "Mann to *O Estado de Paraná*."

41 "Ernst Mann to Comandante da 5a comanda fronteira," April 10, 1979, Folder "Correspondências enviadas por Ernest Mann," FEM.

42 Gasparini et al., "Itaipu e as Sete Quedas," 24, 43.

43 "Escritora prepara ação popular contra Itaipu," *Jornal do Brasil*, September 23, 1982; Secretaria de Estado da Segurança Pública, "DOPS-Sete Quedas-Carlos Pioli," October 5, 1982, Acervo Digital da Universidade Federal do Paraná, accessible online https://acervodigital.ufpr.br/handle/1884/61719 (last accessed April 2023); "Ação contra Itaipu tende ao fracasso," *Folha de São Paulo*, December 27, 1982.

44 Sulema Mendes, "Por Sete Quedas," *Jornal Ilha Grande*, October 16, 1982.

iconic natural feature and tourist attraction.[45] During the 1950s and 1960s, US engineers more seriously considered building a dam that would flood large stretches of the Grand Canyon National Monument. In response to the proposal, environmentalists organized a protest that successfully derailed the project.[46]

There has been less concern with stopping dams that threaten cultural monuments, but authorities have sometimes gone to great lengths to rescue historical treasures from floodwaters. During the 1960s, the Egyptian government and the United Nations spent tens of millions of dollars relocating ancient Egyptian temples so that they would not be destroyed by the Aswan High Dam's reservoir.[47]

By the early 1980s, however, most Brazilians realized that it was too late to save the falls, and they settled for a protest movement whose goal was not to stop the dam or lower its height but merely to bring attention to the disaster to ensure the same ecological errors were not made again in the future. In 1981, the falls' admirers formed the Goodbye Sete Quedas movement, which organized a festival that was to take place at Guaíra Falls in the months before the government closed off access. They called it Quarup, in homage to the Indigenous mourning ritual for a recently deceased chief. On July 24–26, 1982, thousands of people gathered and camped in the former PNSQ for a goodbye ceremony that consisted of speeches, songs, and dance.[48] In his summary of the movement, Mann reflected: "I vote that the country learn from our tears, that in the future the government avoids offering any natural treasures to hydroelectric dams without any effort of resistance from the people, when there exist alternatives to get the same result without destroying nature."[49] For those gathering at the Quarup, the falls could at least serve as a cautionary tale of the costs of building big dams under dictatorship.

Erasing cherished landscapes was just one of the dams' environmental impacts. The government's big reservoirs also engendered profound

[45] See Daniel Macfarlane, *Fixing Niagara Falls: Environment, Energy, and Engineers at the World's Most Famous Waterfall* (Vancouver: University of British Colombia Press, 2020), 42, 155, 183–184.

[46] See Reisner, *Cadillac Desert*, 273–274 and 283–290.

[47] See Reynolds, "Building Past Rockscapes and the Aswan High Dam in Egypt," 192–197.

[48] "Adeus Sete Quedas Festival," *O Estado de Paraná*," March 9, 1982; "Festival 'Adeus Sete Quedas'," *O Estado de Paraná*, May 12, 1982; "Adeus Sete Quedas," *Folha de Londrina*, July 24, 1982; "O que foi o Quarup," *Jornal Ilha Grande*, July 31, 1982; Urban, *Missão (quase) impossível*, 97–103.

[49] "Festival 'Adeus Sete Quedas'."

ecological transformations in the areas they flooded with dramatic consequences for the river's biota and the people interacting with it.

FISHING BOOMS AND BUSTS

The military government's big dams generated massive transformations among fish populations in the areas they flooded. In contrast to some iconic dams known for their detrimental impacts on fishing,[50] none of the dictatorship's dams caused a collapse of fisheries. Some of the most prolific fish in each reservoir did not need to migrate long distances to reproduce, so damming the river had little impact on their reproduction. Moreover, each of these reservoirs had large undammed sections of river immediately upstream, so fish that needed to migrate modest distances to reproduce could do so. Downstream fish populations were harder hit, but even there, fisheries did not collapse, with the exception of the stretch downstream of Sobradinho, where a series of dams impeded migration. Though some fish disappeared from the reservoirs because sluggish water did not suit them, other species filled this new ecological niche and proliferated in the reservoirs. Rather than collapse outright, fisheries in these new reservoirs unfolded along a changing pattern through the mid-1990s: an initial boom, and then overfishing and diminished returns thereafter.

Protecting fish from dams has a long history in Brazil. In the 1910s and 1920s, many of the private companies in the Southeast that erected the country's earliest dams built fish ladders, which are graded cascades that allow migratory fish to surmount dams. In the late 1920s, the state government in São Paulo – where most of these early dams were built – codified the practice into law and mandated that anyone building a dam there also had to build a fish ladder.[51]

Two subsequent national fishing laws softened this requirement. The 1938 Fish Code mandated that dam builders protect fish, but listed both ladders and pisciculture stations as suitable options.[52] The dictatorship

[50] On the Columbia River dams, see Richard White, *The Organic Machine: The Remaking of the Columbia River* (New York: Hill and Wang, 1995).

[51] Memória da Eletricidade, *O meio ambiente e o setor de energia elétrica brasileiro*, 13–16, and Julio Prestes de Albuquerque, *Lei No. 2.250, de 28 de Dezembro de 1927, Estabelecendo medidas relativas á caça e a pesca no território de Estado*, available on the website of the Assembleia Legislativa de São Paulo, www.al.sp.gov.br/repositorio/legislacao/lei/1927/lei-2250-28.12.1927.html (last accessed April 2023).

[52] Getúlio Vargas, *Decreto-Lei No. 794, de 19 de outubro de 1938, aprovando o Código da Pesca*, ACD.

promulgated its own fishing code in 1967, which dropped the reference to ladders altogether and simply stated that dam builders should protect fauna. The new code also created fishing colonies and stipulated that in order to fish legally at reservoirs, fishermen must be registered with one of these colonies.[53] Since it was under no legal obligation to do so, the military government did not build fish ladders on any of its hydropower dams (subsequent administrations added a fish ladder at Itaipu in 2002), and it did little else to mitigate the dams' impact on fisheries.

The generals abstained from building fish ladders because they were in a rush to build their dams quickly and cheaply. Though it turned out that many fish did not need the ladders, such foresight is not what compelled the military regime to eschew them. Indeed, Robert Goodland (introduced in Chapter 3) noted in his environmental impact assessment of Itaipu that researchers knew almost nothing about the migration habits of the Paraná River's fish.[54] The same was true at Sobradinho. As late as the mid-2000s, researchers were still learning about the migration and reproduction habits of important São Francisco River fish.[55] The resilience of fisheries in the first decade after each reservoir formed had more to do with ecology than environmental planning.

Though most reservoirs supported relatively healthy fisheries, they did little to improve the living conditions of those who fished in their waters. Each dam displaced tens of thousands of people without just compensation (see Chapter 4), many of whom turned to fishing to feed their families and earn income. In each case, the ranks of fishermen exploded after reservoirs filled. Increasing competition led to overfishing, which thinned fish populations to such an extent that fishermen had a harder time staying solvent. Most importantly, at each reservoir fishermen worked through a commercial system in which middlemen reaped tremendous profit, while fishermen earned comparatively little.

Fishermen at Sobradinho were the first to experience these changes. For the first six years after the reservoir filled, fisheries boomed. Floodwaters covered organic matter that decomposed and created nutrients that sustained burgeoning schools of fish. There is insufficient data to

[53] Humberto Castelo Branco, *Decreto-Lei No. 221, de 28 de fevereiro de 1967, dispõe sobre a proteção e estímulos à pesca e dá outras providências*, ACD.

[54] Comissão Mista Técnica, "Estudo do Rio Paraná: Relatório especial no. 4."

[55] Alexandre L. Godinho et al., "Migration and spawning of female surubim (*Pseudoplatystoma corruscans*, Pimelodiade) in the São Francisco River, Brazil," *Environmental Biology of Fishes* 80, no. 4 (December 2007): 421–433.

chart the demographic changes among fish species after the reservoir filled, but it is clear that some species found the still waters suitable, perhaps even preferable to the undammed river.[56] One report claimed that the reservoir was producing around 20,000 metric tons of fish annually during its early years, and other reports put the figure at 30,000–50,000 tons annually during the height of the boom in the early to mid-1980s.[57] With about 14,000–15,000 people fishing on the reservoir at that time, a rough estimate for an individual fisherman's average catch during the height of the boom, in round numbers, is a range of 2,000–3,600 kg of fish per year.

Soon thereafter, production began a steep decline, and by the mid-1990s, the reservoir's fishery was a ghost of its former stature. Dom José Rodrigues (introduced in Chapter 4) – who in addition to being a bishop was also then the president of the Pastoral Council of Fishermen – claimed that the reservoir produced less than 4,000 metric tons of fish in 1991, and local newspapers reported that annual yields were around 5,000 tons in the mid-1990s.[58] Sobradinho's fishing boom had turned to bust.

This crash had dire consequences for fishermen, who were now finding it almost impossible to earn an income from fishing. In one case, over the course of a week, five local fishermen together caught just 68 kg of fish, or about 14 kg each. If yields remained that low for the entire year, their annual yield would be around 700 kg per year each. Even fishermen catching nearly twice that amount claimed that it was not profitable enough to sustain them. Another group of nine fishermen spent five days camping and fishing on the reservoir, collectively catching around 210 kg of fish, or about 23 kg each.[59] Such yields sustained over the course of a year would amount to an annual catch of about 1,200 kg per fisherman, which is almost double what their peers caught and near the lowest estimate for the reservoir's early years, but still roughly half the figure for the boom years (using the middle and high estimate). Even though this

[56] "Sobradinho: O milagre dos peixes," *Veja*, June 20, 1979.

[57] For the low number, see World Bank and World Fisheries Trust, *Migratory Fishes of South America: Biology, Fisheries, and Conservation Status* (Washington, DC: World Bank, 2003), 217. For the higher range, see "Encontro de pescadores em Juazeiro de 01 a 04/08/1990: 'o pescador e o meio ambiente'," *Caminhar Juntos* no. 156, October 1990, 2, BDJR and "A morte do Velho Chico," *Jornal do Commercio*, April 2, 1995, 6–7.

[58] "Encontro de pescadores em Juazeiro," 2, and "A morte do Velho Chico," 6.

[59] "A morte do Velho Chico," 6.

second group was much more successful than their colleagues, their catch still did not earn them enough to pay their expenses.

Diminishing yields brought great hardship because fishermen operated within an exploitive system in which middlemen captured most of the profits from fishing. On the São Francisco, fishermen sold their catch to merchants who had the capital to purchase ice and vehicles, which allowed them to store and transport fish to regional markets. These middlemen charged fishermen high prices for ice and other supplies and paid them little for their catch.[60] Because they earned little from the sale of each fish, fishermen counted on large catches to stay solvent.

Natural causes were partly responsible for the bust. The reservoir's initial nutrient boom dissipated as the breakdown of organic matter ran its course, and the regional climate further curbed food access for some species. Sobradinho is located in a hot tropical climate where the temperature is consistent most of the year. Because the temperature of the air above the reservoir was more or less constant, Sobradinho's water remained stratified, and nutrients did not circulate. Nutrients abounded at the bottom of the reservoir where they sustained bottom-dwelling fish, but the reservoir's still and stratified water was not able to support many of the fish that had once prospered in the river, whose moving water circulated nutrients in spite of consistent air temperatures.

Overfishing also contributed to the bust. The number of people fishing in the stretch of the river that Sobradinho flooded had mushroomed after the reservoir filled. The Brazilian government estimates that in the mid-1980s, there were about 2,000 fishermen on the reservoir.[61] The actual number of people fishing was likely much greater considering the socio-economic conditions in the area. Sobradinho's floodwaters displaced tens of thousands of people who were given little indemnity, many of whom resettled on infertile land surrounding the reservoir or in shantytowns on the outskirts of nearby urban centers (see Chapter 4). With insufficient resources to farm, many turned to fishing to feed their families and earn supplemental income. In the early 1980s, Rodrigues estimated that there

60 Agrosig Engenharia and Meio Ambiente Eireli, "Relatório 10: Programa de monitoramento do Rio São Francisco durante o período de vazão reduzida," March 2018, 6, ACHESF, and Comissão Pastoral dos Pescadores, "Situação dos pescadores do Lago de Sobradinho," *Caminhar Juntos* no. 61, November 1981, 11, BDJR.

61 This figure comes from the São Francisco and Parnaíba Valley Development Company [Companhia de Desenvolvimento dos Vales de São Francisco e do Parnaíba, CODEVASF] and is referenced in Agrosig Engenharia e Meio Ambiente Eireli, "Relatório 10: Programa de monitoramento do Rio São Francisco," 6.

were between 14,000 and 15,000 people fishing on Sobradinho's reservoir.[62] Though perhaps exaggerated, these figures are probably closer to the true number of fishermen at that time.

Weather and limited incomes were also factors in the bust, confining these crowds of fishermen to small stretches of the reservoir, thereby limiting fishing grounds and exacerbating overfishing when the ecological conditions that had engendered the boom faded. The area surrounding Sobradinho was characterized by strong winds, which created large waves. Since most fishermen could only afford small, unmotorized wooden boats that could not handle these waves, a large percentage of those fishing were restricted to the shoreline and the relatively shallow areas in the middle sections of the reservoir with smaller waves.[63] Furthermore, since Sobradinho's reservoir had flooded an area of more than 4,214 sq. km, fish had a large space to spread out across and could avoid capture in areas that fishermen could not access due to strong winds.[64]

As catches fell, fishermen began using illegal equipment, such as small mesh nets that could catch smaller, and thus younger, fish.[65] Catching young fish prevented them from reaching adulthood and reproducing, thus further diminishing fish populations. The agencies charged with regulating the fishing industry on the reservoir were underfunded and were not able to curtail illegal fishing techniques, which further exacerbated the pressures of overfishing.[66]

Fishermen from outside the area added to the pressure that locals put on the fishery and were often the target of blame for predatory practices.[67] Outsider fishermen were a common phenomenon at each of the dictatorship's big reservoirs, and some groups traveled from dam to dam to take advantage of the propitious fishing conditions during the first few years after reservoirs filled.[68] At each fishery, outsiders were the minority,

62 Comissão Pastoral dos Pescadores, "Situação dos pescadores do Lago de Sobradinho," 11, and "Assembleia de pescadores em Pilão Arcado," *Caminhar Juntos* no. 102, April 1985, BDJR.

63 Comissão Pastoral dos Pescadores, "Situação dos pescadores do Lago de Sobradinho," 7, and "Relatório 10: Programa de monitoramento do Rio São Francisco," 6.

64 World Bank and World Fisheries Trust, *Migratory Fishes of South America*, 207.

65 Comissão Pastoral dos Pescadores, "Situação dos pescadores do Lago de Sobradinho," 11; "A morte do Velho Chico," 6.

66 "Sobradinho: O milagre dos peixes," 108; "A morte do Velho Chico," 6; World Bank and World Fisheries Trust, *Migratory Fishes of South America*, 223.

67 "A morte do Velho Chico," 6.

68 Angelo Antonio Agostinho et al., "Características econômicas e sociais das atividades pesqueiras no reservatório de Itaipu," presented at the Regional Symposium on Fish

but these groups often had access to greater capital and better equipment – such as bigger boats – which enabled them to have an outsized impact on catches.

Though overfishing and environmental changes may account for the decline in fish catches after the mid-1980s, they alone cannot account for the economic disaster that befell Sobradinho's fishermen. Exploitive middlemen made declining yields ruinous for fishermen. Fishing was only profitable during the boom years when catches were high; when yields began dropping in the late 1980s, fishermen could no longer afford ice and other supplies that middlemen sold at inflated prices. Fishing had been a temporary salvation for those displaced by Sobradinho, but by the mid-1990s it was no longer a sustainable means of earning a living.

STILL WATERS IN THE SOUTH

The situation at Itaipu unfolded along similar lines. Most of the fishermen on the reservoir during the 1980s and 1990s were Brazilian. Land on the Paraguayan side was heavily forested (see Chapter 3), which made accessing the shoreline difficult. Moreover, with the exception of the Avá Guaraní (see Chapter 4), the Paraguayan side of the reservoir was relatively unpopulated during those decades.

The stretch of the Paraná River that became Itaipu's reservoir had not been an important fishery for non-Indigenous Brazilians.[69] Most who moved into Western Paraná starting in the 1960s (see Chapter 4) invested in farms rather than fishing. Some families fished for subsistence, but few earned a livelihood fishing.[70] Indeed, Goodland had argued that the absence of commercial fisheries was one of the characteristics that made Itaipu such an appealing dam site in terms of its environmental footprint.[71]

Commercial fishing along the Paraná began in earnest in the late 1970s. Tourism growth in Guaíra and Foz de Iguaçu increased the demand for fish in hotels and restaurants, and the road improvements that allowed more tourists to visit the region also enabled fishermen to

Management at Latin American Dams, which was part of the Food and Agriculture Organization of the United Nations' conference in Havana, Cuba, October 24–28, 1994, 19, RI-UEM.

69 Agostinho, "Represamentos: Impactos sobre a ictiofauna e sua mitigação," 1989, BIB-FO.

70 Agostinho et al., "Características econômicas e sociais das atividades pesqueiras,"16.

71 Comissão Mista Técnica, "Estudo do Rio Paraná, Relatório especial no. 4," S-1.

expand their clientele.[72] This nascent fishing industry was small by national standards, but it was growing fast in response to demand from tourists and regional markets.

The dam threatened to upend this fledgling industry, and, starting in 1976, Itaipu began studying the river's aquatic biota in order to anticipate some of the changes.[73] The company's reports predicted that there would be a short surge in fish in the months after the reservoir filled, after which populations would stabilize, with catches remaining greater than during the period before the reservoir.[74]

Fish populations did in fact boom in the reservoir. Though the absence of fish catch data prior to reservoir filling makes it impossible to determine the reservoir's true impact, fish abounded after the reservoir filled. Some species, however, did not. Of the 102 species that Itaipu's biologists identified in the river in the late 1970s, about a fifth were missing from the reservoir. Most of these fish were still present downstream, and the ones that disappeared altogether did not have sizeable populations to begin with.[75]

Guaíra's fishing grounds exemplify this ecological transformation. There, the environmental and socioeconomic conditions set in motion by Itaipu led to the decline of the treasured cascudo preto (*Rhinelepis aspera*), a catfish that feeds on microscopic plant and animal remains on the bottom of the river and resides in submerged rock piles, such as the area around the flooded Guaíra Falls.[76] It was one of the most abundant species in the area before the reservoir filled and in the months after,[77] but its populations crashed thereafter due to overfishing and sedimentation.[78] Guaíra Falls is at the uppermost end of the reservoir, where silt-laden

[72] Agostinho et al., "Características econômicas e sociais das atividades pesqueiras," 16.

[73] See Consórcio para Estudos de Meio Ambiente (COMAM) and Itaipu, "Estudos para avaliação do impacto ecológico aquático, plano de trabalho" 1978, BIB-FO; COMAM e Itaipu, "Acréscimo dos trabalhos impacto ecológico aquático, proposta técnica" 1979, BIB-FO.

[74] Ibid. [75] Agostinho, "Represamentos: impactos sobre a ictiofauna e sua mitigação."

[76] On cascudo, see Agostinho, "Relatório final do projeto: Avaliação do estoque do Cascudo Preto, *Rhinelepis aspera*, Cascudo Abacaxi, *Megalancistrus aculeatus* e Cascudo Branco, *Hypostomus sp.* no Rio Paraná, nas imediações da cidade de Guaíra," Vol. 2, Processo 98.10.12000-1, NDJFP.

[77] Agostinho, "Relato sobre a situação da pesca profissional no reservatório de Itaipu, na área de abrangência da Colônia de Pescadores Z-13, Guaíra," December 1989, Vol. 2, Processo 98.10.12000-1, NDJFP.

[78] "Situação da pesca no reservatório Itaipu," *Jornal do Pescador*, 1992, Vol. 6, Processo 98.10.12000-1, NDJFP.

water from the upper stretches of the Paraná and its tributaries deposits its sediments over the rocky areas at the bottom of the reservoir, destroying the habitat of the cascudo.[79]

However, while cascudo populations declined, the armado (*Pterodoras granulosus*) populations soared. The armado (also a type of catfish) did not reside in the area's submerged rocks, so it was unaffected by sedimentation. Catches of armado made up for the declining cascudo stocks and enabled fishermen in Guaíra to maintain catches, at least through the mid-1990s.[80]

During the late 1980s and early 1990s, annual fish production across the entire reservoir was around 1,500 metric tons (1,500,000 kg).[81] During that period, there were about 830 fishermen on Itaipu's waters each year, which meant that individual fishermen were catching upwards of 1,800 kg of fish from the reservoir annually.

Commercial fisheries thus endured the changes unleashed by the reservoir, but abundant fish did not translate to economic stability for fishermen. Similar to what unfolded at Sobradinho, hundreds of displaced people became either full- or part-time fishermen after Itaipu's reservoir filled. Biologist Angelo Agostinho estimates that 84 percent of the roughly 830 fishermen on the reservoir in the late 1980s were those recently displaced from nearby farms by agricultural mechanization and Itaipu's floodwaters.[82] Expelled from their homes and jobs with little to no indemnity (see Chapter 4), many displaced people turned to fishing to feed their families.[83]

79 Agostinho, "Relato sobre a situação da pesca profissional no reservatório de Itaipu"; Agostinho et al., "Características econômicas e sociais das atividades pesqueiras no reservatório de Itaipu," 16.

80 "A produção de pesca na área de Guaíra." For more on changing fish demographics in other parts of the reservoir, see Johnson, "Temples of Modern Pharaohs," 289–292.

81 This catch worked out to about 12 kg/hectare/year, which, according to Brazilian fish specialists, is incredibly low compared to small reservoirs in Northeastern Brazil that have an annual catch of about 152 kg/hectare/year, or reservoirs in Africa that have about 100 kg/hectare/year. However, the catch is high by regional standards, as most other reservoirs on the main stem and tributaries of the Paraná average less than 5 kg/hectare/year. Agostinho et al., "Características econômicas e sociais das atividades pesqueiras no reservatório de Itaipu,"15–17.

82 Ibid., 31.

83 Agostinho estimated that in 1993, about 40 percent of the fishermen on Itaipu's reservoir were fulltime fishermen, 36 percent were small farmers who also fished, 21 percent were day laborers on farms who also fished, and 3 percent were urban laborers who fished. Ibid, 33–34.

These novice fishermen confronted similar problems to those at Sobradinho. Fishermen at Itaipu worked within a commercial system in which middlemen reaped most of the profits and fishermen earned comparatively little. In 1993, fishermen on Itaipu's reservoir earned an average of about US$24 a month (Agostinho reported his findings in dollars rather than cruzeiros). Though this was more than a farmhand usually earned before the reservoir formed, it was far from the standard of living that people expected the dam to generate.

The most important factor accounting for low earnings was the middlemen who set prices. They lowballed fishermen, selling the fish for three to six times what they bought them for. These profits enabled middlemen to invest in tools that fishermen needed – such as fuel, ice, and refrigeration – which they sold at marked-up prices. Agostinho estimated that the average gross income for fishermen between 1987 and 1993 was about US$100 a month, though their expenses cut this sum by three quarters.[84] Fishermen did well to adapt to the dramatic ecological changes that unfolded beneath the reservoir's still waters, but enduring socioeconomic inequalities limited their abilities to improve their standard of living.

Tucuruí also brought dramatic changes to its river's fisheries, which unfolded along a similar pattern as other reservoirs: There were dramatic demographic changes among the river's fish and a decrease in biodiversity, but net increases in fishing productivity in the reservoir. Downstream communities, in contrast, saw steep declines in the years after the reservoir filled.[85] One exceptional feature of Tucuruí's fishery is the presence of high methylmercury concentrations among people for whom the reservoir's fish is their main source of protein. Methylmercury is a highly toxic organic form of mercury, and the anoxic bottom layers of reservoirs are propitious environments for the bacteria that transform mercury into methylmercury, which fish consume. Mercury occurs naturally in soils and can also be introduced into reservoirs through atmospheric deposition and human activity, such as placer gold mining. Researchers are still assessing the factors responsible for Tucuruí's high concentrations.[86]

[84] Ibid., 85–88. [85] See WCD, *WCD Case Study Tucuruí*, xii.
[86] Gabriela P. F. Arrifano et al., "Large-Scale Projects in the Amazon and Human Exposure to Mercury: The Case Study of the Tucuruí Dam," *Ecotoxicology and Environmental Safety* 147 (2018): 299–305.

The boom-and-bust nature of these fisheries reflects a global pattern for large reservoirs in the tropics and subtropics. It is standard for newly filled reservoirs to generate short-lived nutrient booms that support temporarily prosperous fisheries.[87] For example, Kariba's reservoir, which filled between 1959 and 1963, supported a flourishing fishery that briefly sustained thousands of Gwembe Tonga, whom the dam had displaced, before collapsing.[88] The fishery recovered and expanded decades later, but its early boom and bust were emblematic of a global pattern that also characterized the Brazilian military government's big reservoirs.

Changes in fish populations also reflect the global trend for reservoirs in the tropics. Chinese reservoirs also saw an overall decline in species diversity and the diminishing or extinction of some migratory fish, but a corresponding proliferation of slow water-adapted, or lentic, species.[89] The scholarly consensus is that reservoirs are generally less productive than river fisheries, but fast-growing populations of lentic fish enable tropical reservoirs to boast reasonably productive fisheries nonetheless.[90] Such fisheries have been especially productive when governments stock them with well-adapted fish, and fishermen practice pisciculture, the latter of which has helped sustain Brazilian reservoir fisheries since 2003.[91]

MALARIA ALONG RESERVOIR MARGINS

Flooding Guaíra Falls also eliminated a natural barrier for *Anopheles darlingi*, Brazil's most efficient malaria-transmitting mosquito.

[87] For Africa, see T. Petr, "On Some Factors Associated with the Initial High Fish Catches in New African Man-Made Lakes," *Archives of Hydrobiology* 75, no. 1 (1975): 32–49.

[88] Thayer Scudder, "The Kariba Case Study," Social Science Working Paper 1227, 10, California Institute of Technology Division of the Humanities and Social Sciences, 2005.

[89] Yiguang Zhong and Geoff Power, "Environmental Impacts of Hydroelectric Projects on Fish Resources in China," *Regulated Rivers: Research and Management* 12 (1996): 81–98.

[90] Gerd Marmulla, ed., *Dams, Fish, and Fisheries: Opportunities, Challenges, and Conflict Resolution*, FAO Fisheries Technical Paper No. 419 (Rome: FAO, 2001); Paul A.M. van Zwieten et al., eds., *Review of Tropical Reservoirs and their Fisheries: The Cases of Lake Nasser, Lake Volta, and Indo-Gangetic Basin Reservoirs*, FAO Fisheries and Aquaculture Technical Paper No. 557 (Rome: FAO, 2011).

[91] For short overviews of pisciculture at Sobradinho and Itaipu, see Marco Aurélio Rotta et al., "Analyzing the Supply Chain for Fish in Sobradinho Lake, Bahia, Brazil," *World Aquaculture* (2016): 53–57, and Itaipu, "More Fish in Our Waters," www.itaipu.gov.br/en/the-environment/more-fish-our-waters-0 (last accessed April 2023).

An. darlingi inhabit a large area of South America and are found across nearly all of Brazil except for the arid Northeast and the area southeast of the Iguaçu River. *An. darlingi* are not the most common mosquitoes in Brazil, but they are responsible for about 80 percent of all malaria transmission in the country because their preference for human blood and their propensity to rest on the interior walls of houses make them highly efficient transmitters of malaria.[92]

The stretch of the Paraná River between Guaíra Falls and Foz de Iguaçu was largely *darlingi*-free during the decades preceding the construction of Itaipu. *An. darlingi* inhabit sizeable bodies of stagnant water and were therefore rare amid the fast-moving waters of the narrow canyon that stretched between Guaíra and Foz de Iguaçu.[93] Because there were so few *An. darlingi* along this stretch of the Paraná River, malaria was all but absent there in the decades prior to the completion of Itaipu.[94]

Government measures were also helping to stymie malaria in the area. In 1957, the federal government launched a massive campaign to eradicate malaria, based primarily on spraying DDT in the interior of homes, which worked effectively against *An. darlingi*. In 1959, it began to target Paraná.[95] The campaign – led, after 1970, by the Superintendency of Public Health Campaigns [Superintendência das Campanhas de Saúde Pública, SUCAM] – did not succeed at eradicating malaria in Brazil, but by the late 1970s and 1980s, it had largely erased the scourge

[92] José Consolim et al., "O Anopheles (Nyssorhynchus) darlingi ROOT, 1926 e a malária no lago de Itaipu, Estado de Paraná, Brasil: Uma revisão de dados," *Arquivo Biológico Tecnológico* 34, no. 2 (June 1991): 263–286, 263, and Maria Eugênia Moreira Costa Ferreira, "Ocorrência de malária na área de influência do reservatório de Itaipu, margem esquerda, Paraná, Brasil," (PhD Diss., Universidade de São Paulo (USP), 1996), 68, 71–73.

[93] Consolim and Joaquim T. Galvão, "Sobre os anofelinos do Rio Paraná: Densidade e regime do rio," *Arquivo Biológico Tecnológico* 16 (1973): 174–181, AAMS.

[94] This is the consensus among most researchers studying malaria and Itaipu, though a lacuna of official data makes it difficult to verify the claim. Ana Lucia Falavigna-Guilherme et al., "Retrospective Study of Malaria Prevalence and Anopheles Genus in the Area of Influence of the Binational Itaipu Reservoir," *Revista Institucional de Medicina Tropical de São Paulo* 74, no. 2 (March–April, 2005): 81–86, 83.

[95] Simone do Rocio Ferreira and Ennio Luz, "Malária no Estado de Paraná—Aspetos históricos e prognose," *Acta Biológica Paranaense* 32 (2003): 129–156, 147, and Sean Michael Griffing et al., "A historical perspective on malaria control in Brazil," *Memória Instituto Oswaldo Cruz* 110, no. 6 (2015): 701–718, 708.

from Paraná.[96] In 1976, the government declared the disease eradicated from the districts surrounding Itaipu.[97]

In the late 1980s, however, the districts adjacent to Itaipu's reservoir became the site of a malaria epidemic. The government registered the first cases of autochthonous malaria in 1986, and in 1988 the number of cases rose to epidemic proportions.[98] The outbreak reached its peak in 1989, with upwards of 1,000 autochthonous cases and more than 16,500 imported cases.[99] The documentary record does not identify how many people died during the epidemic. It is near certain that some lives were lost, but it is likely that mortality rates were low given that the majority of cases were treated and were *Plasmodium vivax*, a debilitating albeit less deadly species of the malaria-causing parasite compared to the more lethal *Plasmodium falciparum*.[100]

The 1988–1990 epidemic was the result of two major environmental transformations: soaring populations of both *darlingi* mosquitoes and people emigrating to the region who were carrying malaria plasmodium. When Itaipu's reservoir filled in 1982, it opened a pathway for *An. darlingi* to move southward. The preferred habitat of *darlingi* mosquitoes is large bodies of stagnant fresh water between 68 and 82.5 degrees Fahrenheit with aquatic plants that can shelter their larvae from predators.[101] The margins of Itaipu's reservoir were ideal real estate for *An. darlingi*, which multiplied dramatically in the still water along the reservoir margins.[102]

96 Juvêncio Mazzarollo, "Surto de malária na fronteira," *Oeste* no. 36, April 1989. The Brazilian government has done a poor job of preserving the records of this campaign, in part because in 1990, the federal government dissolved SUCAM and decentralized malaria control, charging state health departments and individual districts with the responsibility of preventing or controlling malaria in their jurisdictions. Most of the records were lost during this transition, though some reports can be found in the BFUNASA.

97 Ferreira and Luz, "Malária no Estado de Paraná," 153.

98 Mazzarollo, "Surto de malária na fronteira."

99 Ferreira, "Ocorrência de malária na área de influência do reservatório de Itaipu," 49–67; Falavigna-Guilherme et al., "Retrospective Study of Malaria Prevalence," 82.

100 Ibid. *Falciparum* has an average case fatality rate of roughly 25 percent for previously unexposed people who go untreated, and *vivax* has a case fatality rate of roughly 5 percent for previously unexposed people who go untreated and about 1–2 percent for those who are treated.

101 Consolim and Galvão, "Sobre os anofelinos do Rio Paraná."

102 Consolim et al., "O Anopheles (Nyssorhynchus) darlingi ROOT, 1926 e a malária no lago de Itaipu"; José Consolim et al., "Culicídeos (Díptera, Culicidae) do Lago de Itaipu, Paraná, Brasil. I. Município de Foz de Iguaçu," *Acta Biológica Paranaense* 22 (1993): 83–90; Ueslei Teodoro et al., "Culicídeos do Lago de Itaipu, no Rio Paraná, Sul do

Itaipu's reservoir also slowed the water in the lower stretches of its tributaries, creating a base for mosquitoes to spread into the surrounding countryside. During the 1988–1990 epidemic, many of the autochthonous cases occurred more than five kilometers from Itaipu's reservoir, which is roughly the limit of the flying range of *An. darlingi*. Yet, none of these cases were more than five kilometers from a major tributary.[103]

In addition to creating a propitious aquatic environment for *An. darlingi*, Itaipu's reservoir perhaps also created more suitable air temperatures for the mosquito. *An. darlingi* only tolerate air temperatures above 75 degrees Fahrenheit and humidity between 70 and 80 percent. The reservoir possibly increased both air temperature and humidity along its edges, allowing *darlingi* mosquitoes to survive the colder months that had previously slashed their numbers.[104]

The second big environmental transformation was the reintroduction and circulation of the malaria plasmodium in areas where *darlingi* populations were booming. Itaipu played a prominent role: About a tenth of those displaced by its reservoir (see Chapter 4) chose to go to agricultural colonization schemes or to work in informal mining camps, called garimpos, in Rondônia and Mato Grosso, two of the most malarious states in Brazil at the time.[105] Much of the work there was seasonal because floods during the rainy season made farming and mining impossible. Some found the working conditions there so grueling that they gave up and returned to Western Paraná. Others who stayed in Rondônia or Mato Grosso full time made seasonal trips to Western Paraná to visit relatives over the holidays. Whatever their circumstances, those traveling

Brasil," *Revista Saúde Pública* 29, no. 1 (1995): 6–14; Anthony Érico Guimarães et al., "Prevalência de anofelinos (Díptera: Culicidae) no crepúsculo vespertino em área da Usina Hidrelétrica de Itaipu, no Município de Guaíra, Estado do Paraná, Brasil," *Memória Instituto Oswaldo Cruz* 92, no. 6 (1997): 745–754.

103 Ferreira, "Ocorrência de malária na área de influência do reservatório de Itaipu," 170–191.

104 Maria Eugênia Moreira Costa Ferreira, "'Doenças tropicais': O clima e a saúde coletiva. Alterações climáticas e a ocorrência de malária na área de influência do reservatório de Itaipu, PR," *Terra Live* 19 (2003): 179–191. For more on the potential changes to the microclimate around the reservoir, see Leila Limberger and Sandra Elisa Contri Pitton, "Percepção climática de moradores lindeiros ao reservatório da UH de Itaipu," in Souza, ed., *Estudos Regionais: Estrutura, Agentes e Processos* (Cascavel: Editora Unioeste, 2012), 401–430.

105 Ferreira estimates that during the 1989 epidemic, roughly 2/3 of the imported malaria cases came from Rondônia and Mato Grosso. See Ferreira, "Ocorrência de malária na área de influência do reservatório de Itaipu," 49–67, 170–191.

between Amazônia and Western Paraná kept the malaria plasmodium circulating in the region.

Neighboring Paraguay was the other region that provided a steady supply of malaria to Western Paraná. Malaria had long been endemic in Paraguay, and the government there had fewer resources than Brazil to fight the scourge, which meant that it was often prevalent in the municipalities that bordered Itaipu's reservoir.[106] The treaty that the Paraguayan and Brazilian military governments signed in 1966 that had committed them to building a binational dam also included clauses granting Brazilians favorable access to land in the Paraguayan borderlands. The result was that after 1973 there was a sustained migration of Brazilian farmers to the lands on the Paraguayan side of Itaipu's reservoir.[107] The Avá Guaraní (introduced in Chapter 4) traditionally migrated freely between the two countries and did so even more after the reservoir flooded their land. Lastly, a host of commercial actors, such as illegal drug distributors, used the reservoir as part of their overland routes between Paraguay and Brazilian cities.[108] Like those returning from Amazônia, migrants from Paraguay kept reintroducing malaria plasmodium to the margins of Itaipu's reservoir, which was becoming home to denser swarms of *darlingi* mosquitoes. The combination of increasing numbers of people hosting the malaria parasite and efficient mosquito vectors set off the 1988–1990 malaria epidemic.

A final contributing factor to the epidemic was SUCAM's relaxing of its malaria control procedures in the preceding years. During its eradication campaign, SUCAM had sprayed houses in the area with DDT and monitored for malaria. But after it had declared the area malaria-free in the late 1970s, it relaxed its malaria control programs and turned its attention and resources to fighting other diseases.

During the epidemic, malaria sickened people of all classes, but it was poor communities living and working along the margins of the reservoir who suffered disproportionately. The Avá Guarani at Ocoí were among the earliest to be affected during the outbreak, and they saw some of the

106 For more on the history of malaria in Paraguay, see Servicio Nacional de Erradicación del Paludismo (SENEPA), "Informe nacional de eliminación: Paraguay país libre de malaria," 2017, ASENEPA-AP.

107 See Jacob Blanc, "Enclaves of Inequality: Brasiguaios and the Transformation of the Brazil-Paraguay Borderlands," *Journal of Peasant Studies* 42, no. 1 (2015): 145–158.

108 Interview with Mario Pilleco, Foz de Iguaçu, November 6, 2018. Mario Pilleco worked for both SUCAM and the Secretaria de Saúde de Paraná (SESA) in Foz de Iguaçu for 26 years between 1984 and 2010.

highest malaria rates in the region. In their 1986 letter to the president of the World Bank, the Avá Guarani complained about mosquitoes and malaria, which had sickened about 30 percent of the people living in Ocoí. They wrote: "here we do not have water without malaria; we are drinking water full of bugs and malaria."[109] In 1989, the percentage of Avá Guarani at Ocoí infected with malaria rose to 50 percent.[110]

Malaria had disproportionally impacted the Avá Guarani because they lived meters from the edge of the reservoir, where mosquitoes bred, and built their houses using tree branches and other wooden material that created holes through which mosquitoes could pass freely. Impoverished after Itaipu flooded their land, many Avá Guarani suffered malnutrition, making them more susceptible to other diseases, such as malaria, and had little money to buy prophylactic medicine or treatments.[111]

Fishermen and poor families in Foz de Iguaçu also suffered disproportionately for similar reasons. Fishermen were constantly exposed to malaria-carrying *darlingi* mosquitoes while working on the reservoir and living in porous houses alongside it. In Foz de Iguaçu, roughly 80 percent of the malaria cases came from Bela Vista, in the northeast corner of Vila C.[112] It is the neighborhood closest to the reservoir, and some houses are just meters from the water's edge. Vila C is also one of the city's poorest neighborhoods. When it built the dam, Itaipu Binacional assembled three neighborhoods next to the construction site to house workers and segregated them by social class (Vilas A, B, and C). Vila C housed unskilled and semiskilled workers. Itaipu had planned to demolish the neighborhood upon completing the dam, so it built houses using basic materials that left giant gaps, similar to the housing conditions at Ocoí and fishing villages.[113] Many residents living there also suffered from malnutrition and could not afford prophylactic medicine or treatments.

[109] According to the Avá Guarani, malaria had infected 43 of the 147 people living in Ocoí that year, 27 adults, and 16 children. "Avá Guarani to Barber Conable," September 18, 1986, Folder BR.PR.GN5d Guarani Itaipu, ACIMI, Brasília.

[110] According to Ferreira, malaria infected 153 of the 300 people living in Ocoí that year. Ferreira, "Ocorrência de malária na área de influência do reservatório de Itaipu," 203.

[111] Mazzarollo, "Surto de malária na fronteira."

[112] Ferreira, "Ocorrência de malária na área de influência do reservatório de Itaipu," 170–191.

[113] John Howard White, "Itaipú: Gender, Community, and Work in the Alto Paraná Borderlands, Brazil and Paraguay, 1954–1989," (PhD Diss., University of New Mexico, 2010), 165.

The government moved swiftly and successfully to tame the epidemic. SUCAM's office in Foz de Iguaçu, believing malaria to have been eradicated from the area, had been focusing all its attention and resources on dengue prevention. But when malaria cases started surging after 1986, it jumped into action and restarted a series of measures to control the scourge. Its approach was twofold. First, it sent teams of workers to spray DDT on the interior walls of homes alongside the reservoir.[114] Second, it sent teams of both employees and volunteers to identify sick people and administer treatment, which typically consisted of a fourteen-day course of chloroquine or primaquine.[115]

The campaign was a success. Between January and May 1990, the number of autochthonous cases dropped to 192, and only 45 more were reported during the rest of the year, bringing the total number of autochthonous cases in 1990 to 237. In 1991, there were fewer than 60 autochthonous cases, and between 1992 and 1993, fewer than 20.[116]

SUCAM's efforts were bolstered by economic improvements in the region, which obviated the need for seasonal labor in malarious areas, helping curb malaria rates in subsequent decades.[117] During the 1990s and 2000s, malaria fluctuated between zero and 150 cases in the municipalities along the reservoir, and in most years, there were well under 100 cases.[118] In 1995, there was a small outbreak of 141 cases in Foz de Iguaçu and Santa Terezinha, and in 2002, there was another surge of 94 cases at the Ocoí reserve, an epidemic for a population who then numbered about 578 people.[119] These comparatively minor outbreaks

114 SUCAM, "Informe No. 010/89 – ASI/SUCAM," April 21, 1989, Folder 28, Record Group BR DFANBSB IS.INF.ECE – Erradicação e Controle de Endemias, AN-BSB, and Interview with Mario Pilleco.

115 Interview with Maria do Carmen, Foz de Iguaçu, September 27, 2018. Maria do Carmen worked for both SUCAM and SESA in Foz de Iguaçu for more than 35 years starting in 1979.

116 Ferreira, "Ocorrência de malária na área de influência do reservatório de Itaipu," 49–67

117 Interview with Mario Pilleco. This claim is difficult to confirm with quantitative data because censuses do not account for seasonal migration patterns.

118 Falavigna-Guilherme et al., "Retrospective Study of Malaria Prevalence"; Ferreira and Luz, "Malária no Estado de Paraná"; Marta Bértoli and Maria da Luz Ribeiro Moitinho, "Malária no Estado do Paraná, Brasil," *Revista da Sociedade Brasileira de Medicina Tropical* 34, no. 1 (2001): 43–47; Enéas Cordeiro de Souza Filho, "Epidemiologia da malária no Estado de Paraná, Brasil, 2002 a 2008," master's thesis, Universidade Federal de Paraná, 2010, 61.

119 On the 1995 outbreak, see Bértoli and Moitinho, "Malária no Estado do Paraná," 45. On the 2002 outbreak, see Souza Filho, "Epidemiologia da malária no Estado de Paraná," 61.

notwithstanding, the government had largely controlled malaria in the communities along the reservoir margins.[120]

The municipalities on the Paraguayan side of the reservoir struggled harder to control the disease. During the 1970s and 1980s, malaria rates were high across Paraguay, and after 1982, Itaipu's reservoir became one of the country's biggest epicenters. Officers of the National Service of Malaria Eradication [Servicio Nacional de Erradicación del Paludismo, SENEPA] often remarked that the communities within a five-kilometer radius of the reservoir had some of the highest rates in the region (departments of Alto Paraná and Canindeyú), which was in turn the most malarious area of the country.[121]

The 1988–1990 epidemic hit harder and lasted longer in Paraguay. During the first six months of 1990, Paraguayan communities along Itaipu's margin hosted 2,113 cases of autochthonous malaria.[122] Like their Brazilian counterparts, SENEPA workers sprayed houses with DDT and sent teams of employees and volunteers to undertake public health surveys and provide treatment to infected people. The disease was so deeply entrenched in Paraguay that SENEPA efforts were unsuccessful in the short term. But as socioeconomic conditions improved in subsequent decades and the agency gained greater resources, it stepped up its campaign and succeeded in making Paraguay malaria-free in 2013, a designation that the World Health Organization certified five years later.[123]

Though devastating for the people affected, Itaipu's malaria epidemic was mild compared to many dam-associated outbreaks. Irrigation projects in the late nineteenth and early twentieth centuries engendered outbreaks orders of magnitude greater than Itaipu's. Between the 1860s and 1910s, the British imperial government built irrigation systems in

[120] For more on malaria among the Avá Guarani at Ocoí in the 2000s, see Souza Filho, "Epidemiologia da malária no Estado de Paraná"; Roderlei de Araujo, "Composição da fauna de Anopheles (Diptera: Culicidae) da Reserva Indígena do Ocoy, foco ativo de malária no município de São Miguel do Iguaçu, Paraná," master's thesis, Universidade Federal de Paraná, 2011; Araujo et al., "Composição da fauna de Anopheles (Diptera: Culicidae) da Reserva Indígena do Ocoy, foco ativo de malária no município de São Miguel do Iguaçu, Paraná (Brasil)," *Acta Biológica Paranaense* 43 (2014): 85–112.

[121] SENEPA, "Informe de las actividades desarrolladas por SENEPA en el área especial, año 1990," 5, ASENEPA-AP; Acordo sanitário Brasil Paraguai, "Ata da 8ª reunião do comitê de coordenação," October, 1990, ASENEPA-AP.

[122] Acordo sanitário Brasil Paraguai, "Ata da 8ª reunião do comitê de coordenação."

[123] For more on the eradication campaign, see "Informe nacional de eliminación: Paraguay país libre de malaria," 2017, ASENEPA-AP.

Northwestern India that caused malaria rates to surge, sickening and killing thousands of people annually.[124] Irrigation systems in Puerto Rico were similarly devastating.[125]

Dam-associated malaria outbreaks diminished after the 1940s – when DDT and other factors helped curb malaria – and even more so at hydropower dams, which do not include networks of irrigation canals that can spread mosquito populations. Even so, outbreaks occasionally occurred, and some took heavy tolls. In the late 1990s, the reservoir of the Bargi Dam on the Narmada River in India was the site of a malaria outbreak that killed hundreds and sickened thousands.[126]

Moderate by such standards, Itaipu's malaria epidemic was nevertheless noteworthy given that many hydropower reservoirs built between the 1950s and 1980s did not lead to outbreaks at all. For example, neither Kariba nor Aswan's reservoirs, two of Africa's biggest, led to malaria surges,[127] and many other reservoirs in the tropics and subtropics have not generated malaria. In some cases, malaria did not spread because parasite species were absent; in other cases, ecological conditions were not suitable for local vectors, or potential human hosts were scarce.

Two factors combined to render malaria at Itaipu comparatively mild. First, the disease ecology was not very severe to begin with. *An. darlingi* thrived on Itaipu's margins, but the reservoir was at the southernmost edge of their habitat, and they were thus more sensitive to seasonal weather shifts than in other areas, such as the Amazon. Moreover, malaria was not endemic, and it therefore took a steady stream of infected

[124] Elizabeth Whitcombe, "The Environmental Costs of Irrigation in British India: Waterlogging, Salinity, and Malaria," in David Arnold and Ramachandra Guha, eds., *Nature, Culture, Imperialism: Essays on the Environmental History of South Asia* (New York: Oxford University Press, 1995), 237–259; Mark Harrison, "'Hot Beds of Disease': Malaria and Civilization in Nineteenth-Century British India," *Parassitologia* 40 (1998): 11–18, 14; Ira Klein, "Development and Death: Reinterpreting Malaria, Economics, and Ecology in British India," *The Indian Economic and Social History Review* 38, no. 2 (2001): 147–179, 170–172.

[125] Matthew P. Johnson, "Swampy Sugar Lands: Irrigation Dams and the Rise and Fall of Malaria in Puerto Rico, 1898–1962," *Journal of Latin American Studies* 51, no. 2 (May 2019): 243–271.

[126] N. Singh et al., "Malaria and the Narmada-River Development in India: A Case Study of the Bargi Dam," *Annals of Tropical Medicine and Parasitology* 93, no. 5 (1999): 477–488.

[127] WCD, *WCD Case Study Kariba Dam Zambia and Zimbabwe, Final Report* (Cape Town: WCD, 2000), 120–121; M. A. Abu-Zeid and F. Z. El-Shibini, "Egypt's High Aswan Dam," *International Journal of Water Resources Development* 13, no. 2 (1997): 209–218, 216.

migrants to continually reintroduce the parasite. This happened more often in the tumultuous years following the reservoir's formation. Once the regional economy improved, people found less need to migrate seasonally between malarious areas and thus did not circulate the disease as much.

The second factor was the work of SUCAM. Itaipu played an important role in financing the campaigns to combat malaria in the region, but the company itself did not provide the infrastructure, training, or manpower to fight malaria. SUCAM was an efficient and organized agency that was effective in carrying out its malaria control campaigns. Most of the residents in the region were receptive to the agency's work and cooperated. It was therefore able to spray home interiors with DDT without resistance and to use local social networks to effectively identify cases and treat them in a timely manner. While both factors proved effective at curbing malaria at Itaipu, neither was in play at Tucuruí, where ecological changes unleashed by the reservoir threatened to create an even bigger epidemic.

LAUGHING WITH CLOSED MOUTHS

Malaria rates were soaring in Amazônia during the military regime. The scourge had long been endemic to the region, as the wet and humid rainforest provided excellent habitat for malaria-carrying *An. darlingi* mosquitoes.[128] However, prior to the 1960s, there was little malaria circulating among the isolated Indigenous communities that inhabited the forest, and few population centers were big enough to host the parasite in endemic or epidemic form. The demographic situation changed after 1964, when the military government started sending hundreds of thousands of nonresistant Brazilians from the Northeast and Southeast into the forest to populate it (see Chapter 1). Everywhere that nonresistant Brazilians settled, malaria threatened to severely sicken them.[129]

Tucuruí was no exception. Tens of thousands of Brazilians, most of whom were recent migrants, labored to erect the giant dam. As was the case at the region's agricultural villages and mines, clearing the forest for

[128] On malaria in Amazônia, see Maria Auxiliadora Bessa Barroso and Raul Diniz Souza Amorim, *Malária no Amazonas: Registros e memória* (Manaus: Editora Valer, 2017).

[129] For national context, see Joseli Oliveira-Ferreira et al., "Malaria in Brazil: An Overview," *Malaria Journal* 9 (2010).

the dam's construction site created cluttered landscapes dotted with pools of stagnant water after rains – ideal habitat for *darlingi* mosquitoes. The combination of mosquito swarms and a large number of nonresistant laborers created the perfect conditions for malaria to spread.

The municipality of Tucuruí had little malaria before construction began, but thereafter malaria rates soared. The municipality's average annual parasite index (API) – which measures the number of confirmed new cases, expressed per 1,000 individuals – was roughly 14 during the decade prior to construction. During the construction years (1975–1984), the average API shot up to 82 and fluctuated between 38 and 150 annually.[130] Though malaria was sickening Tucuruí's workers at record rates for that locale, the rate was not high by the standards of other Amazonian colonization projects.[131] But Tucuruí had one unique feature that distinguished its malaria potential from that of farming and mining projects: Once finished, the dam would create a gigantic reservoir whose vegetation-covered shoreline would create ideal real estate for *Anopheles* mosquitoes.

Tucuruí's reservoir had the potential to engender a malaria epidemic. Goodland had warned about the malaria threat in his environmental impact study. In addition to reservoir margins, he worried that uncleared forests and water weeds would harbor mosquitoes and that annual drawdowns for power production would expose breeding grounds that could produce prolific swarms of mosquitoes. Goodland had seen the rising malaria rates firsthand when he carried out his impact study, and he anticipated that the reservoir would only exacerbate the problem: "even before the reservoir construction, malaria already was seriously increasing in the area, and must be expected to worsen."[132] Malaria was a likely consequence of building the reservoir, and Goodland had urged the government to prepare for it.

130 Rosa Carmina de Sena Couto, "Hidrelétricas e saúde na Amazônia: Um estudo sobre a tendência da malária na área do lago da UHE Tucuruí-Pará-Brasil," (PhD Diss., Escola Nacional de Saúde Pública (ENSP)/Fundação Oswaldo Cruz (Fiocruz), 1996). I consulted this rare dissertation in Record Group ENSP, Box 188, Folder 1, AFIOCRUZ. Couto reproduced much of the information from her dissertation in "Malária o custo social da hidrelétrica de Tucuruí-PA, Brasil," in Couto et al., eds., *Saúde, trabalho e meio ambiente: políticas públicas na Amazônia* (Belém: Editora Universitária UFPA, 2002), 107–121.

131 SUCAM, Distrito regional do Pará, "Estratégia adotada no combate à malária em áreas de alta transmissão no Estado do Pará," March 1984, 11, BFUNASA.

132 Goodland, *Environmental Assessment of the Tucuruí Hydroelectric Project*, 151–53.

Eletronorte heeded Goodland's advice and hired scientists from the National Institute of Amazonian Research [Instituto Nacional de Pesquisas da Amazônia, INPA] and SUCAM to study the relationship between malaria and the reservoir. Dr. Wanderlei Tadei, the head scientist of these studies, shared Goodland's concern and built on his conclusion that controlling these swarms would be impossible given the ecological characteristics of the region – heavy rains created swampy lands that made overland transportation difficult, and many houses were built either without walls or with material that was unsuitable for DDT spraying. Tadei recommended that Eletronorte maintain a 1.5-kilometer belt of lightly forested land between the reservoir edge and lakeside communities to prevent malaria transmission.[133]

Eletronorte ignored Tadei's advice and instead settled displaced people in a haphazard manner that included placing some communities adjacent to the reservoir. The ecological results were predictable, and in 1986, Tadei noticed a rise in *darlingi* mosquitoes that portended a dramatic increase in malaria rates. *An. darlingi* populations continued to rise over the following year, with ominous implications. Suddenly, however, they stopped.

The reason was that another mosquito had displaced them. Tucuruí's reservoir had created especially good habitat for *Mansonia titillans* mosquitoes, which do not transmit malaria, though they are inefficient transmitters of some rare arboviruses. These mosquitoes, found widely across Central and South America, thrive under similar conditions as *darlingi* mosquitoes, and they are especially prolific in environments that have a lot of water weeds (called macrophytes), because their larvae attach to the submerged parts of aquatic vegetation.[134] The Tocantins River is naturally rich in nutrients that it picks up from eroding riverbanks on its course from the highlands of Central Brazil to the Atlantic Ocean. When the Tucuruí Dam plugged the river, nutrient-rich water accumulated and fed macrophytes, which expanded prolifically across the surface of

[133] Dr. Wanderlei Pedro Tadei, "Estudos de ecologia e controle ambiental na região da UHE de Tucuruí, convênio ELN/CNPq/INPA, relatório setorial, segmento doenças endêmicas, estudo das espécies de anófeles, vetores de malária," January–June, 1984, BINPA. Many of the reports from this working group are available at both BINPA and AELN.

[134] Ralph Harbach, "Mansonia, Blanchard, 1901," Mosquito Taxonomic Inventory, 2014. https://mosquito-taxonomic-inventory.myspecies.info/node/11703 (last accessed April 2023).

the reservoir.[135] *Mansonia* mosquitoes found even more habitat in the remains of the flooded trees (called *paliteiros* in Portuguese) that Eletronorte failed to clear prior to filling the reservoir (see Chapter 3).

In addition to habitat, these mosquitoes also found abundant supplies of blood meals in wild animals that had been captured and released as a result of Operation Curupira (see Chapter 3) and people and their domestic animals displaced by the dam and resettled by Eletronorte on the edges of the reservoir (see Chapter 4).[136]

Water turbidity curbed *Anopheles* populations in this otherwise favorable habitat. *Darlingi* mosquitoes thrive in similar conditions to *Mansonia*, with one notable exception: *An. darlingi* cannot tolerate turbid water as well as *Mansonia*.[137] The nutrient-filled water that poured into Tucuruí's reservoir flooded a large amount of organic debris that clouded the water, which discouraged *An. darlingi* from proliferating.[138]

The immense expansion of *Mansonia* habitat after the reservoir finished filling in December 1985 led to a boom in *Mansonia* mosquitoes that quickly displaced *An. darlingi*, thereby curbing malaria transmission. Malaria continued to sicken residents in communities alongside the reservoir, but at rates lower than those during the construction years.[139] Tucuruí's reservoir had not exacerbated the local malaria burden.

135 Wanderlei Pedro Tadei et al., "Evolução das populações de Anopheles e de Mansonia, na área de influência da usina hidrelétrica de Tucuruí (Pará)," *Ciência e cultura* 30 (January 1991): 639. Interview with Wanderlei Pedro Tadei, Manaus, October 7, 2017. Tadei is a malariologist who works for INPA, in Manaus and is one of the foremost experts on malaria and hydroelectric dams in the Brazilian Amazon. He was one of the researchers that the military government hired to carry out malaria studies at Tucuruí before the reservoir filled.

136 "Relatório Comissão de Estudos da Proliferação de Mosquitos a Montante de Tucuruí," December 1989, AISA; Secretaria de Saúde do Pará, "Ata da reunião convocada para definição das ações do poder público quanto a problemática da proliferação de mosquitos a montante da Usina Hidrelétrica de Tucuruí, PA," Belém, 13 de maio, 1991; Sindicato dos Trabalhadores Rurais de Tucuruí (STR-TUC), "Relatório parcial dos problemas da área de proliferação de mosquitos do lago reservatório da Usina Hidrelétrica de Tucuruí," 1992, Folder BR.PA.00.5d (Energético Tucuruí), ACIMI.

137 Michael E. Faran and Kenneth J. Linthicum, "A Handbook on the Amazonian Species of Anopheles (Nyssorhynchus) (Diptera: Culicidae)," *Mosquito Systematics* 13, no. 1 (1981): 1–81, 36; Interview with Wanderlei Pedro Tadei.

138 On water turbidity, see WCD, *WCD Case Study Tucuruí*, 45.

139 Tucuruí and the two other municipalities that border the reservoir, Jacundá and Itupiranga, had an API between 0 and 50 most years between 1985 and 1995, with occasional surges to an API upwards of 80. See Couto, "Hidrelétricas e saúde na Amazônia."

But the *Mansonia* proved to be a devastating plague themselves. They reproduced prolifically, forming giant swarms that attacked communities that had resettled along the reservoir's shore. INPA researchers estimated that the mosquito density in affected communities was at least 600 mosquitoes per person per hour and likely higher.[140] One researcher, working alone, captured more than 1,600 mosquitoes in just four hours.[141]

These dense swarms of mosquitoes ravaged the people living nearby. At the apogee of the crisis in 1991, researchers estimated that the bite rate was as high as 500 mosquito bites per person per hour.[142] It was impossible to live and work under such conditions. One farmer, making light of a grave situation, joked that they tended their crops looking like beekeepers.[143] It was hard to do anything at all, as most people spent all their time under mosquito nets and could hardly open their mouths to eat. One Eletronorte social scientist who was there remembers those years as the time of "laughing with your mouth closed."[144]

By some accounts, animals fared worse than humans. Reports abounded of dogs, cats, pigs, chickens, and other animals dying from mosquito bites, perhaps from excessive blood loss or other causes.[145] In order to avoid bites, some animals became attuned to the biting rhythms of the mosquitoes. When the sun began to set, dogs would run inside and hide under the table until their owners locked up the house.[146]

The response was a mass exodus from the affected area. Of those displaced by Tucuruí, Eletronorte had relocated about 1,500 families (roughly 8,000 people) to areas where they were exposed to *Mansonia* swarms.[147] Most of these families lived adjacent to the reservoir itself, or at least within the flying radius of the *Mansonia* mosquitoes. Some outside that radius were also affected, because the *Mansonia* – in a similar manner as the *An. darlingi* at Itaipu – colonized secondary breeding

140 Tadei et al., "Evolução das populações de Anopheles e de mansonia," 639.

141 Antônio José, "Uma tragédia sem solução: Há cinco anos mosquitos infernizam vizinhos da barragem de Tucuruí," *Jornal do Brasil*, July 8, 1991.

142 Ibid.; "Atingidos por praga querem nova área," *O Liberal*, August 19, 1991.

143 "Moscas ainda infernizam em Tucuruí," *O Liberal*, September 19, 1991.

144 Niviene de Deus Maciel, personal correspondence with author, Brasília, October 2017.

145 José, "Uma tragédia sem solução"; Comissão Pro-Índio de São Paulo, "Os povos indígenas e a hidrelétrica de Tucuruí," July 12, 1989, Folder BR.PA.00.5d (Energético Tucuruí), ACIMI.

146 Interview with Wanderlei Pedro Tadei.

147 Dr. Ruy Saint-Clair Cunha da Secretaria Municipal de Saúde Tucuruí, "Relatório técnico," March 4, 1991, Folder BR.PA.00.5d (Energético Tucuruí), ACIMI.

grounds in pools of water that formed in the newly cleared fields.[148] Some residents found that burning the coconut of the babaçu tree helped expel the swarms, but there was little else they could do to make living conditions there tolerable.[149] Most moved, selling their lots for derisory prices or abandoning them altogether and resettling in the urban center of Tucuruí, where there were far fewer mosquitoes, but otherwise deplorable living conditions. By 1991, the number of households that remained in the affected area alongside the reservoir had dropped to 678.[150] The local government sent a health official who confirmed that the conditions there were simply unlivable due to the incessant mosquitoes that bit both day and night. The situation had become so grim that on March 4, 1991, Tucuruí and neighboring municipalities (which included Parakanã villages) officially declared a state of emergency.[151]

Eletronorte's response during the crisis was lackluster. The mosquito swarms became intolerable shortly after the reservoir finished filling in March 1985, and residents began complaining immediately. In the early years, Eletronorte ignored the complaints, calling the farmers lazy and accusing them of trying to get additional financial support for free.[152] It was not until 1989 that Eletronorte decided to send a team of researchers to study the problem, the results of which validated the residents' complaints. The report noted, as both Goodland and Tadei had, that the ecological conditions were so propitious for mosquitoes that any effort to try and eliminate them would be futile. The researchers recommended instead that Eletronorte simply resettle people outside the reach of the mosquitoes.[153] But even then, Eletronorte refused to act. In 1990, the company undertook a preliminary survey outlining measures that it could take to alleviate the mosquito burden at the Parakanã Indigenous reserve, but it never implemented the program.[154] The local government's formal declaration of a state of emergency did nothing to alter this negligent state of affairs.

148 "Relatório Comissão de Estudos da Proliferação de Mosquitos a Montante de Tucuruí."
149 "Moscas ainda infernizam em Tucuruí."
150 STR-TUC, "Relatório parcial dos problemas da área de proliferação de mosquitos."
151 Prefeitura Municipal de Tucuruí, "Decreto No. 005/91-GP, de 04 de março de 1991, Estado de emergência nas Glebas Parakanã, Tucuruí, Pucuruí, e Andorinha e outras áreas afetadas pelo lago da Usina Hidrelétrica de Tucuruí, pertencentes ao Município de Tucuruí-PA," AISA.
152 STR-TUC, "Proliferação de Mosquitos (histórico de um caso)," undated, AISA.
153 "Relatório Comissão de Estudos da Proliferação de Mosquitos a Montante de Tucuruí."
154 See Eletronorte, "Projeto para monitoramento e avaliação do programa de controle de mosquitos na Reserva Indígena Parakanã," 1991, Document 03022, AELN.

On August 1, 1991, about 300 people began camping along the access road to Eletronorte's headquarters in Tucuruí to protest its desultory response to the mosquito problem. On August 20, the Tucuruí Rural Workers Union [Sindicato dos Trabalhadores Rurais de Tucuruí, STR-TU] opened a case in the local courts against Eletronorte for environmental damages.[155] Still there was no action from Eletronorte. It took a group of displaced people camping outside Eletronorte's headquarters in Brasília for two years between 1991 and 1993 for the company to agree to provide financial support to communities resettling at new sites far from the edge of the reservoir and its mosquito swarms.[156]

In 1993, eight years after it began, the crisis started to diminish. Thereafter, the unstable ecological conditions at the reservoir that had produced the macrophyte boom began to stabilize. The nutrients that had accumulated in the reservoir after it filled were gradually flushed through the turbines and floodgates, and the inflows of nutrient-rich water were not sufficiently concentrated to trigger eutrophication. As a result, by the end of the decade, macrophytes in the reservoir decreased dramatically, as did the *Mansonia* swarms whose larvae depended on their roots for shelter.[157] There were still pockets around the reservoir with dense clouds of mosquitoes, but overall, mosquito clouds were fewer and thinner. More important than the collapse of mosquito populations was the demographic collapse of people living along the edge of the reservoir. Most people had left the area of their own accord already, and after 1993, Eletronorte agreed to help build infrastructure on new sites that encouraged much of the remaining population to move.

CONCLUSION

Between the late 1970s and the late 1980s, the Brazilian government finished most of its big reservoirs, which engendered massive environmental changes that impacted people living near the newly flooded areas. The military regime was in a rush to finish its dams and designated few resources to caring for the nonhuman environment, despite producing a wealth of studies about the environmental impacts that might unfold. Construction proceeded much faster than the planning of environmental

155 STR-TUC, "Relatório parcial dos problemas da área de proliferação de mosquitos."

156 Fearnside, "Social impacts of Brazil's Tucuruí Dam," 483–495.

157 Interview with Wanderlei Pedro Tadei; WCD, *WCD Case Study Tucuruí*, 69; Fearnside, "Social impacts of Brazil's Tucuruí Dam," 490.

safeguards, and in almost all cases, the government and its state-owned companies failed to mitigate ecological impacts that turned disastrous after the floodwaters settled.

The consequences of the dictatorship's disregard for the nonhuman environment surfaced even before the reservoirs filled. In a rush to see the magnificent Guaíra Falls, tourists overwhelmed the precarious infrastructure that the military government neglected to maintain, leading to a deadly accident. The dam builders also ignored the impacts on fish and malaria-carrying mosquitoes. Natural and intergovernmental allies helped soften the damage of fishery collapse and epidemic malaria, but could not prevent overfishing and financial collapse among fishermen, and tenacious swarms of mosquitoes at Tucuruí. Not all of these dams' environmental impacts were disastrous, but those that were shone a bright light on the hollow promises of the military regime's pharaonic environmentalism.

These social and environmental disasters hit just as the military was turning over the reins of government to democratically elected politicians, and the national environmental movement was gathering momentum (see Chapter 1). News of displaced families living under tarps and plum trees in the Northeast (see Chapter 4) and battling mosquitoes in the rainforest reached wide audiences and compelled Brazilian voters and international lenders to reconsider the costs of hydropower.

Despite this damage, the military regime and the civilian administration that succeeded it could boast of the tremendous benefits their dams brought. They argued vociferously that the environmental costs of such progress were lamentable but necessary. It was an enticing argument as these dams indeed supplied the energy that Brazilians needed to accelerate industrialization and economic growth and improve standards of living.

Yet, in its final years the military government had laid the groundwork for a big dam that proved the undoing of this argument. It was the most controversial of the dictatorship's dams and the one that did the most to turn Brazilian voters and international agencies against hydropower. It is to this dam that the next chapter turns.

6

The Notorious Balbina Dam

The Dictatorship's Last and Most Infamous Dam, 1980s–1990s

INTRODUCTION

Even the engineers turned on Balbina. In 1989, the Syndicate of Amazonian Engineers wrote an open letter to Eletronorte repudiating the big dam that the company had built on the Uatumã River, a tributary of the mighty Amazon that runs northeast of Manaus, to sustain industrial growth in the city. Two years earlier, Eletronorte had closed the floodgates, setting in motion a series of anticipated social and environmental catastrophes that environmentalists publicized. Upon witnessing these calamities, the Amazonian engineers concluded that the dam was misguided and argued that the reservoir should be emptied.[1]

The letter surprised Eletronorte. The company was used to getting such criticism from environmentalists, Indigenous communities, and their supporters, but not from engineers, who were the prime beneficiaries of the employment opportunities that dam building generated and part of a professional culture that valued large-scale technological interventions. Eletronorte expected the profession to be united in support of all dam-building projects that Eletrobras and its subsidiaries deemed worthwhile. Eletronorte responded to the syndicate's criticism with a caustic letter correcting what it considered to be misconceptions and attacking the engineers for being disrespectful to the national engineering profession.[2]

[1] The letter of repudiation was also endorsed by the National Federation of Engineers. See Sindicato dos Engenheiros no Estado do Amazonas, "minuta de resposta," June 28, 1989, Folder 1989, AEDS.

[2] "Presidente de Eletronorte Miguel Rodrigues Nunes et al. to Presidente de Federação Nacional de Engenheiros," June 21, 1989, Folder 1989, AEDS.

The syndicate thought it worthwhile to clarify its position. Balbina Dam, not the engineering profession, was the target of its disapproval. The engineers explained: "our repudiation was not against the work of hundreds of salaried engineers and millions of construction workers. It was against the irresponsibility of the government that stood firm in the construction of yet another pharaonic dam that caused destruction and death in Amazônia, a true catastrophe."[3] It was criticism leveled at Balbina alone, a project whose costs they believed far outweighed its benefits. It was an opinion that was becoming increasingly common.

Balbina is widely regarded as an environmental disaster. The Eletronorte president and Amazonas governor who were in charge when the dam was constructed (1981–1989) have since remarked that the reservoir should never have been built.[4] In the 1990s, former science and technology secretary José Goldenberg famously called Balbina "a monument to insanity," a statement that Brazilian president Luiz Inácio Lula da Silva reaffirmed a decade later.[5] Environmentalists had long voiced their opposition to the Brazilian government's large dams, but criticism from such a heterogeneous group of opponents, including the engineering profession itself, was novel. It was the first time in Brazilian history that such a diverse coalition united to denounce a dam.

Balbina inundated 2,350 sq. km of mostly flat tropical rainforest. Sluggish reservoir water means that the powerhouse has produced much less than its installed capacity of 250 megawatts, a pitiful amount of electricity relative to the project's social and environmental costs. The dam inundated the land of the Waimiri-Atroari, an Indigenous community north of Manaus introduced in more detail later on. The reservoir covered land the group valued for its resources and cultural significance, as well as two villages that were then home to about a third of the community. Balbina also engendered a series of environmental problems. Eletronorte decided not to clear the forest in the reservoir area prior to construction, which meant that the floodwaters created ghost forests of drowned trees. Decomposing vegetation turned the reservoir's water acidic and anoxic, which made it unfit for human consumption and thinned fishing catches downstream, thereby creating a subsistence crisis

3 Sindicato dos Engenheiros, "minuta de resposta."

4 "Hoje Eletronorte não construiria Balbina," *A Crítica*, July 18, 1986; "Presidente da Eletronorte abre o jogo e se diz frustrado—Balbina é um pecado," *A Crítica*, March 19, 1989.

5 "No Chile, Lula lembra hidrelétrica de Balbina e defende cuidado ambiental," *O Globo*, April 27, 2007.

for those living along the river. Likewise, decomposing vegetation produced both methane and carbon dioxide underwater, which were released into the atmosphere – contributing to the greenhouse effect – through the surface water of the reservoir and at the turbines and spillway.

Balbina's flood of environmental problems was nothing new. Over the past two decades, the military government had built big dams whose environmental impacts matched those of Balbina. Part of what made the dam unique was that it brought meager benefits that were both temporary and local. Balbina provided a surge of electricity to Manaus, but the city quickly outgrew the dam's power and turned to other sources. Unlike other big dams such as Itaipu and Tucuruí, which powered entire regions, Balbina energized just one city in a part of the country that was not renowned for its manufacturing capabilities.

But it is timing, more than anything else, that explains the dam's unique position as the poster child for the ills of big dams. It was the last of the dictatorship's big reservoirs, and the civilian government that replaced the military regime inherited the project and finished it. The restoration of democratic freedoms emboldened dam critics, and the experience of watching social and environmental disasters unfold at dam sites over the preceding decades had turned many against hydropower. The government built Balbina at a time when there was growing awareness about the deleterious impacts of big dams, and it proceeded with construction despite a spate of evidence that it would bring grave social and environmental consequences that pharaonic environmentalism alone could not mitigate.

CONCRETE IN THE JUNGLE

Manaus was once a city ahead of its time. Between the 1880s and 1910s, the small trade hub became a sprawling cityscape in the middle of the world's largest rainforest. In the second half of the nineteenth century, technological advances in transportation and industry increased the demand for natural rubber, which was found in enormous quantities in the Brazilian Amazon, home to the prolific *Hevea brasiliensis* rubber tree. Production boomed in the forest surrounding Manaus, and the city became a flourishing entrepôt of 50,000 inhabitants.[6] Urban growth

[6] See E. Bradford Burns, "Manaus: Portrait of a Boom Town," *Journal of Inter-American Studies* 7, no. 3 (July 1965): 400–421; Yara Vicentini, *Cidade e história na Amazônia* (Curitiba: Universidade Federal do Paraná, 2004), and Otoni Moreira de Mesquita, *La*

accelerated further after the 1960s – its population numbered more than 2,000,000 in 2020 – turning Manaus into a concrete beachhead in the middle of the jungle.

Manaus' aspirations, like those of any city, were deeply tied to its energy system. During the rubber boom, wealthy merchants reaped tremendous profits from exploiting an organic energy regime based on human labor. Rubber trees were tapped by people, who were paid little for their work and often became indebted to merchants who inflated their prices.[7] The elite merchants who profited from this system used their wealth to infuse the city with energy from a cutting-edge inorganic source, electricity.

Beginning in 1896, Manaus installed electric lighting and electric streetcars. It was among the earliest cities in the world to electrify and the second Brazilian city to do so. Foreign investors were in charge of the company that electrified the city; the state government took over between 1898 and 1908, returning the company to private control thereafter. These municipal power plants used charcoal and wood cut from the nearby forests to generate electricity,[8] which powered a streetcar system that ran through the city and lit a big opera house that hosted famous musicians from around the globe. Together, the manual labor of rubber tappers and electrification had turned Manaus into one of the wealthiest and most opulent cities in the world.

But the boom did not last. In the 1910s, Southeast Asian rubber plantations began producing such copious amounts of rubber that the price collapsed, putting Brazilian rubber producers out of business. The South American Leaf Blight (SALB), a native fungus, hindered the efforts of Brazilian producers to compete. Despite the efforts of some industrialists and the government to overcome the SALB, Brazilian rubber production never returned to its former stature.[9] The economic collapse halted the growth of Manaus, which curtailed electrification and other public services.

During the postwar period, Brazilian politicians decided that industrialization was the best means to revitalize Manaus. Industrial growth, however, required abundant energy, which the city lacked. In the 1950s,

belle vitrine: Manaus entre dois tempos, 1890–1900 (Manaus: Universidade Federal do Amazonas, 2009).

7 Weinstein, *The Amazonian Rubber Boom, 1850–1920*.

8 Dr. Nathalia Capellini, personal correspondence.

9 Dean, *Brazil and the Struggle for Rubber*, 53–66.

the state government took control of the city's electrification system and invested in small expansions to supply the city's fledgling industrial center, but these efforts produced only modest results.[10]

The true industrial boom came only after the military seized power in March 1964. The generals considered stimulating industrialization and population growth in the rainforest a priority for national security (see Chapter 1) and passed legislation to encourage industrial development in Manaus. In 1967, President Humberto Castelo Branco created the Free Economic Zone of Manaus [Zona Franca de Manaus, ZFM], an industrial area with tax exemptions designed to entice companies to build factories there.[11] With the ZFM, industrialization and urban growth began accelerating. The city's population doubled during the next decade, expanding from roughly 310,000 to 640,000 people between 1970 and 1980.[12]

But fuel shortages soon stalled the engine of industrialization. Most of the power plants that electrified Manaus in the 1960s and 1970s were thermoelectric, powered by costly foreign petroleum.[13] Blackouts become more and more common as industrial and population growth put increasing strain on the city's aging electricity infrastructure. Some state senators remembered the 1960s as a decade of darkness, when residents relied on kerosene lamps as much as electric lighting.[14] International political tumult in the subsequent decade only made matters worse. The 1973 energy crisis sent the price of petroleum soaring and made it even more expensive to expand the thermoelectric plants that powered Manaus. To mitigate the high costs of foreign oil, politicians began looking to the region's rivers to save the city's fledgling industrial sector.

State politicians found allies in the federal government. In December 1968, Minister of Mines and Energy José Costa Cavalcanti had promulgated legislation that created a series of hydroelectric studies for the

[10] Senator Jose Lindoso, "Discurso no Senado Federal," September 27, 1977, ASF.

[11] Castelo Branco, *Decreto-Lei No. 288, de 28 de fevereiro de 1967, Altera as disposições da Lei número 3.173 de 6 de junho de 1957 e regula a Zona Franca de Manaus*, ACD.

[12] Instituto Brasileiro de Geografia e Estatística (IBGE) "População nos Censos Demográficos, segundo os municípios das capitais," in *Sinopse do Censo Demográfico 2010*, available at https://censo2010.ibge.gov.br/sinopse/index.php?dados=6&uf=00 (last accessed April 2023).

[13] Assembléia Legislativa de Amazonas, "Carta de Amazonas," June 13, 1979, republished in Senator Raimundo Parente, "Discurso no Senado Federal," June 16, 1979, ASF.

[14] Lindoso, "Discurso no Senado Federal."

Amazon basin.[15] In 1972, the government decided that this first set of studies was inconclusive, and Energy Minister Antônio Dias Leite commissioned a second, more comprehensive set of studies, which had identified the dam site that became Tucuruí (see Chapter 2). These studies also included the area surrounding Manaus, which the government considered as strategically significant as Belém. Between 1973 and 1975, surveyors working for the newly created, state-owned electricity company Eletronorte took boats through the jungle north of Manaus to scout potential dam sites along the Amazon's tributaries.

Engineers soon identified the most propitious location. In April 1976, Eletronorte published its final report on the studies, which identified a site called Balbina, on the Uatumã, as the best of three plausible hydropower sites for electrifying Manaus. Of the three, Balbina was the one that could be built the fastest and cheapest.[16]

The site did have some drawbacks. Unlike Paulo Afonso, Guaíra, or Itaboca Falls, the waterfalls at Balbina were small and of no historical or cultural significance to non-Indigenous Brazilians. Because the rapids were small, damming them would produce little energy. Indeed, engineers estimated that Balbina would only have the capacity to produce at most 250 megawatts of electricity, and seasonal variations in the water flow meant that it would not produce at full capacity for much of the year. With such minimal output, engineers did not expect Balbina to satisfy the demand in Manaus for very long.

Nevertheless, the site appealed to both federal and state politicians and industrialists because it would replace foreign oil and could be built quickly. In 1978, President João Figueiredo publicly promised that the government would build Balbina, and the following year, Eletronorte began building the access road to the construction site. In 1981, Figueiredo declared the land that the reservoir would flood to be government property, and Eletronorte began erecting the dam.[17]

[15] Artur da Costa e Silva, *Decreto No. 63.952, de 31 Dezembro de 1968, Cria no Ministério das Minas e Energia o Comitê Coordenador dos Estudos Energéticos da Amazônia*, ASF.

[16] Eletronorte, "Estudos Amazônia, relatório final, Vol. 4: Aproveitamento hidrelétrico do Rio Uatumã em Cachoeira Balbina, estudos de viabilidade," April 1976, Document 03691, AELN.

[17] João Figueiredo, *Decreto-Lei No. 85.898, de 13 de abril de 1981, Declara de utilidade pública para fins de desapropriação, áreas de terra com benfeitorias, necessárias à formação do Reservatório da Usina Hidrelétrica de Balbina da Centrais Elétricas do Norte do Brasil, S.A. – Eletronorte, no Estado de Amazonas*, ACD.

For Balbina's proponents, the dam was a giant oil field. In the late 1970s, the thermoelectric plants in Manaus, which collectively produced around 205 megawatts, burned over one million barrels of oil every year,[18] and by 1985, nearly three million barrels.[19] Most of this oil was imported and expensive (see Chapter 1), and in 1985, it cost the government about US$80 million to keep the lights on in Manaus. That year, senator and former energy minister César Cals argued, "Balbina is the most important dam in Brazil, equivalent to an oil field ... replacing oil imports is the same as producing domestic oil, it is the same in terms of balance of payments, foreign exchange savings, and energy sovereignty."[20] Balbina's supporters contended that with a projected cost of US$780 million, the savings from slashing foreign oil consumption would pay for the dam within ten years, after which the city could enjoy cheap energy. Balbina's potential monetary rewards were part of its appeal.

So too were its political benefits. Eletronorte's reports did not support the claim that Balbina was the best solution to the energy crisis in technical terms. Early studies showed that expanding the petroleum-powered thermoelectric plants was actually the cheapest option, even with the high price of petroleum. However, the difference was small enough that both the additional thermoelectric plants and the dam had roughly the same price tag.[21] But Balbina had a political advantage over petroleum: Producing electricity using a domestic and renewable source of energy was more popular than perpetuating dependence on foreign petroleum.[22] Nationalism and energy sovereignty had been prominent in Brazil since the 1930s, and both remained widely popular during the military regime and the civilian administration that succeeded it, which encouraged politicians to support domestic energy production. Even Balbina's staunchest critics acknowledged that abandoning foreign oil was a worthy goal.[23] Furthermore, Balbina could be built quickly and

[18] The exact amount was 1.3 million barrels of oil. Author's calculations, the original figure is listed as 480 tons of petroleum daily. See Parente, "Discurso no Senado Federal," June 23, 1979, ASF.

[19] Author's calculations, the original figure is listed as 7,800 barrels of petroleum daily. See Senator Eunice Michiles, "Discurso no Senado Federal," June 20, 1985, ASF.

[20] Senator Alberto Silva, "Discurso no Senado Federal," May 7, 1985, ASF.

[21] Eletronorte, "UHE Balbina-Estudo de viabilidade econômica," 1978, 4–5, Document 01837, AELN.

[22] Eletrobras, "Estudo de caso da usina hidrelétrica de Balbina – Aspectos sócio-ambientais, relatório preliminar," December 3, 1991, 1, 8, 12–19, Document 01922, AELN.

[23] Conselho Indigenista Missionário (CIMI), "Hidrelétrica de Balbina: Luz para os que andam nas trevas?" *Porantim*, January 1979, 4. *Porantim* was published by CIMI out of

with minimal overhead compared to bigger and more powerful dams, such as Cachoeira Porteira, a promising alternative Eletronorte engineers had identified on the Trombetas River in the neighboring state of Pará. It was these political benefits that appealed most to Balbina's supporters, who were happy that the dam's economic benefits dovetailed with these ambitions.

Though the political benefits of Balbina were appealing at all levels of the government, support from the federal government did not always seem assured. Between 1976 and 1985, tensions flared between state and federal politicians because the federal government repeatedly cut funding for the dam when budgets tightened. During this period, politicians from Amazonas returned to the Senate time after time to praise Balbina and to admonish the federal government for slashing funding.

The debate in the Senate over Balbina did not split along party lines, as had the debate over Itaipu and Octávio Marcondes Ferraz's alternative scheme discussed in Chapter 2. There was near unanimous support among industrialists and politicians from Amazonas for Balbina, and a group of bipartisan senators repeatedly defended the dam.[24] Opposition to Balbina – which was rare in the Senate in the 1980s – came from politicians from outside of the state who were concerned about the dam's environmental impact.[25]

Some Amazonian senators tried to persuade the military government to reinstate funding by appealing to one of its worst fears: foreign incursion. Senators claimed that when the city's industrial economy stalled, people began to fear going to bed Brazilians and waking up foreigners. One senator exclaimed: "if Balbina is not built ... Manaus will turn into a port for timber exports, thus opening a portal for international greed."[26]

Other senators played into the fear of political secession, however unrealistic that prospect was. At one Senate hearing, Raimundo Parente read passages from a popular newspaper article that he agreed with: "Amazonas is being treated as if it were not Brazil ... between continuing to be treated like a colony and searching for new directions, the clear

its main headquarters in Brasília, and its regional headquarters, such as Manaus. Copies can be consulted at the ACIMI.

[24] They were Evandro Carreira (MDB-AM), José Lindoso (ARENA-AM), João dos Santos Braga Júnior (ARENA-AM), Joel Ferreira (MDB-AM), Raimundo Parente (ARENA-AM), Eunice Michiles (PFL-AM), and Fábio Lucena (PMDB-AM).

[25] For example, see Silva (PMDB-PI), "Discurso no Senado Federal."

[26] Braga Júnior, "Discurso no Senado Federal," May 9, 1978, ASF.

option is the second alternative. After all, enough humiliation."[27] State politicians emphasized the threat of territorial loss in an effort to shore up federal support for Balbina.

The federal government did not need convincing. The funding cuts reflected the realities of an economic crisis, not an ideological rift. During the 1960s and 1970s, the military government had borrowed a lot of money from international banks to finance big dams and other infrastructure projects (see Chapter 3). Excessive borrowing caused inflation to rise, and by 1980, Brazil's annual inflation rate was 110 percent.[28] Starting in 1981, a global economic recession caused interest rates on Brazilian loans to rise to such high levels that the government could no longer meet its payments and had to restructure its spending to service these debts. Rising inflation and the debt crisis crippled the Brazilian economy and curtailed the money available for building big dams. Given the funding shortages, the federal government prioritized finishing Itaipu and Tucuruí and defunded Balbina.[29]

But funding cuts were not a sign that the federal government had a change of heart regarding the dam. Balbina was planned and executed at the highest echelons of the military government. Though the dam was popular among local politicians, it was the brainchild of the federal government, which had no plans to abandon it.[30] Exchanges during one of the earliest Senate debates about funding – which occurred before the debt crisis hit – highlight the perspective of the federal government that it maintained throughout these Senate hearings. In 1977, Amazonas senator Evandro Carreira complained about cuts for Balbina and implored the government to reinstate funding using arguments about territorial loss that foreshadowed those of other senators. In response, Northeastern senator Virgílio Távora clarified that funds were being cut because of a shortage of money and nothing else: "the federal government knows that the arguments you are making are correct: the government itself makes these arguments."[31] Távora was reminding local senators that the federal government shared their concerns and motivation for erecting the dam. In 1984, Costa Cavalcanti responded to complaints from industrialists in Manaus with the same reminder: The federal government remained

[27] Parente, "Discurso no Senado Federal," June 15, 1985, ASF.
[28] Skidmore, *Brazil: Five Centuries of Change*, 172.
[29] Lucena, "Discurso no Senado Federal," June 13, 1985, ASF.
[30] Eletrobras, "Estudo de caso da Usina Hidrelétrica de Balbina," 12.
[31] Carreira, "Discurso no Senado Federal," November 30, 1977, ASF.

committed to Balbina, but was putting it on hold because it no longer had the funds.[32]

The government eventually found the money to finish Balbina. In 1985, it released the last payment, and construction on the dam's final phase began. Balbina's floodgates closed in October 1987, and the reservoir was full within two years. Manaus at last had its oil field.

DOWN A FAMILIAR PATH

Environmental planning at Balbina followed the same pattern as with its predecessors: acknowledgment of environmental impacts but few concrete measures to mitigate them until construction neared completion. In 1979, Eletronorte published a cursory impact study that recognized the dam would have a huge social and environmental footprint and called for more research on the matter.[33] But such studies were slow to materialize. Construction began in 1981, before any of these environmental studies had begun, and proceeded with little information about the dam's social and environmental impact. Eletronorte did not even have a clear idea of how many people the reservoir would displace.[34]

The dam's most glaring impact was that it would flood the land of the Waimiri-Atroari, an Indigenous community that had inhabited a roughly 83,000 sq. km area north of Manaus for centuries.[35] Eletronorte initially dismissed this concern as inconsequential because its preliminary flyover missions suggested that the region was not inhabited.[36] The company's 1979 cursory environmental impact study had a more sobering assessment. It recognized that the dam would indeed flood the group's land, and

[32] "José Costa Cavalcanti to José Santos Silva Azevedo, Presidente da Associação dos Lojistas da Zona Franca de Manaus," February 15, 1984, Folder 9, FADL, ACME.

[33] Eletronorte, "Diagnóstico preliminar dos efeitos ambientais do projeto Hidrelétrico de Balbina, BAL-10B-0199-RE," September 1979, Document 01576, AELN.

[34] Fundação Nacional do Índio (FUNAI), "Correspondência interno," February 19, 1979, Folder 1979, AEDS; "Eletronorte to FUNAI, Ass.: Eventual interferência do reservatório da UHE Balbina, no Rio Uatumã, com a Reserva Indígena Waimiri-Atroari, no Estado do Amazonas. Apoio da FUNAI aos trabalhos de implantação do projeto," May 15, 1979, Folder 1979, AEDS.

[35] See José Porfírio de Carvalho, *Waimiri Atroari: A história que ainda não foi contada* (Self-Published, 1982), and Renan Albuquerque Rodrigues, "Vidas despedaçadas: Impactos socioambientais da construção da Usina Hidrelétrica de Balbina (AM), Amazônia Central," (PhD Diss., Universidade Federal do Amazonas, Manaus, 2013).

[36] "Eletronorte to FUNAI, Ass.: Interferência do reservatório com a Reserva Indígena dos Waimiris-Atroaris [sic]," September 21, 1976, Folder 1976, AEDS.

the authors worried that the incursion might provoke a violent reaction.[37] However, the report maintained that the reservoir site was then uninhabited, and therefore flooding the land would not damage Waimiri-Atroari culture. Eletronorte was wrong on both counts: The area was in fact inhabited, and flooding it entailed immense cultural loss.

The emptiness that Eletronorte perceived was the reflection of a recent depopulation, not long-standing vacancies, and the areas that Eletronorte surveyed from its planes and helicopters were home to about a third of the remaining Waimiri-Atroari population. These communities were the survivors of a decade-long genocide.[38] In 1968, the military government began constructing a highway, known as BR-174, that ran north from Manaus to Boa Vista, the largest city in the territory of Roraima. This highway was part of the military government's National Integration Program (see Chapter 1) and was designed to help incorporate Roraima into the national fabric. The highway cut through Waimiri-Atroari territory, and as the military's construction crews leveled the rainforest, conflict erupted. The population loss from these clashes was severe. Anthropologists estimate that when construction began on the highway, the Waimiri-Atroari numbered around 3,000 people. However, by 1982, their population had dropped to just 350 people.[39]

Disease and violence were both responsible for the population collapse. Since the Waimiri-Atroari had lived largely isolated from the rest of Brazilian society and their pathogens, new diseases introduced by soldiers building the highway were particularly devastating. Precise records of disease fatalities do not exist, but health officials at the time believed that malaria, measles, and tuberculosis were three of the biggest culprits.[40] Indeed, anthropologists living with the Indigenous community during the 1980s recorded deadly measles outbreaks.[41]

[37] Eletronorte, "Diagnóstico preliminar dos efeitos ambientais," 23–24.

[38] See Comitê Estadual de Direito à Verdade, à Memória e à Justiça do Amazonas, *A ditadura militar e o genocídio do povo Waimiri-Atroari* (Campinas: Editora Curt Nimuendajú, 2014).

[39] Stephen Baines, "Relatório de pesquisa de campo do projeto etnográfico Waimiri-Atroari: Etapa II, 1983," September 9, 1983, Folder 1983, AEDS. For Waimiri-Atroari population estimates between 1905 and 2011, see Comitê Estadual de Direito à Verdade, *A ditadura militar e o genocídio do povo Waimiri-Atroari*, 26.

[40] Rômulo César Saboia Moura, coordinator of the health branch of the Waimiri Atroari Program, "Influência da UHE Balbina sobre a saúde da população indígena Waimiri-Atroari, de julho de 1987 a fevereiro de 1992," 40, Document 00916, AELN.

[41] Doroti Schwade, "Proteção à saúde do índio," 1986, Folder 1986, AEDS; CIMI Norte, "Relatório sobre sarampo na área dos Waimiri-Atroari," June 1981, Folder 1981, AEDS.

Violence was also responsible for many deaths. Specific information about such atrocities is scarce, but there is ample evidence to suggest that the soldiers who built the highway murdered Indigenous people. Such information has become more accessible since 2014, when the National Truth Commission, a government-sanctioned commission formed to investigate human rights abuses during the dictatorship, published its report. Some of the most damning evidence came from schoolchildren. Upon the return to civilian rule in 1985, the Brazilian government sent teachers to the Waimiri-Atroari territory who used drawing as a means to overcome the language barrier. During these classes, students drew scenes of violence, such as military officers firing machine guns at members of their community. One student told the story of an airplane that dropped poisonous chemicals on a village, killing nearly everyone there.[42] Other survivors recounted stories of violence in interviews, such as leader Viana Uome Atroari, who also witnessed aerial attacks.[43] What with disease, gunfire, and – it seems – aerial assaults with lethal chemicals, the incursion of non-Indigenous Brazilians into Waimiri-Atroari territory spelled disaster for their communities.

As their population plummeted, so too did their territorial holdings. The Waimiri-Atroari once inhabited a large area north of Manaus that enabled them to be mobile and engage in hunting, fishing, and shifting cultivation for subsistence. When the military government began building BR-174, it decided to restrict the group to a small reservation. In 1971, the government created the Waimiri-Atroari Reserve, a roughly 16,000 sq. km area with land on both sides of the highway.[44] The territory remained ill-defined over the next decade and was subject to revisions,

[42] The student's comments are vague about the circumstances of the attack, or the chemicals used, mentioning only that "the plane dumped some sort of dust or powder [O avião derramou como que um pó]." The student also mentioned that everyone died quickly and that someone arriving shortly after the massacre found everyone dead "without any signs of violence on their bodies [sem um sinal de violência no corpo]." Comitê Estadual de Direito à Verdade, *A ditadura militar e o genocídio do povo Waimiri-Atroari*, 41–43.

[43] He was vague concerning which weapons were used in these attacks, though he uses the word "bomb" and remembers everything heating up, which suggests either fire or chemicals that burn: "It was some sort of bomb, there in the village … an airplane came and suddenly everything heated up, lots of people died [Foi assim tipo bomba, lá na aldeia … ele veio no avião e de repente esquentou tudinho, aí morreu muita gente]." Ibid. Viana Uome Atroari recounted this story for a TV Brasil documentary, www.youtube.com/watch?v=lbDkac9Nz9k (17:47–18:37) (last accessed April 2023).

[44] Emílio Garrastazu Médici, *Decreto No. 68.907, de 13 de julho de 1971, cria a Reserva Indígena de Waimiri-Atroari, situada no Município de Airão, Estado de Amazonas*, ACD.

some of which were made at the behest of energy and mining companies. In 1976, the National Indian Foundation [Fundação Nacional do Índio, FUNAI] adjusted the southern border at the request of Eletronorte so that Balbina would flood a smaller part of their territory.[45] The most audacious revisions occurred in 1981, when the government discovered a giant tin deposit in the territory. It readjusted the border to exclude the mining site and demoted the territory from an Indigenous reserve to a lower designation that entailed less protection.[46] By the mid-1980s, the Waimiri-Atroari were just 10 percent of their 1960s population and lived on land that was about 20 percent of their traditional territory.

The remaining population was scattered across villages throughout the territory. About a third lived in two villages that Balbina would flood – Taquari and Tapupunã. The land Balbina threatened also had cultural significance. The Waimiri-Atroari cremated their dead and spread their ashes on the Uatumã and its tributaries.[47] Thus, the dam would not only displace much of the remaining population but also destroy their watery cemeteries. When construction started in 1981, Eletronorte had not identified these villages nor acknowledged the threat Balbina posed to their crematory practices.[48]

News of Eletronorte's carelessness spread. In October 1978, the Indigenist Missionary Council [Conselho Indigenista Missionário, CIMI] held a conference about the state of Indigenous communities in the Amazon. Egydio Schwade, CIMI's president, organized the conference. Schwade was a missionary from the interior of Rio Grande do Sul in Southern Brazil who moved to Mato Grosso in the 1960s to assist Indigenous communities. In 1969, he founded Operation Native Amazon [Operação Amazônia Nativa, OPAN], and a few years later he helped found both CIMI and the Pastoral Land Comission [Comissão

45 "Eletronorte to FUNAI, Ass.: Interferência do reservatório," and "FUNAI to Eletronorte," February 10, 1977, Folder 1977, AEDS.

46 Figueiredo, *Decreto No. 86.630, de 23 de novembro de 1981, declara interditada, temporariamente, para fins de atração e pacificação de grupos indígenas, área que discrimina nos Municípios de Novo Airão e Itapiranga, no Estado do Amazonas, e Caracaraí, no Território Federal de Roraima, e da outras providências*, ACD.

47 Porfírio de Carvalho, "Relatório sobre a possível interferência do represamento do Rio Uatumã – UHE Balbina na área indígena Waimiri-Atroari," October 1986, 42–43, BPWA.

48 "Douglas Souza Luz [Eletronorte President] and Newtown Burlamaqui Barreira to Sondra Wentzel [anthropologist]," March 17, 1983, excerpted in Comitê Estadual de Direito à Verdade, *A ditadura militar e o genocídio do povo Waimiri-Atroari*, 59.

Pastoral da Terra, CPT] (see Chapter 4).[49] At the October meeting, a former FUNAI employee who had accompanied Eletronorte on their initial surveys told Schwade about Balbina and the neglectful way that the company was dealing with the Indigenous communities its reservoir would flood.[50] Three months later, in January 1979, CIMI published an article about the dam that lauded efforts to rid Manaus of dependence on foreign petroleum, but expressed concern that trespassing on Waimiri-Atroari land would instigate hostilities.[51]

CIMI's report provoked defensive remarks from Eletronorte, but did little to motivate further study of the problem, alter the timeline of the dam, or modify its design. In 1982, Eletronorte signed an agreement with the National Institute of Amazonian Research [Instituto Nacional de Pesquisas da Amazônia, INPA] to study the environmental impacts of the dam, but such studies were meant only to gather information that would help mitigate detrimental impacts after the fact. The studies were neither for public consumption nor carried out with the intent to modify the design or construction schedule.[52] Eletronorte was proceeding down the familiar path of building a dam without full information about its social and environmental impact and with minimal safeguards.

CIMI kept up the pressure and continued to publish articles about Balbina's impact on the Waimiri-Atroari.[53] In 1980, Egydio left his position as CIMI's executive director, and he and his wife Doroti moved to Itacoatiara, a town on the Amazon River between Manaus and the Uatumã, where they became more deeply involved in helping the Waimiri-Atroari fight Balbina. Doroti was also from a small town in Southern Brazil, and in the 1970s, she became interested in helping Indigenous communities in the North. She heard about OPAN in 1973 and two years later began working for them in the Amazon, where she was known for traveling alone by canoe to remote villages to conduct

49 For short biographies of Egydio Schwade, see "Egydio Schwade, O Amazonense," *Amazonas Atual*, July 9, 2015; "Indigenista e pesquisador Egydio Schwade é o novo cidadão do Amazonas," *Amazonas Notícias*, July 7, 2015.

50 FUNAI, "Correspondência interno."

51 CIMI, "Hidrelétrica de Balbina: Luz para os que andam nas trevas?"

52 These studies are available at BINPA.

53 See CIMI Norte, "Hidrelétrica de Balbina: Um acordo genocida entre Brasil e França," January 5, 1981, Folder BR.AM.WT.5d (Energético), ACIMI; CIMI Norte, "A Hidrelétrica de Balbina na terra dos Waimiri-Atroari," *Cipó: Caderno Informativo do Povo Prelazia de Itacoatiara-AM (Circulação Interna)*, 6, no. 36 (September–October 1981), Folder 1981, AEDS.

outreach to Indigenous communities. She met Egydio while working in the Amazon, and they married in 1978.[54]

In 1983, Egydio and Doroti founded the Waimiri-Atroari Support and Resistance Movement [Movimento de Apoio e Resistência Waimiri-Atroari, MAREWA], and in 1984, they moved to Presidente Figueiredo, a town in rural Amazonas along BR-174 near Balbina and Waimiri-Atroari territory. From their home in President Figueiredo, they spread the word about the plight of the Waimiri-Atroari and called for Eletronorte to slow construction so that more thorough studies could be made and the public could weigh in.[55]

Two developments in 1984 helped their cause. The first was that in January 1984, the Brazilian government approached the World Bank for a loan (see Chapter 3). The military government knew that environmental safeguards would be a prerequisite for loans, so Eletronorte began to take the task of publishing an environmental impact study more seriously.[56] The second development was that in September 1984, Tucuruí's reservoir began to fill and its floodwaters inundated vast stretches of rainforest and unleashed profound environmental consequences (see Chapters 3 and 5). Newspapers reported on these environmental disasters, and both journalists and ecologists almost immediately began commenting that such calamities would be repeated shortly at Balbina.[57] The news galvanized ecologists in Manaus and elsewhere in the Amazon, who began

[54] For short biographies of Doroti Schwade, see José Ribamar Bessa Freire, "Doroti, a doce radical," December 7, 2010, General Collection, ACIMI; Guenter Francisco Loebens, "Intense, passionate and radical embrace of the indigenous cause (in memoriam Doroti Muller Schwade)," December 13, 2010, General Collection, ACIMI; and "Doroti Alice Muller Schwade, missa do 4° ano de falecimento," December 3, 2014, Folder 2010s, AEDS.

[55] See MAREWA, *Resistência Waimiri/Atroari* (Itacoatiara: MAREWA, 1983), Folder 1983, AEDS; "Egydio Schwade to Eletronorte President Douglas Souza Luz and diretor Newton Burlamaqui Barreira," September 1983, Folder 1983, AEDS; MAREWA, "Hidrelétrica de Balbina contra índios e lavradores," August 23, 1984, Folder 1984, AEDS.

[56] World Bank Oral History Program, "Transcript of Interview with Robert Goodland," 6, WBGA.

[57] "Como em Tucuruí, IBDF já prevê riscos ecológicos em Balbina," *A Crítica*, August 12, 1984; Lúcio Flavio Pinto, "Total despreparo," *O Liberal*, August 15–16, 1984; Teodoro G. Meissner, "Balbina vai repetir o caso Tucuruí," *Folha de São Paulo*, September 2, 1984; "Balbina poderá repetir o fenômeno de Tucuruí," *A Crítica*, September 9, 1984; Flavio Pinto, "O INPA e o lago (1–3)," *O Liberal*, September 19–22, 1984; "Meio ambiente: Existe confusão a respeito do progresso," *Correio da Imprensa*, September 28, 1984; "Erro de Tucuruí poderá se repetir na UHE de Balbina," *A Notícia*, December 17, 1984; "IBDF faz alerta sobre o desastre de Tucuruí," *Jornal*

publishing reports about Balbina's potential environmental impact. Critical reporting and activism against Balbina accelerated even further after March 1985, when the Brazilian military government stepped down, terminating the dictatorship.[58]

The pressure pushed Eletronorte to complete its first environmental impact study, which it published in 1985. But even this study did not include a clear outline of the impact on Indigenous people. The report dedicated just a few short paragraphs to the dam's impact on the Waimiri-Atroari, which concluded that it would flood a small area of their land that was uninhabited.[59]

Indigenous people, however, said otherwise. In March 1986, two Waimiri-Atroari leaders wrote the President and other Brazilian politicians about the dams and mines that were destroying their land. Their immediate concern was a much smaller dam that the Taboca mining company was building to power machinery at its tin mine, but the leaders expressed an overall frustration with the repeated incursions into their territory and their associated environmental impacts: "we have not gained anything, not indemnity, not demarcation, and FUNAI keeps helping Taboca and not us. Taboca is taking our riches and deforesting our lands without having negotiated with us ... a greater problem is the Balbina Dam, which will flood the Abonari and Taquari villages and no one is going to move from there."[60] The leaders highlighted that Balbina would in fact flood settlements, though Eletronorte still maintained that no one lived there.

Then, in May 1986, the World Bank approved a big power loan and conditioned the release of the second part on the successful implementation of social and environmental safeguards (see Chapter 3). Eletronorte knew that Balbina was one of the World Bank's biggest concerns. The Bank had written in its May 1986 report about the loan: "the Brazilian

do Comércio, December 19, 1985. I consulted these and other newspaper articles about Tucuruí and Balbina in APF and AEDS.

58 "Balbina: A história que ninguém contou," *A Crítica*, June 24, 1985; Samuel Murgel Branco, "Uma parte da Amazônia vai morrer com Balbina," *Revista Pau Brasil*, July–August, 1986, Folder BR.00.00.5d (Políticas Governamentais Energéticos Balbina), ACIMI. Ecologists also did interviews for international newspapers that spread the word about Balbina. See Catesby Leigh and Jeff Nesmith, "Dam would generate more disease than energy," *Cox Newspapers South America*, December 1985, APF.

59 Eletronorte, "Estudos ambientais do reservatório de Balbina, relatório diagnóstico, BAL-50-1001-RE," 1985, Document 01027, 196–198, AELN.

60 "Viana Uome Atroari and Mário Paroe Atroari to Presidente, Ministério de Minas e Energia et al.," March 10, 1986, Folder 1986, AEDS.

electric sector ... pays only rhetorical attention at higher managerial levels to sociological and environmental variables, as reflected in some of the projects presently under construction and most notoriously Balbina."[61] Domestic critics, such as then-secretary of the environment Paulo Nogueira Neto, also went on record condemning Balbina as Brazil's most infamous dam.[62] To satisfy the World Bank and the growing number of protests from ecologists and Indigenous supporters, Eletronorte sent a team of anthropologists to study the area in October 1986.

To its surprise, Eletronorte's team discovered Taquari and Tapupunã. The report also drew attention to the fact that the dam would flood large areas of land that were essential to community survival despite being uninhabited. The expedition's leading anthropologist José Porfírio Carvalho wrote in his report that "the forest, for us just a simple assemblage of trees, is for them a cornucopia of fruit and animals, as well as a pharmacy."[63] He noted that the trails were well worn, which demonstrated that the area was constantly used for hunting, gathering, and planting. Carvalho also noted that displacement would create trauma, which had the potential to be disastrous for the social fabric of the group.

This information came much too late to alter the design of the dam. By October 1986, Balbina was nearly complete and its reservoir's floodwaters were due to arrive within a year. Because the project was so far along, the government had exempted it from the new law requiring dams to pass an environmental impact assessment (see Chapter 1), and those in charge of granting Balbina its environmental operating license all supported greenlighting it.[64] Eletronorte had invested hundreds of millions of dollars in the dam, which still had widespread support among industrialists and their political allies, and it was not going to halt construction because of new information about its environmental impact.[65]

61 World Bank, "Staff Appraisal Report Brazil Electric Power Sector Loan" May 27, 1986, 53–54, WBGA.

62 "Fim da linha: Secretário da SEMA pede demissão do cargo," *Veja*, July 16, 1986.

63 Porfírio de Carvalho, "Relatório sobre a possível interferência do represamento do Rio Uatumã," 22–23.

64 Eletrobras, "Estudo de caso da usina hidrelétrica de Balbina," 49–51, and "Relatório síntese dos aspectos sócio-econômico e ambientais da UHE-Balbina," December 1997, 44, 47–48, Document 02839, AELN.

65 "Eletronorte não teme bomba ecológica ... mata de Balbina deverá ser afogada," *A Crítica*, November 8, 1985.

The most it would do was compensate the Waimiri-Atroari. In 1985, after a civilian government replaced military rule, FUNAI created a working group to design a program to assist the Waimiri-Atroari.[66] After October 1986, Eletronorte began working with FUNAI on this program, which would provide financial and technical assistance to displaced Indigenous communities as compensation for flooded land.

The program got off to a controversial start. In 1985, at the request of Waimiri-Atroari leadership, FUNAI had sent Egydio and Doroti Schwade to teach at a newly built school in the Yawará village. During their time at the school, children drew pictures and told stories of violence from the military government, and theft and exploitation at the hands of the Taboca mining company. The Schwades were appalled, and they encouraged the Indigenous students to resist these incursions. Around the same time, Eletronorte released its report that confirmed Balbina would displace two entire villages that were home to upwards of a third of the remaining population.

The government deemed the risk of dissent too high. In December 1986, FUNAI expelled the Schwades from the reserve and, in subsequent years, expelled anthropologist Stephen Baines and other researchers who were working in the area.[67] Eletronorte blamed the expulsions on the desires of the Waimiri-Atroari, but that explanation seems unlikely given that the community itself had requested the teachers, who remained popular during their tenure. It is also possible that Eletronorte paid off some leaders to denounce the anthropologists.[68]

Thus, Eletronorte's efforts to design a program that would compensate the Waimiri-Atroari began without input from Indigenous supporters and anthropologists unaffiliated with either FUNAI or Eletronorte. These

[66] Baines, "Relatório Grupo de Trabalho Waimiri-Atroari (30 dias)," September 9, 1985; "Parecer sobre a área indígena Waimiri-Atroari, conforme a proposta do Grupo de Trabalho da FUNAI instituído pela Portaria No. I901/E de 12 de Julho de 1985," 1986, Folder 1986, AEDS.

[67] On the expulsion of Egydio and Doroti, see Schwade, "untitled document," December 7, 1986; CIMI, "Nota a Imprensa," December 22, 1986; and CIMI, "Waimiri/Atroari – Ref. saída (expulsão) de Egydio e Doroti Schwade," December 30, 1986, Folder 1986, AEDS; on the March 1987 expulsion of Silva, see "Marcio Ferreira da Silva to Profa. Cecília Maria Rodrigues de Souza," February 23, 1990; on the 1989 expulsion of Baines, see "Stephen Baines to Profa. Cecília Maria Rodrigues de Souza," February 23, 1990, Folder 1990, AEDS.

[68] For more on the expulsions, see Coordenador do Núcleo de Apoio Waimiri-Atroari Raimundo Nonato Nunes Corrêa, "Relatório a respeito da saída do Sr. Egydio Schwade da área indígena Waimiri-Atroari," December 23, 1986; FUNAI, "Parecer sobre relatório de Raimundo Nunes Correa," December 30, 1986, Folder 1986, AEDS.

allies had been working with the Waimiri-Atroari for years and had developed a strong rapport with the community. Moreover, they were often the only voices encouraging criticism of resource extraction projects on Waimiri-Atroari land. Without such dissenting opinions, the government had a much easier time getting the Indigenous community to endorse mining and dam construction in return for financial compensation.

IN THE FOOTSTEPS OF PHARAOHS

In response to mounting criticisms, Eletronorte executed the traditional showpiece of pharaonic environmentalism: animal rescue missions. By this point, the ostentatious rescue missions were losing their reliability as a legitimate defense against the damaging environmental impacts of big reservoirs.[69] As early as 1979 – before the big rescue missions at Itaipu and Tucuruí – Balbina's architects acknowledged the controversies surrounding such programs.[70] Yet, their report noted that despite being of dubious scientific merit, rescue missions received a lot of international praise. Eletronorte's 1985 environmental impact study admitted that many animals would save themselves and that the rescue missions were losing credibility, but argued there was enough gain to justify rescuing animals at Balbina regardless.[71]

Operation Muiraquitã unfolded in the same manner as other rescue missions. Between October 1987 and June 1988, rescuers sped around Balbina's reservoir on boats and captured roughly 20,000 animals, 30 percent of which were sent to scientific institutions.[72] In addition to Eletronorte's team of rescuers, nearly forty researchers from various

[69] Rogério Gribel et al., "Destinos da fauna de Balbina," *Ciência Hoje* 6, no. 31 (1987): 76.

[70] Eletronorte, "Diagnóstico preliminar dos efeitos ambientais," 12.

[71] Eletronorte, "Estudos ambientais do reservatório de Balbina," 154–160.

[72] The exact dates are October 2 to June 28, Eletronorte, "Operação Muiraquitã, relatório final," July 28, Document 02592, AELN. There was a second, scaled-down phase of this operation that occurred between July 12, 1988, and January 29, 1989. See Eletronorte, "Operação de Resgate Remanescente, relatório 12 de Julho a 31 de agosto 1988," Document 01567, AELN, and Eletrobras, "Estudo de caso da Usina Hidrelétrica de Balbina," 63. The second phase of Balbina's rescue mission overlapped with a much shorter rescue operation at the Samuel Dam in Rondônia, which Eletronorte carried out between November 14, 1988, and March 19, 1989. See Rosa Maria Lemos de Sá, "Effects of the Samuel Hydroelectric Dam on Mammal and Bird Communities in a Heterogeneous Amazonian Lowland Forest," (PhD Diss., University of Florida, 1995), 4–9, 35.

museums and scientific institutes throughout the country, such as the Butantan Institute, participated in the operation. These researchers often went after specific animals that the Eletronorte team was too apathetic to search out and rescue.

The number of animals captured at Balbina was meager compared to Tucuruí's rescue mission, which had captured upwards of 268,000 animals (see Chapter 3).[73] Three variables explain the difference. First, Balbina's reservoir filled slowly; thus, more animals were able to save themselves. Second, the reservoir was shallow, which meant that there were more clusters of treetops protruding from the water that impeded access to trapped animals.[74] Lastly, the rescue mission ended before the job was complete, and the second, scaled-down phase of the operation (which lasted until the end of January 1989) was so underfunded and understaffed that rescuers left many animals stranded on treetops and islands.[75] Three years after the reservoir filled, fishermen reported finding dead animals on most of the islands throughout the reservoir, an indication that rescuers had left countless animals behind.[76]

The rescue operation did little to help many of the animals it captured. Once released, animals struggled to survive. Eletronorte released animals in forested areas on both sides of the river downstream from the dam. Little care went into choosing the sites. Eletronorte had not undertaken a comprehensive study of the local flora and fauna and so had minimal information about the area's ability to support boatloads of additional animals. Moreover, much of the surrounding area was densely forested, which limited rescuer access to small release sites along the riverbanks. Eletronorte did not monitor the animals it released, so the data on survival rates is unreliable. Eletronorte's final report admitted this shortcoming: "the proposed efforts ... for accompanying the released animals were carried out without any scientific methodology and are being overseen by people who do not know the area. These monitoring activities will

73 The distribution of animal species caught at Balbina was also different than at Itaipu and Tucuruí. Operation Muiraquitã's final report includes a breakdown of animal classifications for the 12,687 animals caught and released – which accounted for roughly 70 percent of the total number of animals captured – but not those caught and sent to scientific institutes. Turtles (3,601) were the most frequently captured and released animal, followed by rodents (2,469), arthropods (2,463), deer (1,653), snakes (931), and primates (863). See Eletronorte, "Operação Muiraquitã, relatório final," 13–14.

74 Gribel, "Os custos de Balbina," 1989, 3–4, Folder 1989, AEDS.

75 Eletronorte, "Operação de Resgate Remanescente," 2.

76 Fórum Permanente de Debates da Amazônia (FPDA), "Balbina: 3 anos depois...," Manaus, 1992, 10, Folder 1992, AEDS.

not show conclusive results ..."[77] Thus, Eletronorte's own report admitted that its monitoring program – and therefore the entire rescue operation – was largely superficial. Despite the absence of data on survival rates, anecdotal accounts from Eletronorte's own reports suggest that many released animals died from trauma and competition for resources.

Such dismal results did not stop Eletronorte from boasting. Its final report claimed: "Eletronorte's effort to minimize the environmental impacts from hydroelectric dams, through fauna rescue missions, shows a maturing and awareness of the importance of the environment for the maintenance of wildlife."[78] The rescue mission also caught the attention of the press, and newspapers from as far as São Paulo and the United States reported on the brave efforts of rescuers to save animals.[79] These articles, however, also noted the controversies that surrounded the practice of rescuing animals from floodwaters and its waning popularity among ecologists. Pharaonic environmentalism was proving insufficient to stem the tide of disapproval.

As Balbina's floodgates closed, others had opened: A freshet of scathing publications began pouring out from the desks of environmentalists, anthropologists, religious activists, and others who had long warned of the dam's deleterious impacts.[80] This flood of criticism reached its peak the year that the reservoir finished filling, and though writing waned in

[77] Eletronorte, "Operação Muiraquitã, relatório final," 14. [78] Ibid., 21.

[79] See Marlise Simons, "On Amazon Noah's Ark, All Aboard (Even Wasps)" *New York Times*, March 30, 1988; "Resgatados dezessete mil animais em Balbina," *A Notícia*, July 3, 1988; "Críticas à operação de resgate," *O Estado de São Paulo*, July 31, 1988; and "Operação de resgate esperou saída de animais," *Folha de São Paulo*, August 3, 1988.

[80] "Biólogo critica Balbina," *Jornal do Commercio*, January 19, 1987; Samuel Murgel Branco, "Balbina, demanda energética e ecologia amazônica," *A Crítica*, January 23, 1987; MAREWA, "Balbina: Uma hidrelétrica gerando morte," February 13, 1987, Folder 1987, AEDS; "Balbina vai inundar 2.346 km² de floresta amazônica, o risco é criar um lago podre e morto," *Jornal do Brasil*, June 21, 1987; "Balbina pode provocar desastre ecológico," *O Liberal*, July 9, 1987; CIMI, "Alguns efeitos sociais do projeto hidrelétrico de Balbina," July 1987, Folder 1987, AEDS; "Balbina é uma grande catástrofe," *A Crítica*, August 9, 1987; "Impacto ambiental, Balbina é uma séria ameaça," *Diário de Amazonas*, August 11, 1987; "Balbina, progresso ou catástrofe?" *A Crítica*, August 20, 1987; "Outra Balbina, nunca mais para Amazônia," *A Crítica*, August 26, 1987; Schwade, "Os grandes empreendimentos na Amazônia e seu custo social: Waimiri-Atroari um exemplo ilustrativo," November 21, 1987, Folder 1987, AEDS; "Balbina: Um escândalo ecológico," *Universidade de São Paulo Jornal do Campus* no. 59 (November 1987): 4–5, Folder "Balbina," APF; and MAREWA, *Balbina: Ameaça e destruição na Amazônia* (Manaus: MAREWA, 1987).

subsequent years, a handful of dedicated activists continued to document the reservoir's environmental transformations.[81] These reports publicized the ecological horrors that activists had anticipated and helped turn more people against the dam – including politicians and engineers who once supported it.

ROAD TO REPARATIONS

By the time the floodwaters rose, Eletronorte had realized that pharaonic environmentalism alone would not satisfy its critics. The chorus of disapproval had grown louder after 1985, when the military stepped down, and more concrete measures were needed to compensate those displaced. In 1987, Eletronorte created the Waimiri Atroari Program, funded by both Eletronorte and FUNAI, to indemnify the Indigenous group.[82]

In material terms, the program was successful in helping the Waimiri-Atroari recover from the damage that Balbina and other industrial projects inflicted on their society. The program has its shortcomings, but its benefits stand out as a success story among other derisory compensation measures that the Brazilian government extended to the Indigenous groups its dams displaced. The dictatorship's most controversial dam had the most successful reparation process – thanks to the efforts of activists who raised the alarm about the dam's social and environmental impacts.

The Waimiri-Atroari recuperated territorially. In 1987, the Brazilian government promulgated legislation that reinstated and enlarged the Waimiri-Atroari Reserve, and two years later, the President passed a

81 Fearnside, "Brazil's Balbina Dam"; "Peixes morrem e população passa fome no Rio Uatumã," *A Notícia*, March 9, 1989; "Poluição do Rio Uatumã vai perdurar seis anos," *A Notícia*, March 11, 1989; "Um desastre chamado Balbina," *O Povo Brasileiro*, April 19, 1989; "Caminhada de protesto à poluição de Balbina," *A Crítica*, April 22, 1989; "Grave desastre ecológico: Fome atinge Uatumã," *O Diário*, April 23, 1989; "Engenheiros repudiam Hidrelétrica de Balbina," *O Povo Brasileiro*, May 12, 1989; MAREWA, "Balbina: Massive Destruction of Amazonian Life," May 18, 1989; Gribel, "Eletronorte mata o Rio Uatumã," August 15, 1989; Gribel, "Os custos de Balbina"; Gribel, "Balbina no pais de impunidade," 1989, Folder 1989, AEDS; and Philip Fearnside, *A Hidrelétrica Balbina: O faraonismo irreversível versus o meio ambiente na Amazônia* (São Paulo: IAMÁ, 1990), 49.

82 Eletronorte and FUNAI, "Termo de Compromisso No. 002/87, que entre si celebram a Fundação Nacional do Índio e as Centrais Elétricas do Norte do Brasil S/A, objetivando a implantação de um programa de apoio às comunidades indígenas Waimiri-Atroari tendo em vista a inundação de parte de suas terras imemoriais pela UHE-Balbina," April 22, 1987, Folder BR.AM.WT.5d (Energético), ACIMI.

decree demarcating the land.[83] The new Waimiri-Atroari territory measured 25,856 sq. km, an area roughly the size of the state of Alagoas. The net result of the Brazilian military government's land theft remained a loss. The area the Waimiri-Atroari had inhabited before non-Indigenous Brazilians invaded it was about 83,000 sq. km, meaning they lost upwards of 65 percent of their land to the government and non-Indigenous landowners during the preceding decades.[84] But even such modest territorial gains were notable given that other Indigenous groups displaced by the government's dams lost nearly all their land and received only meager compensation (see Chapter 4). The reserve also helped protect their remaining land from further incursions.

The Waimiri-Atroari population also began to rebound. From its nadir of roughly 350 people in the mid-1980s, the population climbed over the subsequent decades. At the beginning of 2020, the Waimiri-Atroari numbered about 2,200.[85] The group's territorial gains were essential to population growth because the reserve helped shield them from the violence of both the state and individual landgrabbers. The program's public health component was also indispensable. Health officials carried out a series of measures that reduced the death toll of infectious disease. For example, the program's health officials controlled malaria, reducing it from a hyperendemic disease to one that surfaced only in small bouts.[86] Program officials vaccinated Waimiri-Atroari against other diseases, such as measles.[87] Once again, the military government's genocide represented a net loss for the group. Their population remains just a third of what it

[83] José Sarney, *Decreto No. 94.606, Declara de ocupação dos índios Waimiri-Atroari área de terra nos municípios de Novo Airão, Itapiranga, Presidente Figueiredo e Caracaraí, no Estado do Amazonas e no Território Federal de Roraima, e da outras providências*, July 14, 1987; Sarney, *Decreto No. 97.837, Homologa a demarcação administrativa da Área Indígena Waimiri-Atroari, que menciona, nos Estados do Amazonas e Roraima, e dá outras providencias*, June 16, 1989, ASF.

[84] Author's calculations. Eletronorte notes that the 1971 Waimiri-Atroari Reserve set aside an area of about 16,619 sq. km, and MAREWA contends that the traditional Waimiri-Atroari territory was five times the size of the reserve. See Eletronorte, "Estudos ambientais do reservatório de Balbina," 197, and MAREWA, "Hidrelétrica de Balbina contra índios e lavradores," 1.

[85] Since such impressive growth is almost biologically impossible for human populations, it is possible that the estimate for the nadir of 350 people is low or that more people choose to identify as Waimiri-Atroari today than did thirty-five years ago. Waimiri Atroari Program statistics webpage, www.waimiriatroari.org.br/dados (last accessed April 2023).

[86] Moura, "Influência da UHE Balbina sobre a saúde da população indígena Waimiri-Atroari," 114–119.

[87] Porfírio de Carvalho, "Relatório de atividades do Programa Waimiri Atroari," January 2008, 2–3, BELN.

was at the beginning of the twentieth century and about two-thirds of what it was in 1968, the year that construction began on the highway that ripped through their land. Moreover, demographic recovery does little to remediate the trauma that genocide inflicted and the knowledge lost from deceased ancestors. Nevertheless, territorial sovereignty and population recovery are at least the first steps on the long road to reparation.

The program brought other material benefits to the Waimiri-Atroari as well. Coordinators built schools and designed initiatives that encouraged the spread of traditional economic and cultural activities. The program also gave the Waimiri-Atroari access to the material perks of the industrial world, and they became some of the wealthiest people in the region, with a fleet of pickup trucks and a cultural center in Manaus that had a swimming pool.[88] Some celebrated the program for these accomplishments. In 1997, Mário Parwe, one of the Waimiri-Atroari leaders, proclaimed in a promotional pamphlet: "many white people said that the dam was going to kill off fish, game, and also the Waimiri-Atroari, and what happened was exactly the opposite ... the reservoir has life and us as well. This partnership worked out well for the Waimiri-Atroari. Without it, I believe we would have been condemned to extinction."[89] Not all members of the group share this opinion, and taking such laudatory remarks at face value risks obscuring the disastrous history of the Brazilian government's incursion into Waimiri-Atroari land. But the material benefits of the program that Parwe acknowledged are undeniable. For many, it is a commendable example of successful reparation for social and environmental damage.

For some, however, the program's gains came at a steep cost. The Waimiri Atroari Program prohibits contact with outsiders, which means that the program is the only source of information about what is happening within the community. Indigenists and anthropologists who were expelled from Waimiri-Atroari territory in the 1980s, such as Egydio and Doroti Schwade, Stephen Baines, and Marcio Ferreira da Silva, were never permitted to return, which created a rift within the indigenist community. In 1990, the program invited the expelled anthropologists to a conference, and both Baines and Silva declined the invitation because they disapproved of the program and had nothing but criticism to share.[90] Both had personal grudges against FUNAI for expelling them in the middle of their research

[88] FPDA, "Balbina: 3 anos depois...," 19.

[89] Eletronorte, "UHE Balbina Programa Waimiri Atroari, 10 anos – A importância da Usina Hidrelétrica de Balbina para o Estado do Amazonas em 10 anos de operação," 1999, 7, BELN.

[90] "Baines to Souza" and "Silva to Souza."

without clear justification and, in the case of Baines, for publishing false accusations against him. But they were also critical of the program's closed nature. Silva concluded: "In sum, the program does not have independent and critical participants, which, in the end, means there is not any true dialogue."[91] Eletronorte and FUNAI maintain that the decision to ban outsiders came from the Waimiri-Atroari themselves, but their critics charge that prohibiting outside contact is a government strategy to stifle dissent.[92] Though the informational filter remains a point of contention, the material gains that the program has brought are a notable accomplishment given that during the preceding decades, the military government had been in the habit of doing almost nothing to compensate displaced Indigenous communities.

The government's efforts to offer more than lip service to compensate for Balbina's deleterious impacts also extended to the nonhuman environment. In 1989, researchers from INPA recommended setting aside land within the Uatumã Basin for environmental protection. During earlier studies, they had discovered that the area was rich in biodiversity and had a high degree of endemicity. The federal government heeded their advice. In June 1990, it promulgated legislation that created the Uatumã Biological Reserve, which set aside a 5,600 sq. km area for protection along the eastern shore of the Balbina reservoir. In 2002, the government expanded the area to a total of 9,403 sq. km.[93] The reserve prohibited human occupation and use without prior approval of Brazil's environmental protection agency, and it served the dual purpose of slowing sedimentation in the reservoir and compensating for the damage done to the region's flora and fauna.[94] The Uatumã Biological Reserve came on the heels of state legislation that created the Maroaga Cavern Environmental Protection Area, a 3,747 sq. km area within the Municipality of Presidente Figueiredo along the western margin of the reservoir.[95] Unlike the federal reserve, the Maroaga Cavern Protection Area permits human occupation and land use, such as farming and tourism.

[91] "Silva to Souza," 6. [92] FPDA, "Balbina: 3 anos depois...," 19.

[93] Fernando Collor de Mello, *Decreto No. 99.277, Cria, no Estado do Amazonas, a Reserva Biológica do Uatumã e dá outras providências*, June 6, 1990, ACD; Fernando Henrique Cardoso, *Decreto de 19 de setembro de 2002 – Amplia os limites da Reserva Biológica de Uatumã, no Amazonas*, September 19, 2002, ACD.

[94] See Eletronorte, "Usina Hidrelétrica de Balbina, documento ambiental," February 1994, 43–46, Document 03362, AELN; Eletronorte, "Relatório síntese dos aspectos sócio-econômico e ambientais," 29–30, 68–70.

[95] The reserve was expanded form its original area of 2,785 sq. km to 3,747 sq. km in 1994. See Governo de Estado de Amazonas, *Decreto Estadual No. 12.836 de 09 de março de 1990* and Governo de Estado de Amazonas, *Decreto Estadual No. 16.364 de 07 de Dezembro de 1994*, AISA.

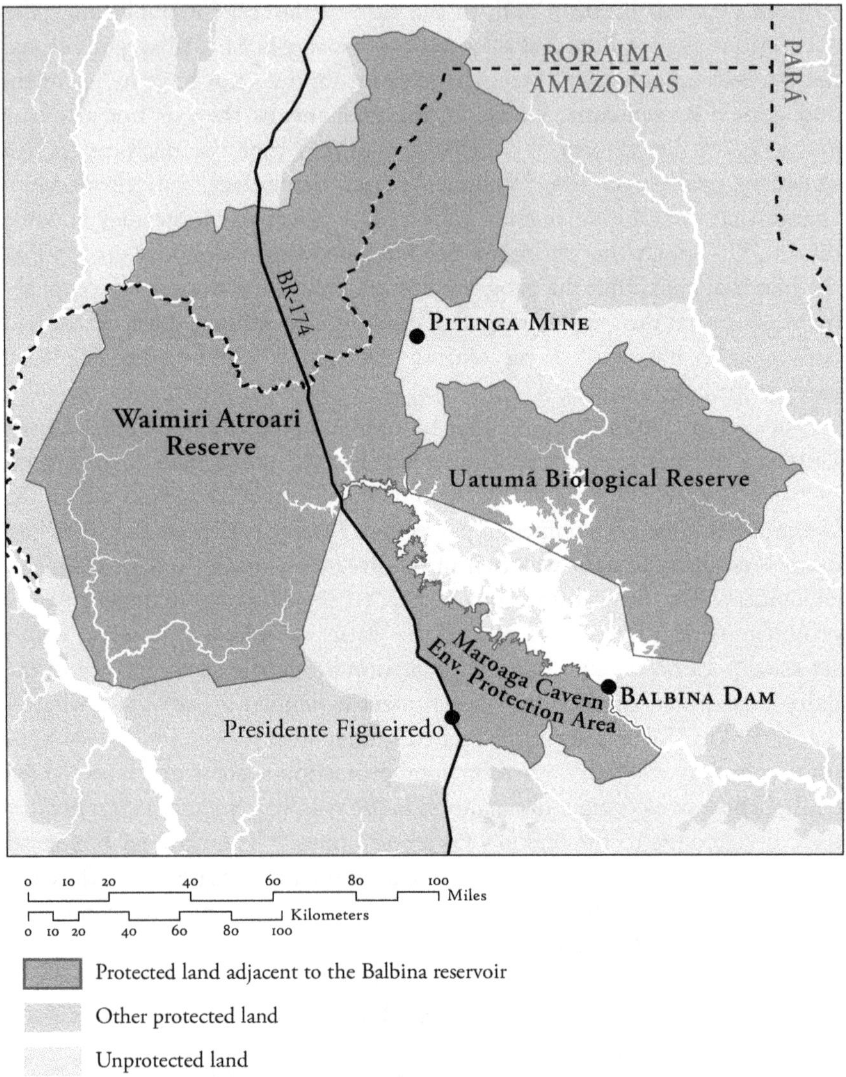

FIGURE 6.1. Protected land surrounding Balbina Dam, 1989–2023. Map by Geoffrey Wallace (G. Wallace Cartography & GIS), based on data from Rede Amazônica de Informação Socioambiental Georreferenciada (www.raisg.org).

The amount of land set aside for conservation is impressive. In total, the government designated roughly 40,000 sq. km of protected land after Balbina's floodgates closed, an area almost equal to the state of Rio de Janeiro (Figure 6.1).[96] The Uatumã Basin is about 70,600 sq. km,

[96] This calculation includes the Waimiri-Atroari Reserve, which is 25,856 sq. km.

meaning that protected land totaled more than half the size of the entire watershed, though the reserves only conserved a quarter of the basin because much of the Waimiri-Atroari Reserve is in another watershed.[97] Of course, legislation does not always ensure preservation – the state reserve around Presidente Figueiredo permits farming and other land use, and illegal occupation is a constant threat in other conservation areas. Nevertheless, these protected areas are a huge boon to environmental defense efforts in the Amazon and help establish a buffer against deforestation by ranchers and farmers. Though such measures can forestall future ecological damage, they did little to mitigate the environmental consequences of the reservoir itself.

GHOST FORESTS

Most of Balbina's undesirable ecological impacts can be traced to some basic features of the land it flooded. First, the dam inundated giant expanses of tropical rainforest, which created vast ghost forests of dead trees. These tracts of lifeless trees extended into the land bordering the reservoir because it raised the water table and waterlogged soils.[98] When trees and plants died and decomposed, they released nutrients into the water and set in motion the process of eutrophication that depleted the reservoir's dissolved oxygen supply. Second, the soil that supported these trees was acidic, and the reservoir water absorbed these acids. Lastly, the dam impounded a slow-moving river and inundated a flat area, which meant that the water turnover rate, or the amount of time it takes for new water to replace old water, was slow. Fast-moving water can flush out nutrients and acids, whereas stagnant water does little to ameliorate undesirable chemical imbalances. Balbina's turnover rate was about one year, a long time compared to Tucuruí, where water turns over every month and a half.[99] Thus, the basic characteristics of the area the reservoir flooded – dead trees, acidic soils, and flat terrain – created anoxic and acidic water that set in motion a series of ecological consequences.[100]

97 Eletronorte, "Relatório síntese dos aspectos sócio-econômico e ambientais," 68–70.

98 Gleicyane Souza Feitosa et al., "Estimativa da zona de deplecionamento da hidrelétrica de Balbina por técnica de sensoriamento remoto," *Anais XIII Simpósio Brasileiro de Sensoriamento Remoto*, Florianópolis, Brasil, 21–26 abril 2007.

99 The turnover rate of one year corresponds to the maximum reservoir height of 50 meters. The turnover rate is about six months when it operates at its minimum level of 46 meters.

100 This paragraph is drawn from Eletronorte, "Parecer sobre a execução de desmatamento do reservatório de Balbina, BAL-50-1000-RE," August 1985, 2–9, Document 02612 AELN; Fearnside, "Brazil's Balbina Dam," 406–410; and Eletronorte, "Relatório síntese dos aspectos sócio-econômico e ambientais," 79–84.

Eletronorte did little to mitigate these environmental impacts. To be sure, there was little it could do about the soils and topographic conditions, but it considered felling the forests to ameliorate the damage from decomposing trees.[101] In 1985, however, the company decided to leave most of the forest intact, felling only a small 55 sq. km area near the dam, as it had done at Tucuruí (see Chapter 3). Eletronorte worried that rapid vegetation regrowth (a feature of tropical rainforests) would quickly diminish the impact of forest clearing, and it decided that the overall ecological benefits would not be worth the high monetary cost of deforesting the area itself.[102]

Eletronorte also failed to entice another company to do it. The forest contained valuable hardwoods, but logging companies were discouraged because Eletronorte had not built roads or other infrastructure in the forest and because the dam's construction timeline was too short for companies to turn a profit given the high overhead of installing infrastructure.[103] Though it took Eletronorte the better part of a decade to finish Balbina, its long timeline was a result of delays that the company could not predict. During construction, it offered companies a period of about two years to log the forest, which they deemed unprofitable.[104]

Changes in water chemistry had the biggest impact downstream, where the largest concentrations of people lived. Most people in the area resided in towns and villages far from the river, and more people – about a thousand or so during the years when the reservoir filled – lived downstream from the dam than alongside the reservoir. Closing floodgates to fill the reservoir reduced the amount of water that reached the river below, and the water that did fall through the dam's turbines was acidic and anoxic. In 1992, researchers found the dissolved oxygen content of the water near the base of the dam to be about 1 mg/l, well below the average for blackwater rivers in the Northern Amazon, which was about 8 mg/l.[105] The caustic water was unpotable and caused itchiness and diarrhea for those exposed to it.[106] Eletronorte dug wells for roughly 200 people living along the stretch of the river closest to the dam, but residents complained that these wells were not deep enough. The chemical

101 Eletronorte, "Diagnóstico preliminar dos efeitos ambientais," 18–20.

102 Eletronorte, "Parecer sobre a execução de desmatamento," and Eletronorte, "Estudos ambientais do reservatório de Balbina," 236–242.

103 Eletrobras, "Estudo de caso da usina hidrelétrica de Balbina," 33–34.

104 Eletronorte, "Diagnóstico preliminar dos efeitos ambientais," 18–20.

105 FPDA, "Balbina: 3 anos depois...," 5.

106 Gribel, "Eletronorte mata o Rio Uatumã."

changes also had deleterious impacts on the river's biota and led to fish die-offs that were so massive they attracted large flocks of heron, a rare sight along the Uatumã and other blackwater tributaries.[107]

These changes engendered socioeconomic changes for the residents along the river. Commerce evaporated as the river dried, and families lost access to essential goods. Healthy fish became so scarce that families had to travel hours by canoe to the headwaters of small tributaries to find big schools. Some families began hunting, but gains were short-lived because the caustic water of the river repelled animals. Most families adapted by raising poultry, a form of subsistence that was not dependent on river water. Some families decided to escape the environmental hazards by moving to Manaus, though researchers (in 1992) found that many of them were unable to find work in the city and were planning on returning to the Uatumã despite the ecological challenges there.[108]

Balbina's downstream impacts diminished over time. So too did reporting about the environmental conditions along the Uatumã, but there are enough accounts to establish a rough timeline. The acute phase of the dam's deleterious downstream impacts seems to have lasted for about five years after the reservoir filled,[109] although depleted oxygen levels persisted for at least ten years and perhaps longer – in the late 1990s, researchers affirmed that anoxic conditions were still rendering the water unpotable.[110]

The environmental impacts along the reservoir itself were less severe. To be sure, the chemical changes damaged the biota, but the giant size of

107 FPDA, "Balbina: 3 anos depois...," 10. 108 Ibid.

109 In 1993, Eletronorte confirmed that the water downstream was still unable to support aquatic life, but the following year some observers noted some signs of life returning to the area, though most evidence came from the reservoir rather than downstream. See Eletronorte, "UHE Balbina, diagnóstico da qualidade da água, fases rio, enchimento e reservatório," 1993, 17, Document 03363, AELN; Sérgio Figueiredo, "Revisitando Balbina: A espera de uma morte anunciada," *Amazonas em Tempo*, April 27, 1994; and Eletronorte, "Relatório síntese dos aspectos sócio-econômico e ambientais," 32. Observations of modest improvements to ecosystem health after 1994 are supported by the fact that scientists had estimated the acute impacts would last at least six years, but in the late 1990s were commenting that the ecosystem around Balbina was recovering faster than anticipated. See José Lauro Thomé, *Um grande projeto na Amazônia: Hidrelétrica de Balbina, um fato consumado* (Manaus: Editora da Universidade de Amazonas, 1999), 78–80.

110 Fearnside, "Revisitando Balbina," *Amazonas em Tempo*, May 4, 1994; Thomé, *Um grande projeto na Amazônia*, 78–80. Thomé also noted that the smell of hydrogen sulfide – the product of decomposing biomass in anoxic water – was still detectible around Balbina in the late 1990s, though he claimed the foul smell was much less pungent than during the years when the reservoir filled.

the reservoir diluted the hazardous elements in the water and softened their impact on fish. Changes in fisheries unfolded along a similar pattern as those at other big reservoirs. New conditions did not suit some species of fish, but others flourished in the novel environment. Local fishermen reported that migratory fish diminished, while carnivorous fish proliferated.[111] Information and statistics about fishing at Balbina are scarce, but multiple reports suggest that throughout much of the 1990s, annual reservoir production was between 500 and 700 tons a year.[112] The boom eventually diminished as nutrients subsided and pressure from fishermen increased. According to one account, by 2012, annual production had dropped to just 70 tons a year.[113]

Environmental factors had as much to do with plummeting production as overfishing. Although undoubtedly real,[114] overfishing had a limited impact for several reasons. First, Balbina's reservoir is giant and shallow and has ample fishing grounds. Second, there were relatively few commercial fishermen at Balbina. During the boom years, about a hundred fishermen consistently fished on the reservoir.[115] Third, since the region around Balbina is relatively unpopulated, demand was not encouraging overfishing. There was greater demand in Manaus, but other fisheries competed for the city's markets. Lastly, the government banned mesh nets and prohibited fishing in the Uatumã Biological Reserve – whose borders included the eastern shore of the reservoir – which means that fish had access to protected areas to regenerate. Overfishing was a concern, but not the cause of the collapse in production. The boom was a response to the changes that unfolded after the reservoir formed, and the bust was a reflection of ecological stabilization.

Balbina's reservoir also did not engender malaria epidemics or intolerable mosquito swarms. Malaria was endemic in the areas surrounding the

111 FPDA, "Balbina: 3 anos depois…," 6–8.

112 This was about 2.1 kg/hectare/year. Eletronorte, "Relatório síntese dos aspectos sócio-econômico e ambientais," 21, 31–32; Geraldo Mendes dos Santos and Arnaldo Braga de Oliveira Jr., "A pesca no reservatório de Hidrelétrica de Balbina (Amazonas, Brasil)," *Acta Amazônica* 29, no. 1 (1999): 145–163.

113 Ana Roberta Pessoa Aguilar Cañas, "Conflitos silenciosos: A pesca amadora no Lago de Balbina, Presidente Figueiredo, Amazonas," master's thesis, Universidade Federal do Amazonas, Manaus, 2012, 51.

114 In May 1996, the state government went so far as to temporarily ban commercial fishing because of conflicts between fishermen. See Eletronorte, "Relatório síntese dos aspectos sócio-econômico e ambientais," 32.

115 FPDA, "Balbina: 3 anos depois…," 6–8, and Eletronorte, "Relatório síntese dos aspectos sócio-econômico e ambientais," 21, 31–32.

dam when construction began; however, the health department successfully controlled malaria at the dam site, and there were no cases of autochthonous malaria during construction. Health department officials tested all workers, administered prophylactic medicine, treated cases, sprayed oil on mosquito breeding grounds, and regularly sprayed houses with DDT. To be sure, malaria sickened many workers, but in each case, workers contracted the disease elsewhere.[116]

Malaria threatened to erupt once the reservoir filled. Both proponents and critics anticipated that the reservoir might engender malaria epidemics, and Eletronorte hired malaria specialists to study the problem in greater detail.[117] But these concerns proved unfounded. During the two years that the reservoir filled, health officials reported that there were small outbreaks of malaria in areas near the reservoir and downstream of the dam.[118] But these outbreaks were small and short-lived, and the region's foremost malaria experts have confirmed that the scourge did not persist near the reservoir.

Two characteristics account for the absence of malaria along the reservoir margins. First, aquatic plants – which harbor malaria-carrying mosquito larvae – did not proliferate in the reservoir because the soils in the basin of the Uatumã are comparatively nutrient-poor.[119] Decomposing trees released enough nutrients to sustain the populations of fish that sheltered in their flooded trunks, but the reservoir was not sufficiently rich in nutrients to support a big macrophyte boom. When the

[116] Interview with Raul Diniz Souza Amorim, Manaus, October 2017; Barroso and Souza Amorim, *Malária no Amazonas*, 356–367.

[117] Eletronorte, "Diagnóstico preliminar dos efeitos ambientais," 10–11; MAREWA, "Hidrelétrica de Balbina contra índios e lavradores," 15; and "Hidrelétrica de Balbina pode provocar epidemias," *A Crítica*, May 26, 1987.

[118] Eletronorte reported just 21 cases in 1988, Eletronorte, "Relatório síntese dos aspectos sócio-econômico e ambientais," 95–96. See also, Moura, "Demanda dos serviços de saúde da área de influência da UHE Balbina antes e durante o enchimento do reservatório, de 1984 a 1989," 86–90, Document 01766, AELN; Moura, "Evolução das condições de vida e saúde da população da área de influência da UHE Balbina durante o enchimento do reservatório – Inquérito epidemiológico durante o enchimento do reservatório," Convênio Eletronorte-Instituto de Medicina Tropical de Manaus, May 1989, 80–84, Document 01767, AELN. For other accounts of these small outbreaks, see "Barragem seria causa de malária," *A Crítica*, October 28, 1989; Schwade, "Malária está pipocando em torno do lago," *Informa Partido de Trabalhadores Presidente Figueiredo* 1, no. 4 (October 1989), Folder 1989, AEDS.

[119] Interview with Wanderlei Pedro Tadei, Manaus, October 2017; Lisardo Osorio Quintero et al., "Ocorrência de espécies de anopheles e outros culicídeos na área de influência da hidrelétrica de Balbina – Cinco anos após o enchimento do reservatório," *Acta Amazônica* 26, no. 4 (1996): 281–296.

reservoir filled, aquatic plants spread briefly in the area near the dam that had been deforested (which Eletronorte removed using machines), but did not otherwise proliferate.[120] Indeed, a team of researchers investigating Balbina's environmental impacts three years after it filled were surprised to find no macrophytes in the reservoir.[121]

Second, there were very few people living along the reservoir and downstream. Health officials recorded about 16,000 people living in the area surrounding Balbina before and after its reservoir filled, but almost none lived along the reservoir itself or within five kilometers of it. Most people in the area lived in the small towns of Presidente Figueiredo, Balbina, and Pitinga, and there were fewer than a thousand people living alongside the river, all of whom were downstream of the dam.[122] Waimiri-Atroari villages did not abut the reservoir, and the Uatumã Biological Reserve protected most of the eastern side from human occupation.[123] The absence of both abundant mosquito habitat and people curtailed the reservoir's influence on malaria, further softening its local impact around the reservoir.

Balbina had more alarming impacts at the global level. The reservoir releases greenhouse gases into the atmosphere that contribute to climate change. In the 2000s-2010s, researchers estimated that the total annual emissions from Balbina had the same warming potential as half the fossil fuels consumed in São Paulo. These scientists estimated that it produces roughly 3 tons of CO_2-equivalent carbon per MWh of electricity, which is tenfold the rate for coal-fired power plants, which produce on average 0.3 tons of CO_2-e carbon per MWh of electricity.[124]

Balbina emits both carbon dioxide and methane through various mechanisms.[125] First, the tops of dead trees that remain above the water line release carbon dioxide into the atmosphere as they decompose

120 See Eletronorte, "Operação de Resgate Remanescente," 3.

121 FPDA, "Balbina: 3 anos depois ...," 10.

122 Moura, "Evolução das condições de vida e saúde da população," 16.

123 On Waimiri-Atroari villages, see Moura, "Influência da UHE Balbina sobre a saúde da população indígena Waimiri-Atroari," 117.

124 Alexandre Kemenes et al., "Methane Release below a Tropical Hydroelectric Dam," *Geophysical Research Letters* 34, no. 12 (June 2007); Kemenes et al., "CO_2 Emissions from a Tropical Hydroelectric Reservoir (Balbina, Brazil)," *Journal of Geophysical Research* 116, no. G3 (September 2011).

125 The following paragraphs are synthesized from Philip Fearnside, "Greenhouse Gas Emissions from Hydroelectric Dams in Tropical Forests," in Jay Lehr and Jack Keeley, eds., *Alternative Energy and Shale Gas Encyclopedia* (New York: Wiley & Sons, 2016), 428–438.

through oxygen exposure. These emissions are compounded by the clusters of dead trees along the shoreline that drowned in waterlogged soils. The release of carbon from rotting ghost forests was a significant source of emissions: Environmental scientist Philip Fearnside estimated that the annual above-water decay emissions alone were 6.4 million tons of carbon in 1990, which is about one and a half times the annual emissions of an average US coal-fired power plant.[126] Emissions from dead treetops were high for about a decade after the reservoir filled and dissipated once the decomposing trees released all their carbon.

The second mechanism through which Balbina emits greenhouse gases into the atmosphere is underwater decay. When organic matter – such as tree trunks and branches – decomposes underwater, it fragments into carbon dioxide and methane compounds, which are released as gases into the atmosphere through bubbling and diffusion. Whether organic matter breaks down into carbon dioxide or methane depends on where in the reservoir decomposition occurs. Water in reservoirs stratifies into two layers: a surface layer that is warmer, lighter, and oxygen-rich because it has contact with the air, and a bottom layer that is colder, heavier, and anoxic.[127] Organic matter that decomposes near the surface – where oxygen is present – becomes carbon dioxide, and organic matter that decomposes at depths where the water is anoxic becomes methane. The bottom layer is able to hold high concentrations of gas because of its temperature and pressure, but methane can still bubble up to the surface. This gas is released into the atmosphere either as methane, or as carbon dioxide if it lingers long enough in the surface layer for bacteria to transform it.

Greenhouse gases are released in a similar manner at turbines and spillways. The penstocks that funnel water to turbines at the powerhouse pull water from the bottom layer of the reservoir, which means the water is infused with methane. Upon its fall to the turbine at the base of the dam, the pressure releases, and methane bubbles form and discharge into the atmosphere in the same way that carbonation under pressure in a soda

[126] Fearnside, "Hydroelectric Dams in the Brazilian Amazon as Sources of 'Greenhouse Gases'," 7–19. Data for the coal-fired power plant comparison was calculated using the US Environmental Protection Agency's Greenhouse Gas Equivalencies Calculator, www.epa.gov/energy/greenhouse-gas-equivalencies-calculator (last accessed April 2023).

[127] In technical terms, the surface layer is referred to as the epilimnion, and the bottom layer as the hypolimnion. The line that separates the two is called the thermocline or oxycline, depending on whether one is referring to the contrast in temperature or dissolved oxygen content. See Fearnside, "Greenhouse Gas Emissions," 6.

bottle does upon unscrewing the top. This process happens so fast that bacteria in the surface layer do not have time to act upon the gas, so most of it is released into the atmosphere as methane rather than carbon dioxide. This process is perhaps the most important mechanism by which Balbina has affected the atmosphere because it releases immense quantities of methane stored in the bottom layer of the reservoir, and methane is a far more potent greenhouse gas than carbon dioxide.

Hardwoods decompose very slowly underwater, so these methane emissions are more gradual compared to the release of carbon dioxide from dead treetops on the surface. Though long-term estimates for underwater decomposition rates at Balbina are not available, Fearnside has suggested that it can take up to 500 years for submerged hardwood trees to decompose in an Amazonian reservoir.[128] Long-term estimates for methane releases from these rotting trees are also unavailable, but Fearnside anticipates methane releases to be fairly constant, at least during the first 50 years after the reservoir formed.[129]

The third mechanism of releasing greenhouse gases is the drawdown area. To produce power, dams release water through penstocks and in doing so draw down the level in the reservoir, creating an exposed area during periods when rainfall is insufficient to refill it. During these dry seasons, weeds and grasses proliferate in the drawdown area only to die when the area is reflooded during the wet season. When this plant matter dies, it decomposes in the reservoir in the same manner as the flooded biomass. However, unlike hardwoods, this soft vegetation breaks down rapidly.

Three factors compound the emissions from the drawdown area at Balbina compared to other dams. First, the reservoir is shallow and covers a flat area, so when water recedes it exposes a large surface area. Researchers estimate that the average annual drawdown area at Balbina is about 660 sq. km.[130] Second, the reservoir is located in a tropical rainforest near the equator, a climate propitious for fast vegetation growth. Lastly, it powered a metropolis desperate for electricity, so during the initial decades of operation, dam operators were constantly drawing down the reservoir to its lowest levels to meet burgeoning demand.

128 Quoted in Fred Pearce, "Water Reservoirs and Greenhouse Emissions," *The Independent*, October 13, 2000.

129 Fearnside, "Hydroelectric Dams in the Brazilian Amazon," 12.

130 Feitosa et al., "Estimativa da zona de deplecionamento."

Not all of these greenhouse gas discharges contribute to the net emissions of the dam. For example, carbon dioxide released from growth in the drawdown area does not count toward net emissions because these plants pull carbon from the atmosphere during photosynthesis, which cancels out the carbon emissions during decomposition.[131] For this reason, Fearnside argues that carbon releases from the drawdown area and continuous inputs from the watershed should be discounted when calculating the net emissions of tropical dams. For him, the only source of carbon dioxide that contributes to the dam's net impact is that of decomposing trees.

Methane, however, is a different story. Plants do not absorb methane during photosynthesis, so the dam alone is responsible for its formation and release into the atmosphere, and all methane discharges count toward the dam's net emissions. Some areas of the Amazon flood seasonally and produce methane naturally, but this was not the case for Balbina. The land the dam flooded was a tropical upland forest that did not flood seasonally, so it would not have had net methane emissions before the reservoir.[132]

In sum, most of Balbina's net greenhouse gas emissions come from carbon dioxide released from decomposing trees during the first decade after the reservoir formed and from methane formed by decomposing biomass in the anoxic bottom layers of the reservoir, which is released at the surface or at the spillway and turbines at the base of the dam. Carbon dioxide emissions were immediate and comparatively brief, whereas methane emissions are expected to continue at fairly constant rates for decades, perhaps centuries, because trees decompose slowly underwater, and rains continually wash plant matter into the reservoir.

In the decades after Balbina's reservoir filled, a rancorous debate erupted about whether it was the exception or the rule for greenhouse gas emissions at tropical dam sites. The debate began in the aftermath of the United Nations Framework Convention on Climate Change (UNFCCC), an international treaty opened for signature during the 1992 United Nations Conference on Environment and Development in Rio de Janeiro that set goals for reducing greenhouse gas emissions. One year later, a group of Canadian researchers published a study hypothesizing that reservoirs might be potential sources of greenhouse gas emissions, and a group of Brazilian researchers responded with a study that

131 See Kemenes et al., "CO_2 Emissions from a Tropical Hydroelectric Reservoir."

132 Ibid.

acknowledged but downplayed the threat.[133] Shortly thereafter, Fearnside published a paper on greenhouse emissions from tropical dams that featured Balbina as the protagonist. It concluded that dams were a much greater source of greenhouse gas emissions than the previous studies claimed.[134] The authors championing more conservative estimates accepted that Balbina had exceptionally high rates of emissions per kilowatt hour because it produced so little power, but they balked when Fearnside suggested that Tucuruí also had higher emissions than coal-fired power plants.[135] Over the next decade, debates raged in the pages of scientific journals about whose estimates were more accurate.[136]

The stakes were high. During the 1990s and 2000s, the Brazilian government was planning to build a host of new dams in the Amazon (see Chapter 7). In 1997, the UNFCCC implemented a new set of tools for curbing emissions in a document called the Kyoto Protocol. Under the protocol's Clean Development Mechanisms, the Brazilian government could receive financial compensation for investing in hydropower, because such projects were considered to be a negligible source of greenhouse gas emissions.[137] Thus, the Brazilian government had a vested

133 John W.M. Rudd et al., "Are Hydroelectric Reservoirs Significant Sources of Greenhouse Gases?" *Ambio* 22, no. 4 (June 1993): 246–248; Luiz Pinguelli Rosa and Robert Schaeffer, "Greenhouse Gas Emissions from Hydroelectric Reservoirs," *Ambio* 23, no. 2 (March 1994): 164–165.

134 Fearnside, "Hydroelectric Dams in the Brazilian Amazon."

135 Luiz Pinguelli Rosa et al., "Are Hydroelectric Dams in the Brazilian Amazon Significant Sources of 'Greenhouse' Gases?" *Environmental Conservation* 23, no. 1 (March 1996): 2–6; Philip Fearnside, "Greenhouse-Gas Emissions from Amazonian Hydroelectric Reservoirs: The Example of Brazil's Tucuruí Dam as Compared to Fossil Fuel Alternatives," *Environmental Conservation* 24, no. 1 (March 1997): 64–75; Philip Fearnside, "Greenhouse Gas Emissions from a Hydroelectric Reservoir (Brazil's Tucuruí Dam) and the Energy Policy Implications," *Water, Air, and Soil Pollution* 133 (2002): 69–96.

136 Philip Fearnside, "Greenhouse Gas Emissions from Hydroelectric Dams: Controversies Provide a Springboard for Rethinking a Supposedly 'Clean' Energy Source," *Climatic Changes* 66 (2004): 1–8; Luiz Pinguelli Rosa et al., "Greenhouse Gas Emissions from Hydroelectric Reservoirs in Tropical Regions," *Climatic Changes* 66 (2004): 9–21; Danny Cullenward and David G. Victor, "The Dam Debate and Its Discontents," *Climatic Changes* 75 (2006): 81–86; Danny Harvey, "The Exchange between Fearnside and Rosa Concerning The Greenhouse Gas Emissions from Hydro-Electric Power Dams," *Climatic Changes* 75 (2006): 87–90; Luiz Pinguelli Rosa et al., "Scientific Errors in the Fearnside Comments on Greenhouse Gas Emissions (GHG) from Hydroelectric Dams and Response to His Political Claiming," *Climatic Changes* 75 (2006): 91–102; Philip Fearnside, "Greenhouse Gas Emissions from Hydroelectric Dams: Reply to Rosa Et al.," *Climatic Changes* 75 (2006): 103–109.

137 Cullenward and Victor, "The Dam Debate and Its Discontents," 81.

interest in research that affirmed its Amazonian dams would be sufficiently emission-free to qualify for financial compensation. Indeed, Luiz Pinguelli Rosa, the head author of most of the studies that pushed for conservative estimates, was the president of Eletrobras in 2003 and 2004. Compounding the contentious nature of the debate was the fact that emission rates vary widely from site to site, which makes generalizing difficult. Indeed, defenders of hydropower as a clean source of energy have pointed out that big reservoirs outside of the Amazon Rainforest, such as Itaipu and those along the São Francisco River, have negligible greenhouse gas emissions.[138]

For reservoirs in the Amazon, most evidence suggests that high emissions are the rule, rather than the exception. There are many points of disagreement about how emissions rates are calculated, but perhaps the most significant is whether or not releases from turbines and spillways are included or omitted.[139] Studies promulgating conservative estimates do not take into account the emissions released as gas bubbles at turbines and spillways.[140] Instead, they measure methane releases from water downstream from the dam, often 50 meters or more from its base. Such measurements underreport emissions because much of the methane is discharged as soon as pressure is released at the turbine or spillway. For example, at Balbina, researchers estimated that 53 percent of the methane passing through its turbines was discharged immediately.[141] The most accurate estimates come from studies that measure the difference in methane concentrations in the water near the penstock intake and near the turbine at the base of the dam. These studies indicate much higher rates of greenhouse gas emissions than those done by authors championing more conservative estimates. Balbina's giant reservoir and small power output make its per-kilowatt-hour emissions exceptionally high, but it is not alone in emitting greenhouse gases at levels equal to or higher than those of fossil fuel-fired power plants.[142]

138 Rosa et al., "Greenhouse Gas Emissions from Hydroelectric Reservoirs in Tropical Regions," 11–16.

139 For an overview of other methodological disagreements about calculating emissions rates, see Fearnside, "Greenhouse Gas Emissions from Hydroelectric Dams in Tropical Forests."

140 Rosa et al., "Greenhouse Gas Emissions from Hydroelectric Reservoirs in Tropical Regions," and Rosa et al., "Scientific Errors in the Fearnside Comments on Greenhouse Gas Emissions."

141 Kemenes et al., "Methane Release below a Tropical Hydroelectric Dam," 3.

142 Kemenes et al., "CO_2 Emissions from a Tropical Hydroelectric Reservoir"; Felipe A.M. de Faria et al., "Estimating Greenhouse Gas Emissions from Future Amazonian Hydroelectric Reservoirs," *Environmental Research Letters* 10 (2015).

CONCLUSION

Between 1981 and 1987, the military dictatorship and the civilian administration that replaced it built Brazil's most infamous dam. Balbina was the culmination of the dictatorship's dam-building campaign – it was a big dam with tremendous political appeal that the government erected as swiftly as it could and with only superficial environmental safeguards. It flooded more than 2,400 sq. km of rainforest that included two Waimiri-Atroari villages, which were home to about a third of their population, survivors of a violent genocide. The ghost forests of dead trees in the reservoir set in motion a series of chemical changes that rendered the water in the reservoir acidic and anoxic, which made it unusable and thinned schools of fish downstream. These same chemical changes produced methane and carbon dioxide in the water, which discharged into the atmosphere at rates equal to or higher than the greenhouse gas emissions of fossil fuel-fired power plants, contributing to climate change.

Balbina's environmental problems alone cannot explain its near-universal disdain. The story of the Waimiri-Atroari is exceptionally tragic in terms of the violent genocide that immediately preceded construction, but their displacement at the hands of Eletronorte was otherwise not different from other Indigenous groups displaced by the military regime's big dams. Moreover, Eletronorte's Waimiri Atroari Program, despite its problems, was the best compensation measure that the government extended to a group displaced by one of its dams. The Waimiri-Atroari have land, health care, and some material wealth, whereas other groups like the Avá Guaraní and Tuxá are still fighting, decades later, for basic compensation (see Chapter 7).

In some respects, Balbina actually fared better than other dams. Unlike Sobradinho, Itaipu, Tucuruí, and Itaparica, which each displaced tens of thousands of non-Indigenous Brazilians, Balbina displaced almost no non-Indigenous Brazilians. The dam inundated sacred Waimiri-Atroari land, but not any cultural sites or natural wonders that non-Indigenous people deemed important. Since there was not a huge cohort of displaced people, fishing booms and busts did not have the same social impact as they did elsewhere. Similarly, the absence of large numbers of people living along the reservoir helped prevent malaria from spreading. Balbina's carbon footprint is much greater than dams in the South, but evidence of its greenhouse gas emissions surfaced long after it had already been widely condemned as an ecological disaster.

This chapter argues that two factors explain Balbina's widespread disapproval. First, Balbina produced very little energy relative to its environmental impact. Its output is minuscule compared to that of Tucuruí and Itaipu, which produce 32 and 56 times as much power, respectively. To be sure, its local benefits were profound. While other dams were built to anticipate future demand and encourage growth, Manaus was desperate for electricity when Balbina was built. In 1992, the dam accounted for 80 percent of the city's electricity.[143] But these benefits were short-lived. By 2004, Balbina supplied just 20 percent of Manaus' electricity and the city once again depended on petroleum and natural gas to keep the lights on.[144] Moreover, these benefits were restricted geographically. More powerful dams such as Tucuruí and Itaipu generated enough energy to power entire regions, whereas Balbina powered just one city in the middle of a rainforest that many came to believe should be valued more for its natural features than its urban landscapes.

The second and most important factor was timing. Eletronorte closed Balbina's floodgates just as Brazilians were reckoning with the damaging results of a host of giant reservoirs built without public input or environmental safeguards. Moreover, it was built at a time when the perception of the Amazon was changing. The environmental movement had encouraged people to see the rainforest as an area to be preserved, and during the 1980s, a movement surfaced to protect it from deforestation.[145]

Balbina and the military government's other environmentally catastrophic dams had political consequences. In response, the newly democratic government passed legislation that required the government to undertake environmental impact studies and get public input before building dams. Shortly thereafter, the popular movement against dams coalesced into a formal organization that began lobbying against big reservoirs. Together, these forces helped curb the government's plans to build more dams, many of which were cancelled altogether or modified significantly so as to reduce their environmental impact. It is to these changing circumstances that the final chapter turns.

[143] "El Nino deixa Eletronorte em estado de alerta," *Amazonas em Tempo*, June 23, 1992.

[144] Mário Adolfo and Rodrigo Araújo, "Hidrelétrica Balbina 15 anos depois," *Amazonas em Tempo*, August 13, 2004; Rodrigues, "Vidas despedaçadas," 52.

[145] See Hecht and Cockburn, *The Fate of the Forest*.

7

Aftermath

The Anti-Dam Movement, Social Injustice, and Climate Change, 1990s–2010s

INTRODUCTION

In April 1989, a diverse group of Brazilian activists met in Goiânia to discuss the coordination of a national movement against dams. For three days, leaders from different local movements that had fought dams during the preceding decades shared horror stories about the social and environmental impacts of government-built reservoirs. Despite the geographic differences that made each case unique, the stories highlighted the consistent government neglect of displaced communities and the deleterious environmental impacts that plagued them. These similarities and shared experiences reminded participants that the reservoirs they had fought against were part of a nationwide dam-building campaign promoted by the federal government and encouraged them to form a national coalition in response.[1] The Goiânia gathering was the first in a series of meetings that resulted, in 1992, in a National Congress for People Impacted by Dams, which created the Movement of People Affected by Dams [Movimento dos Atingidos por Barragens, MAB], an NGO dedicated to helping those displaced by reservoirs and fighting against controversial dams.

The target of the Goiânia meeting was the federal government's latest energy blueprints. In 1987, Eletrobras released a plan that proposed building more than 250 dams over the following two decades, and activists worried that these projects might repeat the disasters that the

[1] Comissão Regional de Atingidos por Barragens (CRAB), "Relatório do Encontro Nacional dos Trabalhadores Atingidos por Barragens – Goiânia, 19, 20, e 21-04-89," Folder 1989, AEDS.

dictatorship's dam-building program had unleashed.[2] The committee's opening report wondered, "what will happen to the displaced people? Is it possible to prevent the repetition of what happened at Sobradinho, Itaipu, Itaparica and Balbina?"[3] The conference organizers believed that creating a united front against these dams would help stop construction outright or at least achieve modifications and equitable indemnity to soften their deleterious impacts. Reacting to recent dam-related catastrophes and emboldened by the fall of the dictatorship, Brazilian activists set out to inaugurate a new era in Brazilian dam building, one in which social and environmental justice carried more weight.

This chapter discusses the changing nature of dam building in Brazil in the 1990s–2010s. It argues that during this period, greater mobilization for social and environmental justice came to characterize the county's dam-building program and that the movement's priorities and achievements were not uniform. Anti-dam activists devoted the most attention to fighting dams in the Amazon Rainforest, where they have been successful in curbing dam construction and forcing modifications to the dams that were built to ameliorate their impacts. Outside of the Amazon, anti-dam activists were less adamant about blocking dam construction outright, but did successfully force modifications and concessions to mitigate reservoir impacts. However, the national anti-dam movement devoted comparatively little attention to fighting for justice on behalf of still-uncompensated Indigenous communities that were displaced by the dictatorship's reservoirs. More than thirty years after being displaced, the Indigenous communities dispossessed by Itaipu and Itaparica are still fighting for the land the government owes them.

Climate-related challenges have been a second defining element of this period. Since the late 1990s, the Brazilian hydropower sector has endured at least three significant droughts that lowered reservoir levels, which curtailed output and led to rolling blackouts. Such episodes could become more common and severe under anthropogenic global warming. Thus, while the Brazilian hydropower sector, excluding the Amazon, has done much to mitigate carbon emissions (see Chapters 3 and 6), the impacts of

[2] Eletrobras, *Plano 2010: Plano nacional de energia elétrica 1987/2010 relatório geral* (Rio de Janeiro: Ministério das Minas e Energia, December 1987).

[3] Central Única dos Trabalhadores and CRAB, "Encontro Nacional dos Trabalhadores Atingidos por Barragens, Goiânia, 19–21 abril, 1989, proposta para debate," Folder 1989, AEDS.

anthropogenic warming threaten to curtail the degree to which reservoirs can produce such valuable low-carbon energy.

THE ANTI-DAM MOVEMENT

The Goiânia meeting was the culmination of a decade of activism across the country. Though it is difficult to pinpoint an exact point of origin for domestic activism against big dams, there is a strong case to be made for both Sobradinho and Itaipu (see Chapter 4). The resistance that arose in response to both reservoirs in the late 1970s inspired other communities to take action against other menacing dams, such as those discussed in the preceding chapters.

The struggle of displaced farmers at Itaipu had a direct influence on the group that organized the meeting in Goiânia, the Regional Commission of the People Affected by Dams on the Upper Uruguay [Comissão Regional de Atingidos por Barragens do Alto Uruguai, CRAB]. In 1979, the military government announced plans to build more than twenty dams on the Uruguay River and its tributaries in the southern state of Rio Grande do Sul. At the time, the government claimed that these dams would cumulatively displace about 15,000 people, but CRAB later estimated that upwards of 200,000 people would be impacted. Those in the affected area were aware of the social and financial calamities then unfolding at Itaipu and other dams, and they worried that such tragedies might be repeated in their communities. In April 1979, activists in these communities attempted to forestall disaster by creating a commission to negotiate with the government for design modifications and equitable indemnity.[4]

Itaipu continued to have an influence on the movement after the formation of the commission. To raise awareness about the dams' potential impacts and build support for their movement, local leaders brought in someone displaced by Itaipu to share their experience and also screened a documentary about the dam and its resettlement troubles.

The commission succeeded in getting most of its demands met. It confronted Eletrosul, the state-owned company building dams on the

[4] This paragraph and the following two are drawn from Franklin Daniel Rothman and Pamela E. Oliver, "From Local to Global: The Anti-Dam Movement in Southern Brazil, 1979–1992," *Mobilization* 4, no. 1 (1999): 41–57, and Carlos Vainer, "Águas para a vida, não para a morte: Notas para uma história do movimento de atingidos por barragens no Brasil," in Henri Acselrad et al., eds., *Justiça ambiental e cidadania* (Rio de Janeiro: Relume Dumará, 2004), 185–216.

Uruguay River, about the basin's first two big dams, Machadinho and Itá. In 1983, CRAB staged a massive protest against Machadinho, and in response, the state legislature recommended that Eletrosul redesign the project. Eletrosul agreed and reduced the size of the reservoir so that it displaced fewer people. Four years later, CRAB persuaded Eletrosul to sign an agreement in which it acknowledged the commission as the legitimate representation of those displaced. It also agreed to indemnify everyone collectively and fairly, and to slow the pace of construction in order to address Machadinho's social problems.

Timing and demographics were both crucial to these achievements. CRAB was Brazil's first dam-resistance movement to begin its activism before construction started. The process of political opening that began in the late 1970s had enabled activism against dams across the entire country, but by then construction on the military regime's most controversial dams was already well underway. In contrast, construction on Machadinho had not started when the commission began its activism. Two demographic factors also helped. The commission was basin-wide, which meant that those fighting against a single dam had the support of other communities near future reservoir sites. Lastly, the commission consisted of mostly white, non-Indigenous farmers, and participants were thus not subjected to the overt and systemic racism that made it even harder for displaced Indigenous communities to wrest concessions from the military government.

After these initial victories, CRAB wanted to maintain pressure on Eletrosul. The commission understood that its agreement was fragile and was not a guarantee that Eletrosul would follow through on its promises. The government had postponed Machadinho and Itá because of funding shortages, and meanwhile, other reservoirs, such as Tucuruí, Itaparica, and Balbina, filled and created social and environmental disasters that alarmed those following the stories. These big and iconic dams captured the attention of the anxious farmers in the Uruguay Basin, who were determined to prevent the stories from being repeated in their communities when the government restarted construction on Machadinho and Itá.

In 1988, CRAB reached out to the national trade union federation Central Única de Trabalhadores, CUT, about organizing a national conference for people displaced by dams. CUT agreed, and in February and March of the following year, it organized regional meetings to prepare for a big national meeting, which the Goiânia-based Pastoral Land Commission [Comissão Pastoral de Terra, CPT] – whose southern branch had been active in CRAB since its beginnings – agreed to host. In Goiânia, activists began planning advocacy work for greater social justice at projected dam sites.

The Brazilian anti-dam movement was part of an international wave of anti-dam sentiment that spread during the late 1980s–1990s. In some cases, such as China, authoritarianism stifled the formation of protest movements.[5] In other cases, such as the United States, rising anti-dam sentiment did not translate into a strong national movement against dams because there were fewer dams to protest, as the country was running out of good dam sites.[6] In democratic countries where dam-building campaigns were still booming, such as India, widespread anti-dam protests emerged. There, in the late 1980s, sociologist turned activist Medha Patkar and others founded the Narmada Bachao Andolan [Save the Narmada Movement] to consolidate activism among groups across the Narmada River's watershed, which was the target of an ambitious government dam-building scheme.[7]

In the late 1980s–1990s, most of the anti-dam movements emerged in parallel with one another, rather than as part of a shared cause. These movements – in Brazil and elsewhere – focused on domestic concerns and sought international support from countries whose lending institutions were funding contentious dams.

However, anti-dam activists soon began collaborating at the international level. In the mid-1990s, a multinational coalition of NGOs published a declaration calling for the World Bank to reassess its position on funding large dams (see Chapter 3). The Bank then carried out a lackluster self-assessment that greenlit further dam construction, and, in response, dam-affected communities from twenty countries met in Curitiba, Brazil, to sign another declaration outlining their shared experiences and demands.[8] In 1998, this international coalition pushed the

[5] See the updated English edition of journalist Dai Qing's *Yangtze! Yangtze! Controversy Over the Three Gorges Dam Project* (Toronto-London: Probe International and Earthscan, 1994). The Chinese government arrested Dai after she published the original edition in 1989, banned the book, and destroyed all the copies it could find.

[6] Reisner, *Cadillac Desert*, 510–511.

[7] See Amita Baviskar, *In the Belly of the River: Tribal Conflicts over Development in the Narmada Valley* (New York: Oxford University Press, 1995); Arundhati Roy, *The Greater Common Good* (Bombay: India Book Distributor, 1999); and Khagram, *Dams and Development*.

[8] Both the 1994 Manibeli Declaration and the 1997 Curitiba Declaration can be found on the website of the International Rivers Network, an influential North American NGO that formed in the mid-1980s to assist dam-affected communities worldwide (last accessed April 2023).

World Bank to co-sponsor, with the International Union for the Conservation of Nature, the World Commission on Dams (WCD), a multidisciplinary team of engineers, activists, scientists, and others, charged with assessing the overall merits of dams worldwide. In 2000, the Commission published a sweeping, and mostly negative, assessment of the world's large dams.[9] Its findings inspired even further anti-dam activism.[10]

In Brazil, by the time of the 1989 Goiânia meeting, activists had more tools at their disposal to help with their fight. In 1986, the government passed the first in a series of laws that made it mandatory for companies to conduct environmental impact studies before building dams and other infrastructure projects. Legislation also required that these studies be presented at public hearings and that input from these hearings be factored into whether or not the government approved projects.[11] In 1988, these environmental safeguards were incorporated into the Constitution.[12] The next year, the government passed legislation requiring companies to pay reparations to still-uncompensated communities impacted by dams and created the Brazilian Institute of Environment and Renewable Natural Resources [Instituto Brasileiro do Meio Ambiente e dos Recursos Naturais, IBAMA], a strengthened environmental protection agency charged with ensuring compliance with these new laws.[13] These legal

9 World Commission on Dams, *Dams and Development: A New Framework for Decision-Making* (Sterling: Earthscan, 2000).

10 Chris Sneddon and Coleen Fox, "Struggles Over Dams as Struggles for Justice: The World Commission on Dams (WCD) and Anti-Dam Campaigns in Thailand and Mozambique," *Society and Natural Resources* 21 (2008): 625–640.

11 O Conselho Nacional do Meio Ambiente (CONAMA), *Resolução Conama No. 001, Dispõe sobre critérios básicos e diretrizes gerais para a avaliação de impacto ambiental*, January 23, 1986; CONAMA, *Resolução Conama No. 006, Dispõe sobre o licenciamento ambiental de obras do setor de geração de energia elétrica*, September 16, 1987; CONAMA, *Resolução Conama No. 009, Dispõe sobre a realização de Audiências Públicas no processo de licenciamento ambiental*, December 3, 1987. These documents are available on Ministério do Meio Ambiente's website (last accessed April 2023).

12 See República Federativa do Brasil, *Constituição da República Federativa do Brasil de 1988* (Brasília: Congresso Nacional, 1988), Título VIII, Da Ordem Social, Capítulo VI, Do Meio Ambiente, Art. 225. Accessed through ACD.

13 Nelson Carneiro, Lei No. 7.735, *Dispõe sobre a extinção de órgão e de entidade autárquica, cria o Instituto Brasileiro do Meio Ambiente e Recursos Naturais Renováveis e dá outras providências*, February 22, 1989, and José Sarney, *Lei No. 7.990, Institui, para os Estados, Distrito Federal e Municípios, compensação financeira pelo resultado da exploração de petróleo ou gás natural de recursos hídricos para fins de geração de energia elétrica, de recursos minerais em seus respectivos territórios, plataformas continental, mar territorial ou zona econômica exclusiva, e dá outras providências*, December 28, 1989, ACD.

protections were powerful tools for the anti-dam movement, and determined activists were successful in using them to forestall further construction of disastrous reservoirs.

Many dams were never built. Consider the case of Cachoeira Porteira, a big dam planned on the Trombetas River northeast of Manaus. Engineers identified Cachoeira Porteira as a propitious dam site during the series of studies the government funded between 1973 and 1976 that had produced Tucuruí and Balbina (see Chapter 2).[14] Indeed, the dam contended with Balbina as the best choice to power Manaus, and some of Balbina's critics even endorsed it as a safer alternative to Balbina because it would produce much more power and had a smaller reservoir.[15] However, the government chose Balbina because it was much closer to the city and could be built faster and cheaper (see Chapter 6). Cachoeira Porteira came back into the spotlight in the late 1980s, when it became clear that Manaus would outstrip Balbina's energy supply almost immediately. The government released plans stating that it would begin construction in 1990 and finish the first stage in 1995.[16]

But Eletronorte never built the dam. In 1988, it published the dam's environmental impact study, the first in Pará to be open to public comment. The report detailed the detrimental impacts that the dam would have on local people, which included both Indigenous groups and *quilombos*, Afro-Brazilian refugee communities established by those who had escaped slavery.[17] In 1989, these groups organized resistance and publicized their disapproval of the dam.[18] The worsening economic

[14] See Eletronorte, "Estudos Amazônia, relatório final, Vol. V: Aproveitamento Hidrelétrico do Rio Trombetas em Cachoeira de Porteira, estudos de viabilidade," April 1976, Document 05188, AELN.

[15] Fearnside, "Brazil's Balbina Dam".

[16] Eletronorte, "Estudos Hidrelétricos de Cachoeira Porteira, relatório final de viabilidade," August 1987, Documents 02480 and 02481, AELN.

[17] Eletronorte, "Aproveitamento Hidrelétrico de Cachoeira Porteira, relatório de Estudos de Impacto Ambiental (EIA)," January 1988, Document 02167, AELN, and Eletronorte, "Aproveitamento Hidrelétrico de Cachoeira Porteira, Relatório de Impacto Ambiental (RIMA)," January 1988, Document 02140, AELN.

[18] See Comissão Pro-Índio de São Paulo, "O que é o aproveitamento Hidrelétrico de Cachoeira Porteira?" 1989, Folder "Grandes Projetos," Document TA-0056-0020, ACPT; Associação das Comunidades Remanescentes de Quilombos do Município de Oriximiná, "Sumário da reunião sobre hidrelétricas, June 30-July 2, 1989," Folder "Hidrelétrica Cachoeira Porteira," Document PA-0000-1076-00010, ACPT; "Relatório do I Seminário sobre as Barragens de Cachoeira Porteira, October 13-15, 1989," Folder "Atingidos por Barragens," Document TA-0085-0060, ACPT.

crisis further discouraged Eletronorte from building the dam. Soon thereafter, funding returned, but the government never approved the project's environmental license.[19] In 2019, the government relaunched the project as part of the Barão do Rio Branco infrastructure project proposal, but as of April 2023, no further news has surfaced, and it remains to be seen whether or not the dam will be built.

Eletronorte also never built the other dams planned for the Trombetas River and its tributaries. Cachoeira Porteira was the first of nine dams planned for the river, none of which have yet to be constructed.[20] The same is true of many other Amazon rivers that the government had planned to plug with dams. For example, the government had planned at least six dams for the Xingu River southwest of Belém, but just one, Belo Monte, was built.[21] In total, the Brazilian government planned to build roughly ninety dams in the Amazon, of which about sixteen dams, or 18 percent, were built (Figure 7.1). If the dams on the stretches of the Tocantins and its tributaries in the Cerrado – a large tropical savanna that borders the Amazon – are excluded, the figure drops to sixty-nine dams planned for the Amazon since the dictatorship, of which ten were built, which means that the government built just 14 percent of the dams it planned.[22] Budget squeezes during the 1980s and 1990s partly accounted for the slowdown. But after budgets loosened, the protests of Indigenous communities, environmentalists, and their allies continued to stall dams, curbing the government's dam-building ambitions significantly.

The government modified the dams it did build to mitigate their environmental impacts. Consider the infamous Belo Monte Dam, Brazil's most

[19] For more on Cachoeira Porteira, see Teixeira, *Energy Policy in Latin America*, 253–317.

[20] Fearnside, "Hydroelectric Dams in the Brazilian Amazon as a Source of 'Greenhouse' Gases," 8.

[21] See Fearnside, "Dams in the Amazon: Belo Monte and Brazil's Hydroelectric Development of the Xingu River Basin," *Environmental Management* 38, no. 1 (2006): 16–27.

[22] Author's calculations. The original list comes from Fearnside, who identified seventy-nine dams from government documents built or planned for the Amazon. Fearnside, "Hydroelectric Dams in the Brazilian Amazon." To this list, I have added sixteen dams not listed by Fearnside and subtracted five that were built during the dictatorship. The added dams are the Jirau and Guajara Mirim dams on the Madeira River; the Jatobá and Chacorão dams on the Tapajós River; the São Manoel, Foz de Apiacas, Colíder, and Sinop dams on the Teles Pires River, a tributary of the Tapajós; the Cachoeira de Caí, Janmanxim, Cachoeira dos Patos, and Jardim de Ouro dams on the Janmanxim River, a tributary of the Tapajós River; and the Estreito, Peixe Angical, São Salvador, and Cana Brava dams on the Tocantins River. The five dams I subtracted were Curacy Nunes, Curuá-Una, Samuel, Tucuruí, and Balbina.

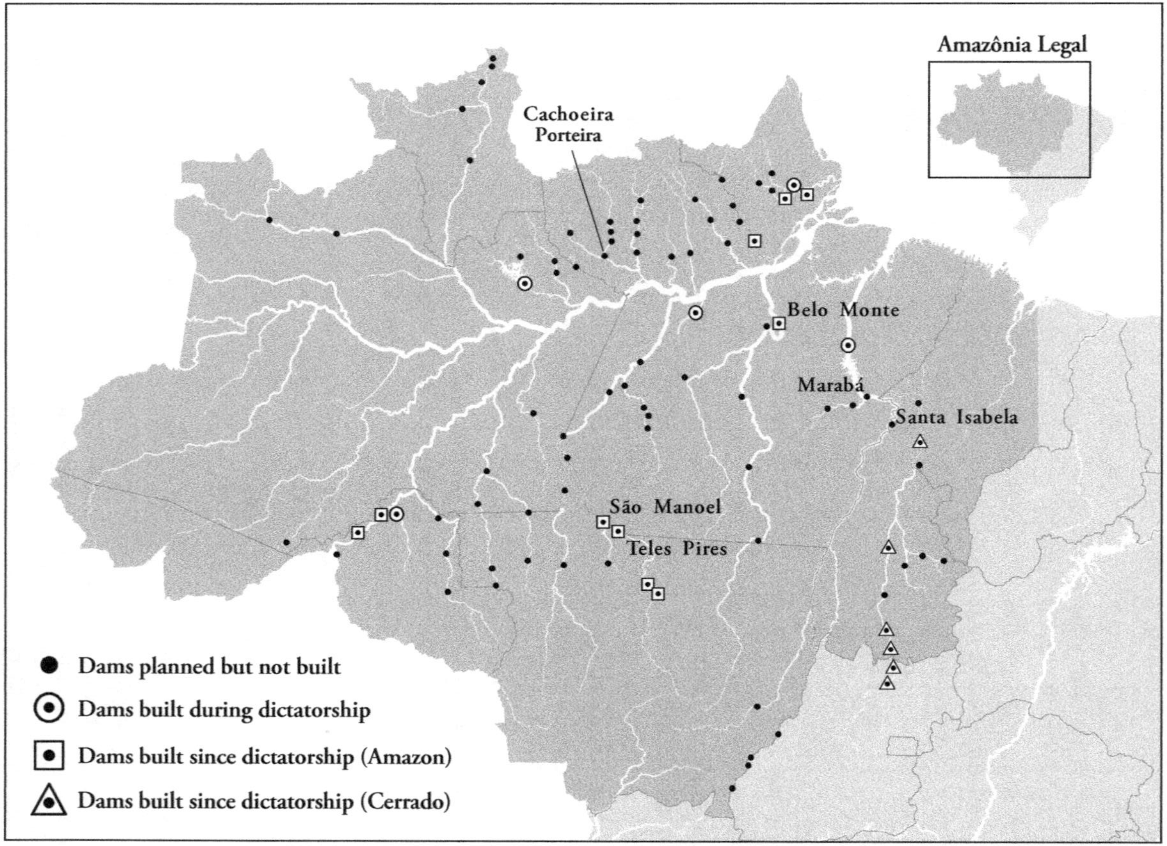

FIGURE 7.1. Hydropower dams in the Amazon Rainforest, 1960s–2023. Locations for proposed dams based on Philip Fearnside (1995), with amendments as discussed in footnote 22 of this chapter. Map by Geoffrey Wallace (G. Wallace Cartography & GIS).

controversial dam since the dictatorship. In 1975, Eletronorte began studying potential dam sites on the Xingu River, a tributary of the Amazon that runs south to north through the state of Pará. Engineers identified at least fourteen good sites, and in 1980, Eletronorte decided to start with two dams, Babaquara and Kararaô, known collectively as the Altamira Complex. Both were located along "the Big Bend,", a short stretch of the powerful river where it dropped more than 90 meters.[23] The Big Bend's energy potential was immense, and engineers estimated that together both dams could produce upwards of 17,000 megawatts of energy, making the Altamira Complex bigger than Itaipu, the most powerful hydropower plant in the world at the time.[24] But the environmental footprint of the project was also colossal. Early studies estimated that together both dams would flood an area of 7,200 sq. km, though some engineers contended that the true figure was closer to 3,600 sq. km, because much of the area to be flooded was underwater during the rainy season regardless.[25] Economic downturn put the project on hold during much of the 1980s, and when engineers returned to the project at the end of the decade, the political landscape had changed. The restoration of democratic freedoms had strengthened the Indigenous and environmental lobbies, which now posed a formidable opposition to big dams.

Environmentalists used direct action and legal tools to force tremendous modifications. The project first made headlines in 1986, and there was such a large outpouring of criticism from Indigenous communities and environmentalists that Eletronorte decided to drop Babaquara altogether. Kararaô's reservoir was just 1,225 sq. km, which means that activism had pressured the government into sparing thousands of square kilometers of land from being flooded. But even Kararaô threatened to flood the land of the Kayapo and other Indigenous groups and to disrupt their subsistence practices. In 1988, Eletronorte began proceeding with the project without releasing drafts of its environmental impact studies for public comment.[26] In February 1989, the Kayapo and their allies organized a meeting with Eletronorte, known as the Altamira Gathering, to

[23] VF Saracura, "History of Studies on Belo Monte Hydroelectric Plant," *Brazilian Journal of Biology* 75, no. 3 (2015): S5–S9.

[24] "Governo quer fazer no Pará hidrelétrica maior que Itaipu," *Folha de São Paulo*, September 28, 1986.

[25] "Lago de Xingu inundará 7,200 quilômetros quadrados," *Folha de São Paulo*, September 28, 1986.

[26] The draft environmental impact and viability studies are available in Eletronorte's archive. See also Eletronorte, "Usina Hidrelétrica de Kararaô, estudo de impacto ambiental

raise their concerns about Kararaô's social and environmental impacts. Discussions were tense, and during one argument, a Kayapo leader ran her machete across the cheeks of Eletronorte's chief engineer. The gathering was widely covered in the press, which raised global awareness about the dam and its deleterious impacts.[27]

The movement gathered enough supporters that politicians found it imprudent to defend the dam without significant alterations. Eletronorte agreed to halt construction until it could better mitigate the dam's social and environmental footprint, and it also agreed to drop the name Kararaô at the request of the Kayapo. The name came from their language, and they disapproved of Eletronorte's appropriation. In the late 1990s, Eletronorte relaunched the project under the name Belo Monte and dramatically reduced its reservoir size, even though downsizing meant big cuts to the dam's output. The smaller reservoir would flood just 500 sq. km of land and would not displace any Indigenous communities.[28] Legal suits by the Public Prosecutor's Office held up construction for almost two decades. In 2011, the government reviewed new environmental impact studies and granted an operating license. In 2016, it completed the dam. After decades of protest, activists had compelled the government to reduce the grandiose Altamira Complex to just a single dam with a small reservoir that did not flood Indigenous land.

The pattern was the same throughout the Amazon. All ten dams that the government built in the Amazon since the dictatorship are considered run-of-the-river dams,[29] a term that refers to dams that do not have a big

preliminar," 1988, Document 00931, and Eletronorte, "Aproveitamento Hidrelétrico de Kararaô, estudos de viabilidade relatório final," January 1989, Document 00997, AELN.

27 See Terence Turner, "The Role of Indigenous Peoples in the Environmental Crisis: The Example of the Kayapo of the Brazilian Amazon," *Perspectives in Biology and Medicine* 36, no. 3 (1993): 526–545; Arsenio Oswaldo Sevá Filho, ed., *Tenotã-Mõ: Alertas sobre as consequências dos projetos do Rio Xingu* (São Paulo: International Rivers Network, 2005); Fearnside, "Dams in the Amazon: Belo Monte"; Terence Turner, "The Kayapo Resistance," in James Spradley and David McCurdy, eds., *Conformity and Conflict: Readings in Cultural Anthropology* (Boston: Pearson, 2008), 391–409.

28 Saracura, "History of Studies on Belo Monte Hydroelectric Plant."

29 This number excludes those on the stretches of the Tocantins and its tributaries in the Cerrado. The ten dams are Belo Monte (440 sq. km reservoir), on the Xingu River; Jirau (250 sq. km reservoir) and Santo Antonio (270 sq. km reservoir) on the Madeira River; São Manoel (150 sq. km reservoir), Teles Pires (150 sq. km reservoir), Colíder (180 sq. km reservoir), and Sinop (340 sq. km reservoir) on the Teles Pires River, a tributary of the Tapajós River; Jari (20 sq. km reservoir) on the Jari River; and Cachoeira Caldeirão (50 sq. km reservoir) and Ferreira Gomes (20 sq. km reservoir) on the Araguari River. The total area flooded by these dams is about 1,870 sq. km, an area smaller than the reservoirs of both Tucuruí and Balbina.

enough reservoir to store water during dry periods and thus depend entirely on rainfall and the natural flow of the river to produce power. In 1989, activists in Goiânia had wondered if they would be able to prevent the repetition of big reservoirs such as Sobradinho and Balbina. After thirty years of fighting, they had succeeded in preventing the construction of such reservoirs in the Amazon.

Activists accomplished this feat without jeopardizing the country's economic growth. Starting in the 2000s, natural gas, biofuels, and nuclear began to account for a much greater percentage of electricity generation than they did previously, and after the mid-2010s, the contributions of wind power grew dramatically. The continued output of hydropower dams – both existing dams and new dams built outside Amazônia – alongside these additional sources of electricity generation has enabled Brazil's economy to grow even as construction on dams in the Amazon basin stalled.

The anti-dam movement has done less to halt dam construction outside the Amazon, but there too, activism has helped ensure that impacts were more benign than those at earlier dams. Some of the most controversial dams were stopped outright. For example, the contentious Marabá and Santa Isabel dams, on the Tocantins and Araguaia rivers, respectively, were both mothballed because of their social and environmental impacts. Other dams were built with modifications and safeguards to mitigate their impacts. Eletrosul modified and built the Machadinho and Itá dams,[30] and the state Companhia Paranaense de Energia's Salto Caxias Dam on the Iguaçu River had a well-timed and funded resettlement program.

Indigenous communities and environmentalists did not always achieve all their goals, and some dams built during this period share at least two disagreeable similarities with the dictatorship's big dams. First, critics have charged that kickbacks from engineering companies were a factor in deciding to build the dams. As with the dictatorship's dams, corruption is difficult to document, though an anti-corruption campaign known as Operation Car Wash [Operação Lava Jato] has exposed a massive corruption scheme between construction firms and Petrobras – the state-owned oil company – that happened during the administration of

[30] On Itá, see Melody Forcelini, "Colossal, Complexa, Imprescindível: Tecnologia de força bruta e Usina Hidrelétrica de Itá/SC," master's thesis, Universidade Estadual do Centro Oeste do Paraná, 2017, and Forcelini and Jó Klanovicz, "A Usina Hidrelétrica de Itá, Brasil e a tecnologia de força bruta na história recente do Rio Uruguai," *Revista Inclusiones* 5 (Enero-Marzo 2018): 175–191.

presidents Luiz Inácio Lula da Silva (2003–2010) and Dilma Rousseff (2011–2016). These construction companies were the same ones building big dams, so it is not unreasonable to think that kickbacks encouraged some politicians to approve controversial dams.

Second, critics have charged the Lula and Dilma administrations with subverting the environmental licensing process at some dam sites, resulting in projects with considerable social and environmental footprints. For example, Belo Monte, despite its tremendous modifications, disrupts the Xingu River fisheries, upon which Indigenous communities and non-Indigenous fishermen depend.[31] Two dams along the Teles Pires River in the Amazon also caused deleterious impacts for the Munduruku, an Indigenous community that traditionally inhabited the land where the dam was built. One of these dams inundated a set of waterfalls that was their culture's equivalent to heaven, as well as other sacred sites.[32] These dams and others built in the Amazon Rainforest also produce greenhouse gases that contribute to the global climate crisis (see Chapter 6). That damage continues despite the improvements in environmental safeguards highlights the importance of upholding and strengthening the environmental licensing process and the heightened potential for harm in the Amazon Rainforest.[33]

Nevertheless, the Lula and Dilma administrations had a much more balanced record on dams and the environment than the military regime. Neither built big reservoirs – all were run-of-the-river dams, which do not

[31] See Brum, "Belo Monte: A anatomia de um etnocídio," *El País*, December 1, 2014; Brum "O pescador sem rio e sem letras: à beira de Belo Monte, uma história pequena numa obra gigante," *El País*, February 16, 2015; Brum, "Belo Monte, empreiteiras e espelhinho," *El País*, July 7, 2015; Brum, "O que Belo Monte delata sobre todos os lados," April 11, 2016; Brum, "Dilma compôs seu réquiem em Belo Monte," May 10, 2016; Brum, "A Veneza de Belo Monte," *El País*, May 15, 2018; Brum, "A notícia é esta: O Xingu vai morrer," *El País*, September 12, 2019; Brum, "Belo Monte, a obra que une os polos políticos," *El País*, December 5, 2019. See also, Instituto Socioambiental, *Dossiê Belo Monte: Não há condições para a Licença de Operação* (São Paulo: Instituto Socioambiental, 2015).

[32] See Sue Branford and Maurício Torres, "The End of A People: Amazon Dam Destroys Sacred Munduruku 'Heaven'," *Mongabay*, January 5, 2017; Branford et al., "Munduruku Standoff against Amazon Dam Builders Potentially Explosive," *Mongabay*, October 17, 2017; and Branford and Torres, "NGOs Denounce Tapajós Basin Intimidation, Violence, Brazil Inaction," *Mongabay*, April 4, 2018.

[33] Outside of the Amazon Rainforest, there is also the case of Barra Grande Dam in Southern Brazil; see Marcos Aurélio Espíndola, "Tecnização da natureza: O desastre ambiental da Usina Hidrelétrica de Barra Grande (SC)," in Eunice Sueli Nodari et al., eds., *Desastres socioambientales em Santa Catarina* (São Leopoldo: Oikos Editora, 2015), 157–181.

flood massive areas to store water – and both kept the most controversial dam sites undammed.

Both administrations also accelerated environmental remediation programs for the big dams built during the dictatorship. Itaipu began cleaning up its environmental image under the Cardoso administration and finished its fish ladder in December 2002. It went much farther under the Lula administration. In 2003, Itaipu launched Cultivating Good Water [Cultivando Água Boa, CAB], a series of initiatives designed to improve local socioeconomic and environmental conditions. In 2010, the Lula administration took another step toward improving the reputation of the country's big dams when it created a national registry of displaced people that could be used to establish the amount of indemnity the government owed.[34] Although CAB has won international praise from the United Nations and the government's registry is a commendable step toward compensating displaced communities, neither program has succeeded in securing environmental justice for displaced Indigenous communities. Indeed, in contrast to the gains at future dam sites, anti-dam activism during the 1990s–2010s did little to help achieve justice for the communities that were displaced during the dictatorship.

THE LANDLESS GENERATION

Many displaced Indigenous communities are still fighting for their land. Shortly after the return to civilian rule, the Brazilian government set aside land for displaced Indigenous groups in the Amazon Rainforest as compensation for flooded land (see Chapter 6). However, the government has still not indemnified Indigenous communities outside the Amazon, such as the Avá Guaraní, displaced by Itaipu, and Tuxá, displaced by Itaparica (see Chapter 4), whose contemporary struggles are not well known nationally or internationally.[35] The Paraguayan government has similarly

[34] *Decreto No. 7.342, Institui o cadastro socioeconômico para identificação, qualificação e registro público da população atingida por empreendimentos de geração de energia hidrelétrica … e dá outras providências*, October 26, 2010, ACD.

[35] On the Tuxá, see the excellent work of two Tuxá scholars, Cruz, "'Quando a terra sair'," and Vieira, "Os índios Tuxá na rota do desenvolvimento," as well as the website of the Consórcio Agropecuário Indígena Tuxá Rodelas https://arquivotuxarodelas.com.br/ (last accessed April 2023). On the Avá Guarani in Brazil, see Gustavo Kenner Alcântara et al., *Avá Guarani: A construção de Itaipu e os direitos territoriais* (Brasília: Ministério Público da União, 2019). Other Indigenous and non-Indigenous groups are also continuing to fight for reparations. Among the most vocal of these groups are landless famers displaced by Sobradinho and fishermen at Guaíra.

neglected the Avá Guaraní living on its side of Itaipu's reservoir. Both governments have dragged their feet for almost four decades, despite having a legal obligation to compensate these Indigenous groups with land of equal quantity and quality to the flooded land and despite repeated promises to comply with this obligation. The result is that an entire generation has grown up without land, which has inflicted severe material hardships, cultural loss, and psychological trauma on these communities. This new generation is tired of the repeated setbacks and is putting pressure on both governments to fulfill their promises.

In August 2015, the Avá Guaraní Tekoha Sauce reclaimed their land.[36] Recall that in 1982, Itaipu's floodwaters displaced thirty-eight Indigenous communities on the Paraguayan side of the reservoir and the government did not uphold its legal obligation. Instead, it sent them to settlements outside their traditional territory that were already occupied by other Indigenous groups. Tekoha Sauce was one of these displaced communities. In 2015, after decades of moving from one overcrowded settlement to another, Tekoha Sauce families decided to return to their homeland, some of which had become part of the reservoir's protective forest belt and remained above the waterline (see Chapter 3). These families decided that if the government refused to honor its legal agreement to grant them land, they would reoccupy it themselves. They chose an unoccupied area on the edge of the forest belt to build houses, a school, and a temple, and to plant crops for subsistence.

Local farmers responded to their occupation with violence. During the previous decades, some of the Paraguayan land that Itaipu had expropriated for its forested belt fell into the hands of wealthy farmers through illicit means.[37] Soy farmer Germán Hutz claimed to own the land that the Tekoha Sauce families returned to. He harassed them and used one of his tractors to flatten their crops. He used his political connections – Vice President Juan Afara was his in-law – to get the government's support. On September 30, 2016, the government sent military police to expel the Indigenous group and set fire to their schools and temples.

[36] Tekoha is an Avá Guaraní term that refers to both land and life. In this context, it means village or community. The following two paragraphs are drawn from Barón, *Los Avá Guaraní Paranaenses*, 111–126, and "INDI [El Instituto Paraguayo del Indígena] analiza conflicto entre Itaipú y Avá Paranaenses," *Ultima Hora*, October 31, 2019. Unless otherwise noted, all newspaper and magazine articles, press releases, court filings, and ordinances cited in this section were accessed through their respective websites.

[37] "La franja de bosques de Itaipú es utilizada para negocios ilícitos," *Ultima Hora*, July 28, 2019.

Forcibly displaced once again, Tekoha Sauce decided to rebuild their community on uncontested land on the edge of the Limoy Biological Reserve in San Alberto, Alto Paraná, and use it as a base to take legal action against Itaipu. Their legal case has gotten stronger since 1982, when Itaipu was arguing that the Avá Guaraní on both sides of the reservoir deserved land, but that other government agencies should foot the bill. In 2008, Paraguay's Truth and Justice Commission published a report that listed Itaipu's botched efforts to resettle and indemnify Indigenous groups among the Stroessner dictatorship's human rights abuses.[38] In 2013, the Paraguayan government built on the findings of this report and passed a resolution formally acknowledging its historical debt.

The Indigenous community is continuing to fight through setbacks. In 2016, the Paraguayan Indigenous Institute [El Instituto Paraguayo del Indígena, INDI] revoked the resolution acknowledging the government's debt to the group, though it reinstated it in 2019 after Itaipu announced it would evict Tekoha Sauce from the biological reserve.[39] In response to Itaipu's threat, both INDI and local NGOs defended the group and initiated a series of meetings to resolve the conflict. Itaipu's directors stated bluntly that what they feared was the precedent. Tekoha Sauce is the first displaced Indigenous community to reopen talks about reparations, and Itaipu fears that other displaced communities will do the same. Nevertheless, Tekoha Sauce activists and their allies are continuing to pressure the government, which is in turn putting pressure on Itaipu to uphold its obligation to the group. Tekoha Sauce even has the support of legislators from Mercosur – a big South American trading bloc – which has helped strengthen the group's position. As of this writing, Tekoha Sauce remains in its temporary settlement and the Paraguayan government has agreed to create a working group to acquire land for them.[40]

[38] "Comisión Verdad y Justicia (CVJ) también señaló deuda de Itaipú con los indígenas," *Ultima Hora*, November 6, 2019.

[39] "INDI analiza resolución que admite deuda de Itaipu con Avá Paranaenses," *Ultima Hora*, November 11, 2019.

[40] "Convocarán a indígenas y al INDI por destierros de Itaipu," *Ultima Hora*, October 5, 2019 "INDI dice apoyar la lucha de los Avá y espera diálogo de parte de Itaipú," *Ultima Hora*, October 11, 2019; "INDI se reunirá con la Itaipú para tratar deuda con Avá Paranaenses," *Ultima Hora*, November 8, 2019; "Harán mesa de trabajo para resolver problemática de comunidad indígena," Cámara de Senadores, Noticias Comisiones, August 10, 2020; "Avá Guaraní Paranaenses, desalojados hace 40 años por el embalse de Itaipu, aún reclaman derechos," *ABC Color*, April 18, 2022.

Avá Guarani communities on the Brazilian side of the reservoir are also putting pressure on their government to uphold its promise. In February 2017, a group of ten families from Tekoha Mokoi Joeguá reoccupied their traditional land in a stretch of Itaipu's forest belt in the municipality of Santa Helena. Six other Brazilian Avá Guarani communities followed their lead and established similar settlements in the area.[41] Recall that Itaipu also flooded 15 sq. km (1,500 hectares) of mostly forested Avá Guarani land in Brazil and that the company compensated the group with an unforested 231 hectare plot. Such derisory landholdings created overcrowded living conditions and dramatically reduced the community's standard of living. Brazilian Avá Guarani communities have been fighting for just compensation for almost four decades, and the reoccupation was a reaction to repeated government negligence.

Setbacks stalling some of the group's biggest advances triggered the direct action. During the 2010s, the Brazilian government created a National Truth Commission to investigate the dictatorship's human rights abuses. In the aftermath of the commission's 2014 final report, the government began taking concrete steps toward Avá Guarani reparations. In 2016, then-Attorney General Rodrigo Janot created a group to investigate Itaipu's historical debt.[42] However, shortly thereafter, the government enacted legislation that threatened to annul the potential gains from the investigation's recommendations. In February 2017, the government permitted an independent working group to review areas that the National Indian Foundation [Fundação Nacional do Índio, FUNAI] was in the process of demarcating. Indigenous communities across Brazil denounced the measure because wealthy agribusinessmen influenced the working group and planned to use it to curtail the demarcation process.[43] Frustrated, Brazilian Avá Guarani decided that nearly four decades without land was long enough and reoccupied their ancestral land.

[41] Conselho Indigenista Missionário (CIMI) Regional Sul, "Avá Guarani retomam parte de antigo Tekoha," *Notícias CIMI*, February 10, 2017, and Júlio Carignano, "Indígenas Avá-Guarani retomam parte de antiga terra tradicional, no Paraná," *Brasil de Fato*, February 17, 2017.

[42] Julio José Araujo and Manuela Carneiro, "Os Avá Guarani e as violações de Itaipu," *Notícias CIMI*, November 11, 2019.

[43] Ministério da Justiça e Cidadania, Gabinete do Ministro, *Portaria No. 80, de 19 de Janeiro de 2019, Cria Grupo Técnico Especializado, GTE, para auxílio em assuntos relacionados a Terras Indígenas*, January 19, 2017, and Centro de Trabalho Indigenista, "Nota de repúdio à Portaria MJ No. 80/2017," January 2017.

Like their Paraguayan counterparts, the group is using the occupied land as a base to mount a legal campaign to get the title for their traditional land. The living conditions are precarious, and the group faces harassment from neighbors, but they remain committed to the occupation, nonetheless. In 2019, the Avá Guarani communities got a huge boost from the government. Raquel Dodge – who succeeded Janot as attorney general – accelerated the investigation into Avá Guarani reparations, and in April, the working group published its final report, which was the first government-commissioned study to acknowledge Itaipu's historical debt to the Avá Guarani. In September, Dodge explicitly recommended reparations.[44]

Once again, setbacks followed these advances. That same month, the President replaced Dodge with Augusto Aras, who rejected the report's acknowledgment of Itaipu's historical debt and archived the recommendation for reparations, putting it on hold for the indefinite future.[45] Shortly thereafter, the Federal Court of Paraná ruled that Itaipu had the legal right to evict the communities from their settlements alongside the reservoir. The case went to the Supreme Court, which overruled the Federal Court of Paraná's decision, and permitted the Avá Guarani to remain in the forest belt for the time being.[46] Despite this victory, the future of Brazilian Avá Guarani communities hangs in the balance. As of this writing, the communities alongside the reservoir remain in their temporary settlements, enduring arduous living conditions while they continue to fight for their land.

The Brazilian government has also reneged on its obligation to set aside land for the Tuxá, for whom the last three decades have been an infuriating nightmare of disingenuous agreements and broken promises that have kept them landless. Recall that in 1988, Itaparica's reservoir displaced the Tuxá, and over the following years, the Companhia Hidrelétrica do Rio São Francisco (CHESF) put off purchasing land for the group despite its legal obligation to compensate them with land of equal quantity and quality. Throughout the 1990s, the Tuxá pressured CHESF to follow through on its agreement, and on various occasions, the

[44] Alcântara et al., *Avá Guarani: A construção de Itaipu e os direitos territoriais*; Raquel Elias Ferreira Dodge, *Ação Civil Originária (ACO) 3300*, Ministério Público Federal (MPF), September 9, 2019.

[45] "STF [Supremo Tribunal Federal] arquiva pedido de indenização bilionária a índios atingidos por Itaipu," *O Globo*, October 31, 2019.

[46] Ministro Dias Toffoli, "Suspensão de Liminar 1.197 Paraná," STF, April 28, 2020; Toffoli, "Suspensão de Liminar 1.218 Paraná," STF, April 28, 2020.

group agreed to renegotiate the terms and timelines of their compensation in order to speed up the process. Despite the renegotiated terms, CHESF continued to stall.[47]

Repeated setbacks pushed the Tuxá to sue the company. In the late 1990s, the federal government considered privatizing CHESF and encouraged it to quickly resolve its responsibilities to Itaparica's displaced communities so that it could better entice potential buyers. Rather than speed up its land acquisition and irrigation projects, CHESF decided it would be cheaper to offer large lump-sum payments to displaced communities. The government floated the idea among the various Indigenous leaders while further stalling land demarcation and irrigation canal construction so as to make payments more appealing. The plan backfired. Having spent a decade without land, the Tuxá were furious that CHESF was now trying to skirt its obligations by offering cash settlements. In 1999, the Tuxá enlisted the Federal Prosecution Service to file a civil action suit against CHESF for reneging on its agreement.[48]

This lawsuit resulted in another series of unfulfilled agreements that set in motion enduring consequences. In 2004 and 2006, the Tuxá and CHESF restructured their settlement. These new accords – called Termos de Compromisso de Ajustamento de Conduta or TACs – maintained the basic elements of previous agreements but introduced at least two notable changes. First, the renegotiated agreement permitted CHESF to substitute its obligation to purchase land for the Tuxá with a one-time payment of R$5,000,000 to FUNAI, which would take charge of acquiring the land. The change did nothing to speed up land acquisition. It took three more years for the executive branch to promulgate the decree establishing the Tuxá Indigenous Reserve, and thereafter, FUNAI stalled on demarcating the land.[49] Second, the Tuxá agreed that CHESF could

[47] The following three paragraphs are drawn from anthropologist Fabiana Lima dos Santos, "Reconhecimento do território do povo Tuxá de Rodelas, Bahia," Informação-E no. 36/2010, MPF, September 2, 2010, AUT.

[48] MPF, Procuradoria da República, Bahia (MPF-PRB), "Ação civil pública com pedido de liminar, contra a CHESF e a União Federal na defesa de direitos dos índios Tuxá, que perderam suas terras tradicionais quando do enchimento da Barragem de Itaparica, situada na fronteira dos Estados de Bahia e Pernambuco, em 1986," Justiça Federal Salvador, June 30, 1999, AUT.

[49] Luiz Inácio Lula da Silva, "*Decreto de 21 de dezembro de 2009, Declara de interesse social, para fins de desapropriação, o imóvel rural que menciona, destinado a assentar famílias da Comunidade Indígena Tuxá de Rodelas, no Município de Rodelas, no Estado da Bahia,*" ACD.

exchange the irrigation projects it had promised for individual payments to Tuxá families.

Disillusionment pressured the Tuxá to accept these new terms, which gave them immediate financial gains but diminished CHESF's responsibility to deliver land and irrigation systems. Many Tuxá did not want to accept the new terms, but others thought that money would be better than nothing. The comments of Tia Dôra, a decade after these negotiations, help to understand this disillusionment. After reflecting upon the trauma of being forcefully displaced by the reservoir's floodwaters, she lamented that the lack of progress despite so many years of fighting for land made her sure that she would "die without ever seeing any land."[50] This same sentiment existed among many Tuxá ten years before; they had already passed two decades without land and were suffering enormous material and psychological hardship as a result.

Since the dictatorship and democratic government alike had proven unwilling to keep promises, the Tuxá reclaimed their land themselves. In January 2010, about 100 families moved to a patch of unoccupied sacred land alongside the reservoir known as D'zorobabé, or Surubabel in Portuguese, and built temporary housing. The government threatened to evict them, but the families remained. Four months into the occupation, FUNAI agreed to resume its demarcation studies. With this concrete step underway, the families decided to disband the following month.[51] Subsequent years proved as disappointing as the previous twenty-two, but the Tuxá never stopped fighting for their land.

The next decade brought big gains and heartbreaking setbacks. On March 17, 2014, President Dilma passed legislation that created a Tuxá Indigenous Reserve and gave FUNAI permission to expropriate more than 40 sq. km for the reserve in an area near Rodelas that is currently covered with coconut monocultures.[52] However, the next year, Gilmar Mendes, a justice on the Supreme Court [Supremo Tribunal

50 Cruz, "'Quando a terra sair'," 116.

51 CIMI, "Povo Tuxá faz retomada de terra tradicional na Bahia," *Notícias CIMI*, March 9, 2010; Conselho Tuxá da Aldeia Mãe, "Manifesto do Povo Tuxá Rodelas," October 5, 2017, full text published in "As caravelas passam ... os Tuxá de Rodelas reafirmam sua identidade ética e seu pertencimento ao território D'Zorobabé," *Notícias da Articulação dos Povos Indígenas do Brasil (APIB)*, October 14, 2017.

52 Dilma Rousseff, *Decreto de 13 de março de 2014, Declara de interesse social, para fins de desapropriação, o imóvel rural que menciona, destinado a assentar famílias da Comunidade Indígena Tuxá de Rodelas, no Município de Rodelas, Estado de Bahia*, March 13, 2014, ACD.

Federal, STF], suspended the decree to placate a group of twenty-one local farmers, who had filed a petition against the decree for the Indigenous reserve and asked the government to choose another area. The farmers claimed that they had also been displaced by Itaparica, but an investigation revealed that only nine had been displaced by the dam and none had been resettled in the disputed area.[53]

The appeal stalled demarcation, but the court eventually upheld the Tuxá's legal defense. In June 2017, the Regional Federal Court of the 1st Region affirmed the 2014 resolution and fined FUNAI for noncompliance.[54] In May 2018, the Justice Department ordered FUNAI to begin demarcating the land and to create a working group composed of anthropologists, environmental scientists, and cartographic professionals. The ordinance gave FUNAI 30 days to organize the group and 180 days to finish its demarcation studies.[55] Like previous assurances, these fines and mandates did not convince the Tuxá that the government would follow through on a promise that was now almost three decades unfulfilled.

In the midst of these legal battles, the Tuxá once again used direct action to reclaim their traditional land. In August 2017, hundreds of families reoccupied D'zorobabé and built a much bigger and more enduring settlement. These Tuxá families have built forty dwellings on about 10 hectares of land alongside the reservoir. Living there is difficult because the community is unable to plant food. Families need irrigation water to raise crops on a scale big enough for subsistence, and before they can build irrigation canals, the government needs to grant them the legal title to their land and provide financial assistance. The constant threat of eviction also compounds the hardship. Local landowners and government officials have repeatedly tried to expel the community, but have thus far failed.[56]

During the 2010s, the Tuxá applied another tool to their fight for land: university education and its mobilization for social and environmental

53 Cruz, "'Quando a terra sair'," 64.

54 "União e Funai são condenadas a pagar R$ 1 mil por atraso em processos de demarcação de terras indígenas na Bahia," Assessoria de Comunicação, MPF na Bahia, July 27, 2017.

55 Wallace Moreira Bastos, *Portaria nº 580, de 8 de maio 2018*, Ministério da Justiça; "Funai institui criação de GT [Grupo de Trabalho] para regularização de área indígena em Rodelas (BA)" *Notícias CIMI*, May 15, 2018.

56 On the threat of eviction, see Conselho Tuxá da Aldeia Mãe, "Manifesto do Povo Tuxá Rodelas"; Ayrumã Flechiá Tuxá and Douglas Mota, "Indígenas na Bahia sofrem ameaça de remoção e lutam na justiça," *Carta Capital*, December 20, 2018; Bruna Caetano, "Após perderem terras para barragem, 490 famílias Tuxá sofrem ameaça de despejo," *Brasil de Fato*, December 21, 2018; D. Aroeira, "Entrevista com liderança indígena do povo Tuxá da Bahia," *A Nova Democracia*, November 2019.

justice. In 2017, the same year that the court affirmed the validity of the decree creating the Tuxá reserve and that the group began to occupy D'zorobabé again, two Tuxá scholars – Felipe Sotto Maior Cruz and Antônio Fernandes de Jesus Vieira – defended their master's theses at the Universidade de Brasília. Both theses documented the violent land grabbing that the group has suffered throughout the centuries, as well as the government's negligence in following through on its obligation to find compensatory territory to replace the land that Itaparica's reservoir flooded. Antônio Fernandes, a Tuxá lawyer, used his master's thesis as an opportunity to write a petition to the Organization of American States (OAS)'s Interamerican Commission on Human Rights (ICHR) that detailed the violations the Brazilian government has committed over the last three decades. Using the courts, direct occupation, and universities, the Tuxá are continuing to fight for justice.

As of this writing, the Tuxá remain landless. In December 2018, when the Justice Department's deadline passed, FUNAI had still not demarcated their land. That year, Brazilians elected an overtly anti-Indigenous president, Jair Bolsonaro (2019–2022), who was openly racist and vocal about rolling back protections for both Indigenous communities and the environment. Bolsonaro made good on his promises, and the Tuxá made no progress during his administration. Some are hopeful that his successor, Lula, will follow through on demarcating land.[57]

In 1990, while the Avá Guaraní and Tuxá were in the midst of fighting for their land, two environmental scientists thousands of kilometers away exchanged letters about their research on the social and environmental impacts of big dams. The correspondence began when Amazonian specialist Philip Fearnside sent Canadian dam specialist David M. Rosenberg his article about the Balbina Dam. Rosenberg responded with a letter outlining some similarities between dams in both countries, including the fact that Indigenous populations often suffered the most damage. His most compelling words were his final reflection: "After 15 years of researching the effects of hydroelectric development, it's clear to me that a lack of social or environmental knowledge is not the problem. It's simply a difference in values held by the proponents of such developments, and the fact that the political and economic power is in their hands."[58]

[57] Bruno Alfano, "Governo vai demarcar 13 novas terras indígenas; saiba quais são," *O Globo*, January 30, 2023.

[58] "David Rosenberg to Philip Fearnside," February 2, 1990, APF.

The story of the Avá Guaraní and the Tuxá's fight for land is one of distorted values among politicians and a shortage of the public outrage needed to hold these officials accountable or elect new ones. Neither misunderstanding nor an absence of knowledge among authorities explains why the government has let four decades pass without giving either group their rightful land. Since the first impact studies, dam-building companies acknowledged that the law obligated them to compensate displaced Indigenous groups with land. These companies and the government agencies assisting them have not followed through because those in power do not care. These politicians care more about maintaining an extremely unequal socioeconomic and racial hierarchy that benefits wealthy white Brazilians at the expense of landless Indigenous communities. These priorities persist among politicians partly because they reflect those of the constituents electing them; many Brazilians harbor prejudice against Indigenous communities. But they also persist because the many Brazilians who do support Indigenous communities are not aware of these conflicts or are not pressuring their representatives to resolve them.

For the cynical, such callousness is standard fare for politicians; for those who expect more from democratic institutions and demand a more just world, it is disgraceful. Electorates have the power to vote politicians into power who reflect their values. I hope that the noise emanating from these pages joins a chorus of voices that thunder loudly enough to pressure politicians to give the Avá Guaraní and Tuxá their land and thereby belatedly serve social and environmental justice.

In addition to louder calls for social and environmental justice, since the 1990s, dam builders and their supporters have also had to contend with climate-related challenges that could perhaps worsen under anthropogenic global warming. Since the mid-1990s, concern about the climate crisis has encouraged many to champion hydropower dams for their carbon-mitigating potential (see Chapters 3 and 6), but climactic changes that are already underway could potentially curtail their impact.

THE SEA BECOMES THE DESERT

In July 2001, a group of families living alongside the margins of Sobradinho found themselves an unlikely new home: the ruins of Remanso, one of the towns the reservoir had covered. For two years, drought had gripped the region, and that year, Sobradinho's reservoir plummeted to one of the lowest levels in its history.[59] Twenty-four years earlier, when the reservoir began filling, Northeastern folk duo Sá e Guarabyra released a hit song about the dam and its social and environmental impacts. The chorus repeated the famous prophecy of Antônio Conselheiro – the preacher whose community the Brazilian government had violently repressed at the Canudos Massacre in 1896–1897. The popular refrain went "the desert will become the sea, and my heart fears that someday the sea will also become the desert."[60] For those watching Sobradinho's reservoir retreat and the ruins of old towns resurface, it must have appeared as though the second part of the prophecy was at last coming true: The sea was becoming the desert.

The drought was countrywide. During the previous rainy season, which normally lasts from November to April, the average streamflow rate for all of the country's major rivers combined was down 20 percent from its historical average. In the Southeast, it was down 30 percent, and in the Northeast, 50 percent. In April 2001, reservoir levels in both the Southeast and the Northeast were averaging just 35–40 percent of their capacity, well short of the 50 percent threshold needed to ensure sufficient water to generate power for the remainder of the year.[61] Engineers had stepped up production at Itaipu – Brazil's most efficient power plant – in order to conserve water in other reservoirs, but it was not enough. The government feared the country was heading toward an energy crisis. In June 2001, it began rationing electricity, mandating that consumers cut consumption by 20 percent and shutting off the power of those who did not comply. Brazilians endured rolling blackouts through the following year, when rains returned and reservoirs refilled.

The drought had set off such a severe energy crisis because it strained an already tight supply and demand balance. Beginning in the late 1980s, Brazil's government began redirecting money away from infrastructure projects and toward paying off the giant debts that the dictatorship had incurred during its construction boom. Between 1988 and 1995, the Brazilian government neither invested in new power plants nor granted

59 Evelson de Freitas, "Em Sobradinho, mar virou sertão," *Folha de São Paulo*, July 29, 2001.

60 Sá e Guarabyra, "Sobradinho," *Pirão de Peixe com Pimenta*, 1977, www.youtube.com/watch?v=EjVbTP7v9nY (last accessed April 2023).

61 Comissão Especial Mista Destinada a Estudar as Causas da Crise de Abastecimento de Energia no País, bem como Propor Alternativas ao seu Equacionamento (hereafter "Comissão Especial"), *A crise de abastecimento de energia elétrica relatório* (Brasília: Congresso Nacional, 2002), 31, General Collection, BCN.

new concessions to private power companies. The slowdown had no immediate consequences because Brazil's debt crisis had engendered a major recession, so demand for electricity had also stalled. However, after 1995, the economy recovered, and thereafter demand for energy soared. Power plant construction was slow to catch up, which made the depleting reservoirs a big threat.[62]

The energy crisis made the administration of Fernando Henrique Cardoso (1995–2003) unpopular and cast doubt on one of his main programs, privatization. Cardoso was a well-known sociologist of the academic left before becoming a politician. He earned his greatest fame as economic minister during the administration of Itamar Franco (1992–1995), who had replaced the impeached Fernando Collor de Mello (1990–1992), the country's first democratically elected president since the end of the dictatorship. In 1994, Cardoso and his team of economists created a new currency and implemented a set of economic policies that successfully curbed the inflation rate – which was an astonishing 2,708.2 percent in 1993 – and reignited economic growth. His success as finance minister made him a popular presidential candidate, and, once in office, he began a massive privatization campaign in order to reduce public debt and further boost the economy. His program aligned with the neoliberalism that governments in Europe and the United States were then promoting, but support for privatizing state companies was hard-won in Brazil because nationalism and a tradition of state ownership in key industries had been entrenched for decades.[63]

Privatizing the electricity sector was already controversial, and the energy crisis was a reminder that such an essential resource might be best left in the hands of the state. The Cardoso administration had privatized some of the sector, but mostly downstream operations, such as distribution. Power generation and transmission had remained in government hands. In 2002, after a big wave of auctions, state-owned companies remained responsible for just 10 percent of electricity distribution, but continued to account for 80 percent of production. The ratio has remained about the same since, though recent administrations have been championing privatization again.[64]

[62] Ibid.

[63] See Bryan McCann, *The Throes of Democracy: Brazil since 1989* (New York: Zed Books, 2008), 34–38.

[64] In 2013, the government was responsible for 15 percent of power distribution and 85 percent of power production. Ramón Espinasa et al., *Energy Dossier: Brazil, Technical Note No. IBD-TN-1121* (Washington, DC: Inter-American Development Bank, November 2016), 19.

The energy crisis also highlighted the consequences of slowing dam building without investing in alternatives and encouraged subsequent administrations to reconsider dams despite their environmental impacts. In 2002, during his campaign, Lula had come out against big dams in the Amazon Rainforest, where all the best remaining dam sites were located. He singled out both Tucuruí and Balbina, criticizing the former for enriching foreign aluminum companies at the expense of local people and repeating former Secretary of Science and Technology José Goldemberg's claim that the latter is a testament to insanity.[65]

However, once in power, his administration approved a series of controversial dam projects in the rainforest. In 2010, Lula endorsed Dilma, his energy minister, as his successor, and after she won the election, her administration began to build the dams the Lula administration had greenlit. She was even more adamant about big dams than Lula had been, and publicly championed building gigantic reservoirs.[66] Though she did not have the political support to build dams requiring massive reservoirs,[67] her administration did build other big and contentious dams, such as those along the Madeira and Teles Pires, and the infamous Belo Monte Dam, as discussed earlier in this chapter.

As the government continues to weigh the costs and benefits of dam building, it will have to consider the degree to which recurrent water shortages might curtail energy output, a challenging task considering the uncertainties that global warming is creating. The 2021 Intergovernmental Panel on Climate Change (IPCC) Report predicts, at 2°C global warming and above, more frequent and enduring drought alongside more frequent and intense flash flooding throughout most of Brazil, with the exception of Southeastern Brazil, where only the latter is expected.[68] It is clear that climate change will exacerbate extreme weather throughout Brazil, but what impact that will have on the country's reservoirs remains uncertain.

65 Luiz Inácio Lula da Silva, "O lugar da Amazônia no desenvolvimento do Brasil," 14–15, Programa de Governo 2002 Coligação Lula Presidente, 2002.

66 André Borges, "Dilma defende usinas hidrelétricas com grandes reservatórios," *Valor Econômico*, June 6, 2013.

67 André Borges, "Governo admite limitações a construir usinas com grandes reservatórios," *Valor Econômico*, August 14, 2013.

68 "Regional Fact Sheet—Central and South America," from Intergovernmental Panel on Climate Change (IPCC), *Climate Change 2021: The Physical Science Basis. Contribution of Working Group I to the Sixth Assessment Report of the Intergovernmental Panel on Climate Change* (New York: Cambridge University Press, 2021).

The outcome will depend in part on human actions. Scientists' predictions for extreme weather associated with climactic changes vary widely depending on the degree of warming, as defined in comparison to pre-industrial averages. In general terms, the more the industrial world is able to curb its emissions – for which Brazilian hydropower dams, except those in the Amazon, are helping – the softer the impact of extreme weather is expected to be. Curbing deforestation in the Amazon Rainforest will also help mitigate the crisis. In addition to being a significant carbon sink – absorbing large amounts of carbon that would otherwise be released into the atmosphere – the rainforest also drives a mobile evapotranspiration process that is an important component of the climate system in Southcentral and Southeastern Brazil.[69] Of course, such actions cannot eliminate the threat of drought altogether. Climate systems have their own vagaries, with or without warming, and a significant amount of warming is already underway in response to past and current emissions.

Recent droughts are perhaps a portent of some of the challenges that Brazil's energy sector might face under increasingly dry conditions. Starting in 2013, drought once again returned to Southeastern Brazil, creating water shortages in São Paulo and other metropolitan areas that reached a nadir during the austral summer the following year. The government instituted rolling blackouts to slow the depletion of its hydropower reservoirs and ramped up production at thermal power plants to make up the deficit. Though severe, the drought was not long-lasting, and rains returned to fill reservoirs in the Southeast shortly thereafter.[70]

Less than a decade later, in 2020–2021, drought pummeled the Southeast once again, with river discharges in the Paraná Basin falling to historic lows and many of the region's major reservoirs dropping below 20 percent capacity.[71] Again, rains returned to the Southeast, but not before generating another wave of rationing reminiscent of the previous two drought-induced energy crises of the past twenty years.

In the semidesert Northeast, the drought that began in 2013 never abated, and it has heavily diminished water supplies. Starting in 2016, Sobradinho's reservoir levels once again plummeted to dangerous lows, and the remnants of inundated towns began reappearing from beneath

[69] Augusto Getirana et al., "Brazil Is in Water Crisis – It Needs a Drought Plan," *Nature* 600 (December 2021): 218–220.

[70] See José Galizia Tundisi, ed., "Dossiê Crise Hídrica," *Revista Universidade de São Paulo* 106 (julho 2015): 8–94. For anglophone coverage, see Jonathan Watts, "Brazil's Worst Drought in History Prompts Protests and Blackouts," *The Guardian*, January 23, 2015.

[71] Getirana et al., "Brazil Is in Water Crisis."

the water.[72] Signs of crisis also appeared downstream. In order to maintain their reserves, upstream reservoirs released little water, drying up Paulo Afonso Falls. The famous falls had been repeatedly reduced to a trickle since Sobradinho and Itaparica were built, but most years annual rains brought them back to life, and a sizeable tourist industry had grown around them.[73] But since 2012, the falls have not returned. Periodic drought has long been a feature of the semiarid Northeast, but the recent dry spell has been one of the longest and most severe in recent history, and it has brought dire consequences for the region's waterways and those who depend on them. As of this writing, the drought has not let up.

In the face of such difficulties, Brazil's energy sector has had to adapt. One of its most helpful tools has been a resilient transmission system. Between 1998 and 1999, Eletrobras and Eletronorte erected the North-South Transmission Line, which connected two regions whose electricity grids were previously isolated from one another. In its earliest phase, this newly united system could not transport enough energy to forestall the rolling blackouts in 2001, but in subsequent decades the government added lines that increased the amount of electricity it could transmit.[74] Brazil now boasts a countrywide transmission system that can move electricity produced in wet areas to regions suffering from drought. This upgraded transmission system is one of the reasons that during the country's most recent droughts, energy rationing was less severe than during the 2000–2001 crisis. Having more dams and gas-fueled power plants also helped, and of course, none of these features were enough to prevent rolling blackouts altogether.

Climate change will demand more adaptation still. It will discourage investment in Amazonian dams because of their greenhouse gas emissions

[72] Dimmi Amora, "Sobradinho chega perto do volume morto e pode voltar a ser sertão," *Folha de São Paulo*, December 26, 2015; Edwirges Nogueira, "Sobradinho, de volta ao sertão," *Agência Brasil*, July 29, 2016; Sumaia Villela, "Vazão do São Francisco é reduzida novamente em maior seca em quase 90 anos," *Agência Brasil*, August 31, 2017; and Ana Lucia Azevedo, "No Lago de Sobradinho, 'mar virou sertão'," *O Globo*, March 22, 2018.

[73] See "Cheia do São Francisco faz renascer as cachoeiras de Paulo Afonso," *Folha Sertaneja*, April 30, 2004; "Cachoeira atrai milhares de turistas," *Folha Sertaneja*, February 28, 2005; "A Cachoeira de Paulo Afonso e sua importância para o turismo," *Folha Sertaneja*, September 30, 2005; "Paulo Afonso: Quase 10 mil visitam as cachoeiras em 4 dias," *Folha Sertaneja*, March 4, 2012. I am grateful to Antônio Galdino, the founder and editor of *Folha Sertaneja*, for letting me consult these articles at his private collection in Paulo Afonso.

[74] Isabela Sales Vieira, "Expansão do sistema de transmissão de energia elétrica no Brasil," master's thesis, Universidade de Brasília, 2009, 23–24.

and could diminish the output of existing dams if droughts and water shortages become more severe and long-lasting. Luckily, fossil fuels and hydroelectric dams are not the only two options for generating electricity, and Brazil has good terrain for alternative energy sources such as solar power and wind. But as the Brazilian government looks ahead to its next energy transition, it cannot forget that it still has a historical debt to settle with displaced Indigenous communities.

CONCLUSION

During the 1990s–2010s, the character of Brazilian dam building changed dramatically. Demands for social and environmental justice became more forceful and compelled the Brazilian hydropower sector to curtail its ambitions. The government had planned to build hundreds of big dams, many in the Amazon Rainforest, but it ended up mothballing most of these projects. For those it did build, it built them without giant reservoirs. Such concessions made this period of dam building much less damaging than its predecessor, though, to be sure, some of these dams had big social and environmental footprints that inflicted trauma on affected communities.

Seeing such episodes within the big picture of improving social and environmental conditions should not excuse the damage that has occurred at dam sites. Rather, it highlights that, since the return to democracy in the mid-1980s, the Indigenous and environmental movements have won some major victories that brought about tremendous improvements for communities affected by dams. Such achievements are a reminder that further gains can be made so long as people ensure that social and environmental justice continue to be potent political forces.

In contrast to the gains that social and environmental justice advocates won at planned and active dam sites, anti-dam activism during this period did almost nothing to help communities displaced during the dictatorship, some of whom are still fighting for just compensation.

Conclusion

So, are dams good or bad? It is a popular question and one that has attracted a lot of attention from scholars and engineering professionals. The scholarly consensus suggests that most mega dams with giant reservoirs entail tremendous social and environmental damage, and that paying the full cost to remediate these impacts would render them less profitable than alternative forms of energy. However, such conclusions do not fit all dams and do little to dissuade those engineers, politicians, and citizens convinced that the widespread benefits of electrification are worth some social and environmental sacrifices at reservoir sites.

This book is more concerned with changing the framing of this question than it is with offering a blanket answer. The question treats dams as uniform and unchanging. But, as this book's case studies illustrate, no two dams are uniform, and the process of building dams can change over time. So rather than encouraging sweeping judgments about the merit of dams, I hope readers form opinions about for whom dams are good and bad and the circumstances that result in a given dam being considered either good or bad. In Brazil, the military regime's dam-building campaign produced a mixed balance sheet.

On one side of the ledger are the social and environmental impacts. This book's five case studies together inundated more than 10,000 sq. km of land and directly displaced about 200,000 people.[1] Among those dispossessed were about 5,000–6,000 Indigenous people, a number that

[1] In total, Brazilian dams have inundated about 34,000 sq. km of land and displaced about 1,000,000 people (200,000 families). See Andréa Zhouri and Raquel Oliveira, "Desenvolvimento, conflitos sociais, e violência no Brasil rural: O caso das usinas

would have been much higher had these communities not suffered immense demographic losses preceding dam construction. Indigenous communities suffered distinct material and spiritual losses because their way of life and culture were tied to the land lost under floodwaters. The specific fates of most displaced individuals are unknown, though it is certain that all endured some degree of hardship and trauma as a result of being dispossessed. Thousands weathered the vagaries of reservoir fisheries, and some had the misfortune to spend the better part of a decade living among ferocious swarms of mosquitoes. The military regime's big reservoirs also flooded immense quantities of flora and fauna, an impact that was most detrimental in the incredibly biodiverse Amazon Rainforest, more than 5,000 sq. km of which was drowned under Tucuruí and Balbina's reservoirs. The full extent of the loss is unknowable because most of the biota was uncatalogued, and it is possible that valuable resources – such as plant-based genetic material used in pharmaceutical drugs – went extinct because of these dams. Lastly, reservoirs drowned cherished waterscapes, such as Guaíra Falls, perhaps the single greatest dam-related loss to the world's environmental patrimony.[2]

On the other side of this ledger are the socioeconomic benefits of electrification. The military dictatorship's big dams added about 45,000 megawatts to the electricity grid, and subsequent construction of transmission lines expanded access to electricity. On paper, all Brazilians now have access to electricity, though in practice low incomes limit access in some communities.[3] Electricity has undergirded economic growth and rising standards of living. Since the late 1980s, millions of people in Brazil have begun earning incomes above the international poverty line, and about ten years have been added to the average Brazilian's life expectancy.[4] Electricity alone is not responsible for either trend, but it is a necessary prerequisite, as jobs, households, and hospitals in an urban and industrial world require it. The military regime's dams also had a net benefit in terms of carbon emissions. Both Tucuruí and Balbina emit at

hidrelétricas," *Ambiente e Sociedade* 10, no. 2 (2007): 119–135, 121, and Capellini, "A grande aceleração e a construção de barragens hidrelétricas no Brasil," 318–319.

2 Here I am in agreement with Patrick McCully, *Silenced Rivers: The Ecology and Politics of Large Dams: Enlarged and Updated Edition* (London: Zed Books, 2001), 33.

3 World Bank Database, "Access to Electricity," https://data.worldbank.org/indicator/EG.ELC.ACCS.ZS (last accessed April 2023).

4 World Bank Database, "Life Expectancy," https://data.worldbank.org/indicator/SP.DYN.LE00.IN (last accessed April 2023). The average life expectancy increased from 65 years in the late 1980s to 75 years in the late 2010s.

least as much CO_2 equivalent as an average coal-fired power plant, but the rest of the dictatorship's dams have collectively spared the atmosphere between 3.6 and 6.2 billion tons of CO_2 emissions (see Chapter 3).

These costs and benefits operate on different timelines. The acute phase of the dams' social and environmental impacts lasted about a decade or so, though trauma often lingered much longer. This has been the case for many displaced Indigenous groups, both because the government has been negligent in compensating them with commensurate land and because some of the spiritual and cultural loss is irreparable. Potential extinctions of rainforest biota are similarly permanent.

The benefits of hydropower will last as long as dam infrastructure holds up, which can theoretically be prolonged with investments in maintenance and dredging, though high costs can deter such efforts. Precise sedimentation data is not accessible, but estimates of the lifespan of these dams before siltation renders them obsolete are somewhere in the ballpark of one hundred to two hundred years.

In terms of aggregate costs and benefits, the dictatorship's hydropower dams are undoubtedly a net benefit for most Brazilians, 86 percent of whom live in cities and profit from electrification. In such accounting, the high percentage of people who profit from hydropower outweighs the smaller percentage of those who suffered its deleterious consequences. For example, for all those who suffered from epidemic malaria and mosquito bites along the margins of Itaipu and Tucuruí's reservoirs, there are millions in metropolitan areas and electrified rural homes who are able to live longer and healthier lives because of electrified hospitals and refrigerators. The post-construction carbon saving for dams outside the Amazon Rainforest is also a tremendous accomplishment given the urgent need to mitigate the looming climate crisis.

Such aggregate accounting is an enticing but ultimately problematic form of assessing the merits of hydropower dams. It overlooks the unequal distribution of costs and benefits and assumes that all disastrous impacts are an inevitable consequence of building dams, which encourages those reaping the benefits of hydropower to accept them. One of this book's goals is to encourage the replacement of such thinking with more nuanced assessments that highlight both the uneven distribution of the costs and benefits of dam building, as well as the choices and contingencies responsible for these outcomes.

This book has three interrelated arguments. The first is that political pressures encouraged the military government to build large reservoirs in environmentally sensitive areas without public debate and before proper environmental studies or mitigation measures were carried out. The military regime's dam-building campaign targeted underpopulated and geopolitically strategic areas in order to meet the objectives of its National Integration Program. Politics also influenced choices about specific dam sites, giving some schemes additional appeal beyond their basic technical and economic merits. Lastly, political pressures encouraged the dictatorship to execute its plans quickly and without public debate. The fledgling regime was looking for projects that would win it immediate political legitimacy and thus hoped to shorten construction timelines as much as possible.

This book's second argument is that the environmentalism the military regime practiced was designed to protect power plant infrastructure and showcase concern for the environment without fundamentally altering project designs or slowing construction. The generals hired Robert Goodland – a renowned ecologist and dam specialist – to study their projects, though they made it clear that his findings would not alter their decision to build at the chosen dam sites. The military government rushed his studies and applied his findings selectively, choosing only to implement programs that would either prolong the life of its power plants or dramatically showcase its environmental consciousness, such as animal rescue missions. These measures were pharaonic environmentalism: programs designed to bolster the image of the military government as a regime that could build durable mega dams and also protect the environment. The generals succeeded at the former but faltered at the latter.

This book's final argument is that the dictatorship's ostentatious environmental protection programs did nothing to forestall a series of social and ecological disasters at dam sites. The military regime disregarded the social costs of its big dams because it felt pressure to build dams quickly and cheaply and because it believed that its pharaonic environmentalism would satisfy its critics. Most of the people the government's dams displaced were marginalized communities, which made it easier to dispossess them without political repercussions. The dictatorship's environmental interventions were similarly ineffective. In some cases, local communities benefited from environmental changes, such as fishermen profiting from booming fisheries in the aftermath of reservoir formation. In other cases, other government agencies helped curb damage, such as the health specialists fighting malaria at Itaipu. However, disregard for people and the environment led to local disasters in enough cases that hydropower soon developed a bad reputation among social justice advocates and environmentalists. The military regime underestimated the organizational power of the Catholic Church and the growing environmental movement, and as stories of catastrophe surfaced, tremendous domestic and international criticism poured in that pressured the government to better ameliorate the social costs of its dams.

These arguments have important implications. Seeing dams as political objects as much as economic ones helps dispel the popular narrative that dam sites are chosen for their technical merits alone, an argument that proponents of high-impact dam sites often use to justify damage. For the narrative's proponents, following unbiased science and economic guidance is the best path to greater material progress, and any social and environmental impacts at sites engineers deem worthy are lamentable but necessary sacrifices for such gains. The corollary is that adjusting projects to mitigate those social and environmental impacts will erase their gains and derail material progress. Similar arguments lie at the heart of debates about economic growth and environmental impacts generally.

Inserting political pressures into the equation discredits this argument in the case of Brazilian dam building and calls it into question elsewhere. If the priorities of wealthy landowners and two intransigent military regimes fighting about their shared border can carry as much or more weight in decision-making processes as technical and unbiased science (see Chapter 2), then so too can efforts to protect marginalized communities and the environment.

It is also important to hold companies accountable for responsible environmentalism. Greenwashing, the term for deceitful claims of sound environmental stewardship, is a common practice among many companies and a roadblock against meaningful environmental reform. The military dictatorship's pharaonic environmentalism was greenwashing with flair, an ostentatious program of superficial environmental protection masquerading as a responsible safeguard. Identifying such practices helps people better recognize them and understand the consequences when such behavior goes unchecked: unmitigated social and environmental damage. Shattering this deception also discourages companies from using it and pressures them to engage in more sincere protection.

Lastly, it is important to remember the military regime's greatest environmental tragedies. In Brazil, authoritarianism worsened the social and environmental consequences of dam building because it foreclosed debate and protest and made it easier for the government to neglect affected communities. The specter of both authoritarianism and anti-environmentalism has not been extinguished – both in Brazil and worldwide – and educating current and future generations about its discontents is the best strategy for ensuring that societies do not have to endure repression and related human rights abuses and environmental injustices.

In Brazil, the stakes have never been higher. In 2018, the country elected an anti-Indigenous and anti-environmentalist president, Jair Bolsonaro (2019–2022), as well as large numbers of right-wing congress members who support his agenda. During his time in office, Bolsonaro expressed open disdain for environmental protection and basic science and celebrated the military dictatorship, calling for its return.

The Bolsonaro administration also reignited the debate over dams. He championed a handful of big dams in the Amazon Rainforest, such as the rebranded Cachoeira Porteira Dam (see Chapter 7), with little regard for safeguards.[5] During his tenure, his allies in the National Congress also began considering a proposal to amend the constitution to permit building dams on Indigenous land (as of April 2023, it is still under consideration).[6] Bolsonaro's blundering and tumultuous single-term presidency did not facilitate a successful dam-building campaign, but other policies created environmental disasters, and big reservoirs might have made the list had his administration been more competent and endured longer.

It is unclear what his successor, Luiz Inácio Lula da Silva, will do with these plans. Lula embraces social and environmental justice and has a comparatively strong record on environmental protection, but during his previous tenure, his administration carried out a handful of controversial dam projects in the Amazon (see Chapter 7). He has not since condemned them, which raises concern among anti-dam activists. Moreover, social and environmental justice opponents and their allies continue to have the majority in Congress, which is still weighing the proposal to greenlight dams on Indigenous land.[7]

The climate crisis gives the story of Brazilian dams additional weight. The overwhelming consensus among scientists is that the industrial world

[5] André Borges et al., "Bolsonaro retoma plano de erguer grandes hidrelétricas na Amazônia," *O Estado de São Paulo*, August 21, 2019; Ministério de Minas e Energia (MME), *Plano Decenal de Expansão de Energia 2029* (Brasília: MME, 2019), 287; and André Borges, "Aneel libera estudos para instalar três megausinas na Amazônia," *Estadão*, January 27, 2023.

[6] See Projeto de Lei 191/2020 and Projeto de Lei 278/2023, www.camara.leg.br/proposicoesWeb/fichadetramitacao?idProposicao=2236765 and www.camara.leg.br/proposicoesWeb/fichadetramitacao?idProposicao=2347078.

[7] See Philip Fearnside, "Lula e as hidrelétricas na Amazônia 3: As barragens planejadas," *Amazônia Real*, February 13, 2023.

will need to undergo a rapid and widespread energy transition to avert the catastrophic impacts of climate change, which includes replacing fossil fuel power plants with low-carbon electricity. Hydropower is a tempting tool for countries with propitious dam sites. In Brazil, however, most of the remaining dam sites are in the Amazon, where reservoirs produce greenhouse gases, which means that hydropower dams cannot be the country's only solution for lowering emissions, and the government will need to invest in other sources and energy conservation.

Outside of Brazil, the country's experience with big dams offers valuable guidance. Throughout the twentieth century, the Brazilian government's electricity grid has been much less carbon-intensive than most of the industrial world. That building these power plants was so fraught with social and environmental controversies illustrates that low-carbon energy has its own environmental footprint. This means that a successful energy transition not only requires curtailing fossil fuel emissions but also ensuring social and environmental justice for communities disproportionally impacted during these transitions. This conclusion should not discourage investments in alternative energy sources but rather encourage governments to ensure that renewable energy systems are installed responsibly.

This book carries an important message of hope. Alongside stories of environmental disaster were tales of people fighting to help those harmed by big reservoirs and to ensure that such projects were not built in the future without proper safeguards. These activists understood that technical considerations alone did not guide decision-making, and they fought to make social and environmental justice a stronger political force to match the political power of dam proponents. The fall of the dictatorship helped their cause, and in subsequent decades, they used legal tools to mitigate some of the government's most harmful dam proposals.

It is important to take stock of such accomplishments because they provide the encouragement needed to overcome contemporary setbacks in social and environmental justice efforts. These stories remind us that things have been grim in the past; people reacted, educated themselves, organized, and pushed for reform that remediated the damage, although they did not always accomplish all their goals. The same can happen again so long as people continue to react and educate themselves. To that end, there is nothing more pressing than discrediting narratives that pit economic growth against social justice and environmental protection, calling out deceitful environmentalism, and remembering past tragedies so that our world's worst environmental disasters are those read about in history books and not the headlines.

I end, once again, with noise. I hope these pages replace in our minds the silence of turning on the lights with the thundering of distant waterfalls, the protests of displaced communities, and other sounds that become background noise that forever remind us of our connections to distant hinterlands and our responsibilities to communities there.

Bibliography

ARCHIVES AND LIBRARIES

Dam Companies and Personal Papers of Engineers and Politicians

ACHESF	Arquivo da Companhia Hidrelétrica do Rio São Francisco [Archive of the São Francisco Hydroelectric Company], Recife, Brazil
BCHESF	Biblioteca da Companhia Hidrelétrica do Rio São Francisco [Library of the São Francisco Hydroelectric Company], Recife
BIB-FO	Biblioteca de Itaipu Binacional [Itaipu Binacional Library], Foz de Iguaçu, Brazil
BIB-AS	Biblioteca de Itaipu Binacional [Itaipu Binacional Library], Asunción, Paraguay
AELN	Arquivo de Eletronorte [Eletronorte Archive], Brasília, Brazil
BELN	Biblioteca de Eletronorte [Eletronorte Library], Brasília
ACME	Arquivo do Centro da Memória da Eletricidade [Archive of the Center for Electricity Memory], Rio de Janeiro, Brazil
FADL	Fundo de Antônio Dias Leite [Antônio Dias Leite Papers]
FJC	Fundo John Cotrim [John Cotrim Papers]
FLL	Fundo Lucas Lopes [Lucas Lopes Papers]
FMT	Fundo Mauro Thibau [Mauro Thibau Papers]
BCME	Biblioteca do Centro da Memória da Eletricidade [Library of the Center for Electricity Memory], Rio de Janeiro
AN-RDJ	Arquivo Nacional [National Archives], Rio de Janeiro
FADL	Fundo Antônio Dias Leite [Antônio Dias Leite Papers]

CPDOC-FGV — Centro de Pesquisa e Documentação de História Contemporânea do Brasil, Fundação Getúlio Vargas [Center for Research and Documentation of Contemporary Brazilian History, Getúlio Vargas Foundation], Rio de Janeiro

FOMF — Fundo Octávio Marcondes Ferraz [Octávio Marcondes Ferraz Papers]

FJM — Fundo de Juracy Magalhães [Juracy Magalhães Papers]

AJPA — Acervo João Paulo Aguiar [João Paulo Aguiar Personal Papers], Recife

NGOs and Personal Papers of Advocates and Scientists

BDJR — Biblioteca Dom José Rodrigues [Dom José Rodrigues Library], Universidade Estadual de Bahia, Juazeiro, Brazil

ACIMI — Arquivo do Conselho Indigenista Missionário [Archive of the Indigenist Missionary Council], Brasília

ACPT — Arquivo da Comissão Pastoral da Terra [Archive of the Pastoral Land Commission], Goiânia, Brazil (accessed online and in person).

APIES — Archivo del Pastoral Indígena de Espirito Santo [Archive of the Indigenous Ministry of the Holy Spirit] (Mariblanca Barón's collection), Nueva Esperanza, Paraguay

ACONAPI — Archivo de la Coordinación Nacional de Pastoral Indígena [Archive of the National Organization of Indigenous Ministries], Asunción

BCEFURIA — Biblioteca do Centro de Formação Urbano Rural Irmã Araújo [Library of the Sister Araújo Urban and Rural Education Center], Curitiba, Brazil

AUT — Acervo Uilton Tuxá [Uilton Tuxá Personal Papers], Paulo Afonso, Brazil

CEPEDAL — Núcleo de Pesquisa e Documentação sobre o Oeste do Paraná [Center for Research and Documentation about Western Paraná], Universidade Estadual do Oeste do Paraná, Marechal Cândido Rondon, Brazil

FEM — Fundo Ernst Mann [Ernst Mann Papers]

FSI — Fundo Sarah Iurkiv [Sarah Iurkiv Papers]

AAMS — Acervo Allan Martins da Silva [Allan Martins Personal Papers], Laboratório Central do Estado de Paraná, São José de Pinhais, Brazil

AISA Arquivo do Instituto Socioambiental [Archive of the Socioenvironmental Institute], São Paulo, Brazil (accessed online).

AEDS Acervo de Egydio e Doroti Schwade (Casa da Cultura do Urubuí) [Egydio and Doroti Schwade Papers (House of Urubuí Culture)], Presidente Figueiredo, Brazil

APF Acervo Philip Fearnside [Philip Fearnside Personal Papers], Manaus, Brazil

Government and Intergovernmental Archives

ACD Arquivo da Câmara dos Deputados [Archive of the Chamber of Deputies], Brasília (accessed online and in person).

ASF Arquivo do Senado Federal, Brasília [Archive of the Senate], Brasília (accessed online and in person).

BCN Biblioteca do Congresso Nacional [Library of the National Congress], Brasília

AN-BSB Arquivo Nacional [National Archives], Brasília (accessed online)

BN-RDJ Biblioteca Nacional [National Library], Rio de Janeiro (accessed online)

BNP-AS Biblioteca Nacional del Paraguay [National Library of Paraguay], Asunción

NDJFP Núcleo de Documentação da Justiça Federal do Paraná [Center for Documentation at the Federal Court of Paraná], Curitiba

BFUNASA Biblioteca da Fundação Nacional de Saúde [The Library of the National Health Foundation], Brasília

ASENEPA-AP Archivo del Servicio Nacional de Erradicación del Paludismo, Sector Alto Paraná [Archive of the National Malaria Eradication Service, Alto Paraná Sector], Ciudad del Este, Paraguay

AFIOCRUZ Arquivo da Fundação Oswaldo Cruz [Archive of the Oswaldo Cruz Foundation], Rio de Janeiro

BINPA Biblioteca do Instituto Nacional de Pesquisas da Amazônia [Library of the National Institute for Amazonian Research], Manaus

BPWA Biblioteca do Programa Waimiri Atroari [Library of the Waimiri Atroari Program], Manaus

US-LOC US Library of Congress, Washington DC, United States (accessed online and in person)

WBGA World Bank Group Archive, Washington DC (accessed online)

Select Public, University, and Research Institute Libraries

BPP Biblioteca Pública do Paraná [Public Library of Paraná], Curitiba

DDP Divisão de Documentação Paranaense [Paraná Documents Division]

BFUNDAJ Biblioteca da Fundação Joaquim Nabuco [Library of the Joaquim Nabuco Foundation], Recife

RI-UEM Repositório Institucional da Universidade Estadual de Maringá [Institutional Repository of the State University of Maringá], Maringá, Brazil (accessed online).

INTERVIEWS

Malaria specialist Raul Diniz Souza Amorim, Manaus, October, 2017
Malaria specialist Wanderlei Pedro Tadei, Manaus, October, 2017
CPT-Western Parná coordinator Werner Fuchs, Curitiba, August, 2018
Guaíra writer Edson Galvão, Guaíra, September, 2018
Guaíra historian Ana Menél, Guaíra, September, 2018
Malaria specialist Maria do Carmen, Foz de Iguaçu, September, 2018
Malaria specialist Mario Pilleco, Foz de Iguaçu, November, 2018
Sobradinho's head engineer, João Paulo Aguiar, Recife, March, 2019
Paulo Afonso historian Antônio Galdino, Paulo Afonso, April, 2019
Paulo Afonso tour guide Antônio Souza, Paulo Afonso, April, 2019

NEWSPAPERS AND MAGAZINES

A Crítica (Manaus, Brazil)
A Notícia (Manaus, Brazil)
A Nova Democracia (Rio de Janeiro, Brazil)
ABC Color (Asunción, Paraguay)
Acción (Asunción, Paraguay)
Agência Brasil (Brasília, Brazil)
Amazonas Atual (Manaus, Brazil)
Amazonas em Tempo (Manaus, Brazil)
Amazonas Notícias (Manaus, Brazil)
Amazônia Real (Manaus, Brazil)
Brasil de Fato (Curitiba, Brazil)
Brasil de Fato (São Paulo, Brazil)
Caminhar Juntos (Juazeiro, Brazil)
Carta Capital (São Paulo, Brazil)
Ciência Hoje (Rio de Janeiro, Brazil)
Correio da Imprensa (Cuiabá, Brazil)

Cox Newspapers South America (United States)
Dialogo Indígena Misionario (Asunción, Paraguay)
Diário de Amazonas (Manaus, Brazil)
Diário de Pernambuco (Recife, Brazil)
El País (Madrid, Spain)
Folha de Londrina (Brazil)
Folha de São Paulo (Brazil)
Folha Sertaneja (Paulo Afonso, Brazil)
Inter-Press Service (Rome, Italy)
Jornal do Commercio (Recife, Brazil)
Jornal do Commercio (Rio de Janeiro, Brazil)
Jornal de Curitiba (Brazil)
Jornal Ilha Grande (Guaíra, Brazil)
Jornal do Brasil (Rio de Janeiro, Brazil)
Jornal do Pescador (Maringá, Brazil)
Luta Indígena (Xanxerê, Brazil)
Memoria Institucional Butantan (São Paulo, Brazil)
Mongabay (Menlo Park, United States)
New York Times (United States)
New Yorker (United States)
Newsletter of the Cary Arboretum of the New York Botanical Garden (United States)
Notícias CIMI (Brasília, Brazil)
Notícias da Articulação dos Povos Indígenas do Brasil (Brasília, Brazil)
O Estado de São Paulo (Brazil)
O Estado de Paraná (Curitiba, Brazil)
O Globo (Rio de Janeiro, Brazil)
O Liberal (Belém, Brazil)
O Povo Brasileiro (Manaus, Brazil)
Oeste (Foz de Iguaçu, Brazil)
Popular Mechanics (New York, United States)
Poeira (Curitiba, Brazil)
Porantim (Manaus, Brazil)
Revista Brasileira de Energia Elétrica (Rio de Janeiro, Brazil)
Revista Manchete (Rio de Janeiro, Brazil)
Revista Pau Brasil (Brazil)
The Conversation (Waltham, United States)
The Guardian (London, United Kingdom)
The Independent (London, United Kingdom)
Tribuna da Bahia (Salvador, Brazil)
Tribuna Liberal (Asunción, Paraguay)
Ultima Hora (Asunción, Paraguay)
Universidade de São Paulo Jornal do Campus (Brazil)
Valor Econômico (São Paulo, Brazil)
Veja (São Paulo, Brazil)
World Aquaculture (Sorrento, United States)

ONLINE DATABASES

(Links valid as of April, 2023)

International Energy Agency

Brazil: <https://www.iea.org/countries/brazil>
United States: <https://www.eia.gov/totalenergy/data/annual/>
Electricity, Installed Capacity: <https://www.eia.gov/international/data/world/electricity/electricity-capacity>

World Bank Database

Hydropower: <https://data.worldbank.org/indicator/EG.ELC.HYRO.ZS>
Electricity Access: <https://data.worldbank.org/indicator/EG.ELC.ACCS.ZS>
Life Expectancy: <https://data.worldbank.org/indicator/SP.DYN.LE00.IN>
Urbanization: <https://data.worldbank.org/indicator/SP.URB.TOTL.IN.ZS>

Greenhouse Gas Emissions

SEEG, Sistema de Estimativas de Emissões e Remoções de Gases de Efeito Estufa [Greenhouse Gas Emission and Removal Estimating System]: <http://plataforma.seeg.eco.br/total_emission#>

US EPA, Greenhouse Gas Equivalencies Calculator: <https://www.epa.gov/energy/greenhouse-gas-equivalencies-calculator>

PUBLISHED PRIMARY SOURCES

Government Documents

Brasil, República Federativa do. *Constituição do Brasil de 1967*. Brasília: Congresso Nacional, 1967.

Constituição da República Federativa do Brasil de 1988. Brasília: Congresso Nacional, 1988.

Comissão Especial Mista Destinada a Estudar as Causas da Crise de Abastecimento de Energia no País, bem como Propor Alternativas ao seu Equacionamento. *A crise de abastecimento de energia elétrica relatório*. Brasília: Congresso Nacional, 2002.

Comisión Nacional sobre la Desaparición de Personas. *Nunca más: Informe de la Comisión Nacional sobre la Desaparición de Personas*. Buenos Aires: EUDEBA, 1984.

Eletrobras. *Plano 2010: Plano nacional de energia elétrica 1987/2010 relatório geral*. Rio de Janeiro: Eletrobrás and Ministério das Minas e Energia, December 1987.

Geisel, Ernesto. *II Plano Nacional de Desenvolvimento (1975–1979)*. Brasília: Congresso Nacional, 1974.

Instituto Brasileiro de Geografia e Estatística (IBGE). *Censo demográfico de 1940*. Rio de Janeiro: Serviço Gráfico do IBGE, 1940.

Censo demográfico de 1960. Rio de Janeiro: Serviço Gráfico do IBGE, 1960.
Sinopse do Censo Demográfico 2010. Rio de Janeiro: Serviço Gráfico do IBGE, 2010.
Médici, Emilio Garrastazu. *I Plano Nacional de Desenvolvimento (PND), 1972/74*. Brasília: Congresso Nacional, 1971.
Ministério do Interior. *Relatório da Delegação do Brasil à Conferência das Nações Unidas sobre Meio Ambiente Humano, Estocolmo, 1972, Volume 1*. Brasília: Ministério do Interior, 1972.
Relatório da Delegação do Brasil à Conferência das Nações Unidas sobre Meio Ambiente Humano, Estocolmo, 1972, Volume 2. Brasília: Ministério do Interior, 1972.
Ministério de Minas e Energia. *Plano Decenal de Expansão de Energia 2029*. Brasília: Ministério de Minas e Energia, 2019.

River Basin, Dam, and Energy Surveys

Derby, Orville A. "Reconhecimento Geológico do Vale de S. Francisco," in *Relatório de W. Milnor Roberts, Engenheiro Chefe da Comissão Hidráulica sobre o Exame do Rio São Francisco desde o Mar até a Cachoeira de Pirapora (extensão de 2,122 quilômetros), feito em 1879–1880*. Rio de Janeiro: Typographia Nacional, 1880.
Espinasa, Ramón, Augusto Bonzi Teixeira, and Fernando Anaya. *Energy Dossier: Brazil, Technical Note No. IBD-TN-1121*. Washington DC: Inter-American Development Bank, November 2016.
Goodland, Robert. *Sobradinho Hydroelectric Project Environmental Impact Reconnaissance*. Millbrook: Cary Arboretum of the New York Botanical Garden, September 1973.
Environmental Assessment of the Tucuruí Hydroelectric Project, Rio Tocantins, Amazonia. Brasília: Eletronorte, 1977.
Goodland, Robert and Richard Pollard. *Chixoy Development Project: Environmental Impact Reconnaissance*. Millbrook: Cary Arboretum of the New York Botanical Garden, 1974.
Halfeld, Henrique Guilherme Fernando. *Atlas e relatório concernente a exploração do Rio de São Francisco desde a Cachoeira da Pirapora até ao Oceano Atlântico, 1852–1854*. Rio de Janeiro: 1860.
Marcondes Ferraz, Octávio. *Relatório preliminar sobre o aproveitamento do Salto de Sete Quedas (Guaíra) Rio Paraná*. São Paulo: Escritório Técnico O.M.F., December 1962.
Sampaio, Theodoro. *O Rio São Francisco e a Chapada Diamantina, 2ª Edição*. Salvador: Progresso Editora, 1955 (first published 1906).

Social and Environmental Movement Reports

Centro de Defesa dos Direitos Humanos do Submédio São Francisco. *Itaparica: A Dor de um Povo Gerando Energia*. Petrolândia: Centro de Defesa dos Direitos Humanos do Submédio São Francisco, 1992.

Comissão Pastoral da Terra (CPT). *O mausoléu do faraó: A Usina de Itaipu contra os lavadores do Paraná*. Goiânia: CPT, 1978.
Ilhas de resistência: Os lavadores no Vale do Rio São Francisco. Goiânia: CPT, 1979.
Evaristo Arns, Paulo. *Brasil: Nunca mais*. São Paulo: Arquidiocese de São Paulo, 1985.
Mazzarollo, Juvêncio. *A taipa da injustiça: Esbanjamento econômico, drama social e holocausto ecológico em Itaipu 2ª Edição*. Curitiba: CPT Paraná, 2003 (first published 1980).
Movimento de Apoio e Resistência Waimiri-Atroari (MAREWA), *Resistência Waimiri/Atroari*. Itacoatiara: MAREWA, 1983.
Balbina: Ameaça e destruição na Amazônia. Manaus: MAREWA, 1987.

Books and Memoirs

Barros, Wanderbilt Duarte de. *Parques Nacionais do Brasil*. Rio de Janeiro: Ministério de Agricultura, 1952.
Brunetto, Elio J. and Jonas D. Muraski, eds. *Memórias de Sete Quedas*. Concórdia: Equiplan Serviços Gráficos Ltda., 1983.
Cardozo, Efraim. *Los derechos del Paraguay sobre los Saltos del Guairá*. Asunción, 1965.
Debernardi, Enzo. *Apuntas para la historia política del Itaipu*. Asunción: Editora Gráfica, 1996.
Gibson Barboza, Mario. *Na diplomacia, o traço todo da vida*. Rio de Janeiro: Editoria Record, 1992.
Hermes, Gabriel. *No País dos Dólares*. Belém, 1952.
O Baixo Amazonas. Rio de Janeiro: Editora Leitura, 1966.
Howarth, David. *The Shadow of the Dam*. London: Macmillan Press, 1961.
Porfírio F. de Carvalho, José. *Waimiri Atroari: A história que ainda não foi contada*. Self-published, 1982.
Quintella, Wilson. *Memórias do Brasil Grande: A história das maiores obras do país e dos homens que as fizeram*. São Paulo: Editora Saraiva, 2008.
Rebouças, André. *Excursão ao Salto do Guaíra, o parque nacional*. Rio de Janeiro, 1876.
Robins, Eric and Ronald Legge. *Animal Dunkirk: The Story of Lake Kariba and "Operation Noah" the Greatest Animal Rescue Since the Ark*. London: Herbert Jenkins, 1959.
Walsh, John and Robert Gannon. *Time is Short and the Water Rises*. New York: Tower Publications, 1967.

Speeches

Nehru, Jawaharlal. "Temples of the New Age Speech, July 8, 1954." In *Jawaharlal Nehru's Speeches Vol. 3, March 1953-August 1957*, 1–4. New Delhi: Ministry of Information and Broadcasting, 1958.

DISSERTATIONS AND THESES

Araujo, Roderlei de. "Composição da fauna de Anopheles (Diptera: Culicidae) da Reserva Indígena do Ocoy, foco ativo de malária no município de São Miguel do Iguaçu, Paraná." Master's thesis, Universidade Federal de Paraná, 2011.

Barbosa, Maria Helena. "Memórias do regime militar em Guaíra." Undergraduate thesis, Universidade Estadual do Oeste de Paraná, Marechal Cândido Rondon, 2003.

Bishop, Elizabeth. "Talking Shop: Egyptian Engineers and Soviet Specialists at the Aswan High Dam." PhD Diss., University of Chicago, 1997.

Cañas, Ana Roberta Pessoa Aguilar. "Conflitos silenciosos: A pesca amadora no Lago de Balbina, Presidente Figueiredo, Amazonas." Master's thesis, Universidade Federal do Amazonas, Manaus, 2012.

Capellini, Nathalia. "Historiciser les barrages en Amazonie Brésilienne: Environnement, conflit et politique dans la planification et la construction de Tucuruí." PhD diss., Université Paris-Saclay, 2019.

Carvalho, Maria Lucia Brant de. "Das terras dos Índios a Índios sem terras, O Estado e Os Guarani do Oco'y: Violência, Silêncio e Luta." PhD diss., Universidade de São Paulo, 2013.

Couto, Rosa Carmina de Sena. "Hidrelétricas e saúde na Amazônia: Um estudo sobre a tendência da malária na área do lago da UHE Tucuruí-Pará-Brasil." PhD Diss., Escola Nacional de Saúde Pública (ENSP)/Fundação Oswaldo Cruz (Fiocruz), 1996.

Cruz, Felipe Sotto Maior. "'Quando a terra sair,' os índios Tuxá de Rodelas e a Barragem de Itaparica: Memórias do desterro, memórias da resistência." Master's thesis, Universidade de Brasília, 2017.

Daou, Ana Maria Lima. "Políticas de estado e organização social camponesa: A barragem de Sobradinho." Master's thesis, Museu Nacional and Universidade Federal do Rio de Janeiro, 1988.

Dávalos, Nelson Eduardo Bernal. "Vulnerabilidade socioambiental e os impactos do reassentamento e dos eventos climáticos extremos sobre o povo indígena Tuxá de Rodelas-Bahia." PhD diss., Universidade de Brasília, 2021.

Ferreira, Maria Eugênia Moreira Costa. "Ocorrência de malária na área de influência do reservatório de Itaipu, margem esquerda, Paraná, Brasil." PhD diss., Universidade de São Paulo, 1996.

Forcelini, Melody. "'Colossal, Complexa, Imprescindível': Tecnologia de força bruta e Usina Hidrelétrica de Itá/SC." Master's thesis, Universidade Estadual do Centro Oeste do Paraná, 2017.

Goodland, Robert. "An Ecological Study of Cerrado Vegetation of South-Central Brasil." PhD diss., McGill University, 1969.

Johnson, Matthew P. "Temples of Modern Pharaohs: An Environmental History of Dams and Dictatorship in Brazil, 1960s-1990s." PhD diss., Georgetown University, 2021.

Lima, Vilma Fiorotti. "Guaíra nos tempos de Ernst Mann." Undergraduate thesis, Universidade Estadual do Oeste de Paraná, Marechal Cândido Rondon, 1994.

Magalhães, Antônio Carlos. "Os Parakanã: Quando o rumo da estrada e o curso de águas perpassam a vida de um povo." Master's thesis, Universidade de São Paulo, 1982.

"Os Parakanã: Espaços da socialização e suas articulações simbólicas." PhD diss., Universidade de São Paulo, 1994.

Magalhães, Sônia Barbosa. "Lamento e dor: Uma análise sócio-antropológica do deslocamento compulsório provocada pela construção de barragens." PhD diss., Universidade Federal do Pará, 2007.

Marques, Juracy. "Cultura material e etnicidade dos povos indígenas do São Francisco afetados por barragens: Um estudo de caso dos Tuxá de Rodelas, Bahia, Brasil." PhD diss., Universidade Federal da Bahia, 2008.

Menezes, Luciano Silva de. "A Barragem de Itaparica e os atingidos de Petrolândia-PE." Master's thesis, Universidade do Estado da Bahia, Juazeiro, 2017.

Nasser, Nássaro Antônio de Souza. "A Economia Tuxá." Master's thesis, Universidade Federal da Bahia, Salvador, 1975.

Nasser, Elizabeth Cabral. "A sociedade Tuxá." Master's thesis, Universidade Federal da Bahia, Salvador, 1975.

Ribeiro, Maria de Fátima Bento. "Itaipu, a dança das águas: Histórias e memórias de 1966 a 1984." PhD diss., Universidade Estadual de Campinas, 2006.

Rodrigues, Renan Albuquerque. "Vidas despedaçadas: Impactos socioambientais da construção da usina hidrelétrica de Balbina (AM), Amazônia Central." PhD. diss., Universidade Federal do Amazonas, Manaus, 2013.

Sá, Rosa Maria Lemos de. "Effects of the Samuel Hydroelectric Dam on Mammal and Bird Communities in a Heterogeneous Amazonian Lowland Forest." PhD diss., University of Florida, 1995.

Salomão, Ricardo Dantas Borges. "Etnicidade, processos de territorialização e ritual entre os índios Tuxá de Rodelas." Master's thesis, Universidade Federal Fluminense, 2006.

Santos, Ana Paula dos. "Lago de memórias: A submersão das Sete Quedas." Master's thesis, Universidade Estadual de Maringá, 2006.

Silva, Edcarlos Mendes da. "Desterritorialização sob as águas de Sobradinho: Ganhos e desenganos." Master's thesis, Universidade Federal da Bahia, 2010.

Silva, Mara Dhulle dos Santos. "Guaíra: A cidade em tempos de ditadura civil-militar (1964–1985)." Master's thesis, Universidade Estadual do Oeste de Paraná, Marechal Cândido Rondon, 2016.

Siqueira, Ruben Alfredo de. "Do que as águas não cobriram: Um estudo sobre o movimento dos camponeses atingidos pela barragem de Sobradinho." Master's thesis, Universidade Federal de Paraíba, 1992.

Soares, Daniella Feteira. "Paisagem e memória: Dos Saltos de Sete Quedas ao Lago de Itaipu." Master's thesis, Universidade Federal do Rio de Janeiro, 2001.

Souza, Maria Emília Medeiros de. "Mymba-Kuera." Undergraduate thesis, Universidade Estadual de Oeste de Paraná, Foz de Iguaçu, 1987.

Souza Filho, Enéas Cordeiro de. "Epidemiologia da malária no Estado de Paraná, Brasil, 2002 a 2008." Master's thesis, Universidade Federal de Paraná, 2010.

Vieira, Antônio Fernandes de Jesus. "Os índios Tuxá na rota do desenvolvimento: Violações de direitos." Master's thesis, Universidade de Brasília, 2017.

Vieira, Isabela Sales. "Expansão do sistema de transmissão de energia elétrica no Brasil." Master's thesis, Universidade de Brasília, 2009.
White, John Howard. "Itaipú: Gender, Community, and Work in the Alto Paraná Borderlands, Brazil and Paraguay, 1954–1989." PhD diss., University of New Mexico, 2010.

SECONDARY SOURCES

Aarão Reis Filho, Daniel. *Ditadura e democracia no Brasil: Do golpe de 1964 à Constituição de 1988*. Rio de Janeiro: Zahar, 2014.
Aarão Reis Filho, Daniel, Marcelo Ridenti, and Rodrigo Patto Sá Motta, eds. *A ditadura que mudou o Brasil: 50 anos do golpe de 1964*. Rio de Janeiro: Zahar, 2014.
Abd-El Monsef, Hesham, Scot E. Smith, and Kamal Darwish. "Impacts of the Aswan High Dam after 50 years." *Water Resources Management* 29 (2015): 1873–1885.
Abu-Zeid, M.A. and F.Z. El-Shibini. "Egypt's High Aswan Dam." *International Journal of Water Resources Development* 13, no. 2 (1997): 209–218.
Acker, Antoine. "A Different Story in the Anthropocene: Brazil's Post-Colonial Quest for Oil (1930–1975)." *Past and Present* 249 (November 2020): 167–211.
Acselrad, Henri and Maria das Graças da Silva, "Conflito social e mudança ambiental na barragem de Tucuruí." In *Conflitos Ambientais no Brasil*, edited by Henri Acselrad, 175–193. Rio de Janeiro: Relume Dumará, 2004.
Aditjondro, George and David Kowalewski. "Damning the Dams in Indonesia: A Test of Competing Perspectives." *Asian Survey* 34, no. 4 (April 1994): 381–395.
Alcântara, Gustavo Kenner, João Akira Omoto, Julio José Araujo Junior, and Luciana Maria de Moura Ramos. *Avá Guarani: A construção de Itaipu e os direitos territoriais*. Brasília: Ministério Público da União, 2019.
Alves, Maria Helena Moreira. *State and Opposition in Military Brazil*. Austin: University of Texas Press, 1988.
Amaral, Adzamara. *Sento-Sé: Memória de uma cidade submersa*. São Paulo: Lisbon International Press, 2020.
Araújo, Maria Lia Corrêa de, Magda de Caldas Neto, and Ana Eliza Vasconcelos, eds. *Sonhos submersos ou desenvolvimento? Impactos sociais da Barragem de Itaparica*. Recife: Massangana, 2000.
Araujo, Roderlei de, Allan Martins da Silva, Enéas Cordeiro de Souza Filho, and Ennio Luz, "Composição da fauna de Anopheles (Diptera: Culicidae) da Reserva Indígena do Ocoy, foco ativo de malária no município de São Miguel do Iguaçu, Paraná (Brasil)." *Acta Biológica Paranaense* 43 (2014): 85–112.
Arco Blanco, Miguel Angel del and Santiago Gorostiza. "'Facing the Sun': Nature and Nation in Franco's 'New Spain' (1936–51)." *Journal of Historical Geography* 71 (2021): 73–82.
Armiero, Marco and Wilko Graf von Hardenberg. "Green Rhetoric in Blackshirts: Italian Fascism and the Environment." *Environment and History* 19, no. 3 (2013): 283–311.

Aroucha, Edvalda Pereira Torres Lins. "Ecologia das mulheres do São Francisco." In *Ecologias do São Francisco*, edited by Juracy Marques, 258–365. Self-Published, 2006.

Arrifano, Gabriela P.F., Rosa C. Rodríguez Martín-Doimeadios, María Jiménez-Moreno, Vanesa Ramírez-Mateos, Núbia F.S. da Silva, José Rogério Souza-Monteiro, Macus Augusto-Oliveira, Ricardo S.O. Paraense, Barbarella M. Macchi, José Luiz M. do Nascimento, and Maria Elena Crespo-Lopez. "Large-Scale Projects in the Amazon and Human Exposure to Mercury: The Case Study of the Tucuruí Dam." *Ecotoxicology and Environmental Safety* 147 (2018): 299–305.

Baer, Werner. *Industrialization and Economic Development in Brazil*. Homewood: Irwin, 1965.

Barón, Mariblanca. *Los Ava Guaraní Paranaenses: Un etnocidio sistemático*. Series Biblioteca Paraguaya de Antropología Vol. 107. Asunción: Centro de Estudios Antropológicos de la Universidad Católica, 2017.

Barros, Antonio Teixeira de. "Brazil's Discourse on the Environment in the International Arena, 1972–1992." *Contexto Internacional* 39, no. 2 (May-August 2017): 421–442.

Barroso, Maria Auxiliadora Bessa and Raul Diniz Souza Amorim. *Malária no Amazonas: Registros e memória*. Manaus: Editora Valer, 2017.

Baviskar, Amita. *In the Belly of the River: Tribal Conflicts over Development in the Narmada Valley*. New York: Oxford University Press, 1995.

Beeson, Mark. "The Coming of Environmental Authoritarianism." *Environmental Politics* 19, no. 2 (March 2010): 276–294.

Berga, Luis. "The Role of Hydropower in Climate Change Mitigation and Adaption: A Review," *Engineering* 2 (2016): 313–318.

Bértoli, Marta and Maria da Luz Ribeiro Moitinho. "Malária no Estado do Paraná, Brasil." *Revista da Sociedade Brasileira de Medicina Tropical* 34, no. 1 (2001): 43–47.

Blanc, Jacob. "Enclaves of Inequality: Brasiguaios and the Transformation of the Brazil- Paraguay Borderlands." *Journal of Peasant Studies* 42, no. 1 (2015): 145–158.

"The Last Political Prisoner: Juvêncio Mazzarollo and the Twilight of Brazil's Dictatorship." *Luso-Brazilian Review* 53, no. 1 (2016): 153–78.

Before the Flood: The Itaipu Dam and the Visibility of Rural Brazil. Durham: Duke University Press, 2019.

Brain, Stephen and Viktor Pál, eds. *Environmentalism under Authoritarian Regimes: Myth, Propaganda, Reality*. New York: Routledge, 2019.

Brannstrom, Christian. "Was Brazilian Industrialisation Fuelled by Wood? Evaluating the Wood Hypothesis, 1900–1960." *Environment and History* 11 (2005): 395–430.

Brüggemeier, Franz-Josef, Mark Cioc, and Thomas Zeller, eds. *How Green Were the Nazis? Nature, Environment, and Nation in the Third Reich*. Athens: Ohio University Press, 2005.

Buckley, Eve E. *Technocrats and Politics of Drought and Development in Twentieth-Century Brazil*. Chapel Hill: University of North Carolina Press, 2017.

Burns, E. Bradford. "Manaus: Portrait of a Boom Town." *Journal of Inter-American Studies* 7, no. 3 (July 1965): 400–421.

Campos, Pedro Henrique Pedreira. *Estranhas Catedrais: As empreiteiras brasileiras e a ditadura civil-militar, 1964–1988*. Rio de Janeiro: Editora da Universidade Federal Fluminense, 2014.

Capellini, Nathalia. "A grande aceleração e a construção de barragens hidrelétricas no Brasil." *Varia História* 34, no. 65 (2018): 315–346.

Capellini, Nathalia and Carlos Gomez Florentin, "Hydroelectric Dams and the Rise of Environmentalism under Dictatorship in Brazil and Paraguay (1950–1990): The Case of Itaipu." In *Environmentalism under Authoritarian Regimes: Myth, Propaganda, Reality*, edited by Stephen Brain and Viktor Pál, 51–74. New York: Routledge, 2018.

Chatterjee, Elizabeth. "The Poor Woman's Energy: Low-Modernist Solar Technologies and International Development, 1878–1966." *Journal of Global History* 18 (November 2023): 439–460.

Clements, Frank. *Kariba: The Struggle with the River God*. London, Methuen and Co., 1959.

Collier, David, ed. *The New Authoritarianism in Latin America*. Princeton: Princeton University Press, 1979.

Colson, Elizabeth. *The Social Consequences of Resettlement: The Impact of Kariba Resettlement Upon the Gwembe Tonga*. Manchester: Manchester University Press, 1971.

Comitê Brasileiro de Barragens (CBDB). *A História das Barragens no Brasil, Séculos XIX, XX, e XXI: Cinquenta Anos do Comitê Brasileiro de Barragens*. Rio de Janeiro: CBDB, 2011.

Comitê Estadual de Direito à Verdade, à Memória e à Justiça do Amazonas. A ditadura militar e o genocídio do povo Waimiri-Atroari. Campinas: Editora Curt Nimuendajú, 2014.

Comissão Nacional de Verdade (CNV). *Relatório Volume I*. Brasília: CNV: 2014.

CNV *Relatório Volume III: Mortos e Desparecidos Políticos*. Brasília: CNV: 2014.

Consolim, José and Joaquim T. Galvão. "Sobre os anofelinos do Rio Paraná: Densidade e regime do rio." *Arquivo Biológico Tecnológico* 16 (1973): 174–181.

Consolim, José, Ennio Luz, Nilson Jorge de Mattos Pellegrini, and Pedro Barroso Torres. "O Anopheles (Nyssorhynchus) darlingi ROOT, 1926 e a malária no lago de Itaipu, Estado de Paraná, Brasil: Uma revisão de dados." *Arquivo Biológico Tecnológico* 34, no. 2 (June 1991): 263–286.

Consolim, José, Nilson Jorge de Mattos Pellegrini, and Ennio Luz. "Culicídeos (Díptera, Culicidae) do Lago de Itaipu, Paraná, Brasil. I. Município de Foz de Iguaçu," *Acta Biológica Paranaense* 22 (1993): 83–90.

Cordell, Geoffrey A. "Biodiversity and Drug Discovery - A Symbiotic Relationship." *Phytochemistry* 55 (2000): 463–480.

Couto, Rosa Carmina de Sena. "Malária o custo social da hidrelétrica de Tucuruí-PA, Brasil." In *Saúde, trabalho e meio ambiente: políticas públicas na Amazônia*, edited by Rosa Carmina Couto, Edna Ramos de Castro and Rosa Acevedo Marin, 107–121. Belém: Editora Universitária UFPA, 2002.

Cronon, William. *Nature's Metropolis: Chicago and the Great West*. New York: N.W. Norton & Co., 1991.

Cullenward, Danny and David G. Victor. "The Dam Debate and Its Discontents." *Climatic Changes* 75 (2006): 81–86.

Cunha, Euclides da. *Os sertões: campanha de Canudos*. Rio de Janeiro: Laemmert Editores, 1902. Translated in English as *Rebellion in the Backlands*. Chicago: University of Chicago Press, 1944.

Dai Qing. *Yangtze! Yangtze! Controversy Over the Three Gorges Dam Project*. Toronto-London: Probe International and Earthscan, 1994.

Daou, Ana Maria Lima. "Efeitos sociais da Barragem de Sobradinho: A experiência dos camponeses do Brejo de Dentro." *Revista Geográfica* 120 (1994): 27–40.

Dean, Warren. *Brazil and the Struggle for Rubber: A Study in Environmental History*. New York: Cambridge University Press, 1987.

With Broadax and Firebrand: The Destruction of the Brazilian Atlantic Forest. Berkeley: University of California Press, 1995.

Dias Leite, Antônio. *Energia do Brasil, 3ª edição*. Rio de Janeiro: Editora Nova Fronteira, 2014. Earlier edition translated in English as *Energy in Brazil: Towards a Renewable Energy Dominated System*. London: Earthscan, 2009.

Ding, Xiangli. "'The Yellow River Comes from Our Hands': Silt, Hydroelectricity, and the Sanmenxia Dam, 1929–1973." *Environment & History* 27, no. 4 (November 2021): 665–694.

Doyle, Timothy and Adam Simpson. "Traversing More than Speed Bumps: Green Politics under Authoritarian Regimes in Burma and Iran." *Environmental Politics* 15, no. 5 (2006): 750–767.

Duarte, Regina Horta. *A biologia militante: O museu nacional, especialização científica, divulgação do conhecimento e práticas políticas no Brasil, 1926–1945*. Belo Horizonte: Universidade Federal de Minas Gerais, 2010. Translated in English as *Activist Biology: The National Museum, Politics, and Nation Building in Brazil*. Tucson: University of Arizona Press, 2016.

"'Turn to Pollute': Poluição atmosférica e modelo de desenvolvimento no 'milagre' brasileiro (1967–1973)." *Revista Tempo* 21, no. 37 (2015): 64–87.

Dudley, Marianna. "When's a Gale a Gale? Understanding Wind as an Energetic Force in Mid- Twentieth Century Britain," *Environmental History* 26, no. 4 (October 2021): 671–695.

Eaglin, Jennifer. *Sweet Fuel: A Political and Environmental History of Brazilian Ethanol*. New York: Oxford University Press, 2022.

Espíndola, Marcos Aurélio. "Tecnização da natureza: O desastre ambiental da Usina Hidrelétrica de Barra Grande (SC)." In *Desastres socioambientales em Santa Catarina*, edited by Eunice Sueli Nodari, Espíndola, and Alfredo Ricardo Silva Lopes, 157–181. São Leopoldo: Oikos Editora, 2015.

Espósito Neto, Tomaz. *Itaipu e as relações brasileiro-paraguaias de 1962 a 1979: Fronteira, energia e poder*. Saarbrücken: Novas Edições Acadêmicas, 2017.

Estrela, Ely Souza. "Assentamentos indígenas no Médio São Francisco: O caso dos Tuxá e Pankarú." *Revista Travessia* 39, no. 1 (2001): 42–46.

Sobradinho: a retirada de um povo. Salvador: Editora da Universidade do Estado da Bahia, 2010.

Evenden, Matthew. *Fish versus Power: An Environmental History of the Fraser River*. New York: Cambridge University Press, 2004.

Falavigna-Guilherme, Ana Lucia, Allan Martins da Silva, Edson Valdemar Guilherme, and Dina Lúcia Morais. "Retrospective Study of Malaria Prevalence and Anopheles Genus in the Area of Influence of the Binational Itaipu Reservoir." *Revista Institucional de Medicina Tropical de São Paulo* 74, no. 2, (March-April, 2005): 81–86.

Faran, Michael E. and Kenneth J. Linthicum. "A Handbook on the Amazonian Species of Anopheles (Nyssorhynchus) (Diptera: Culicidae)." *Mosquito Systematics* 13, no. 1 (1981): 1–81.

Faria, Felipe A.M. de, Paulina Jaramillo, Henrique O. Sawakuchi, Jeffrey E. Richey, and Nathan Barros. "Estimating Greenhouse Gas Emissions from Future Amazonian Hydroelectric Reservoirs." *Environmental Research Letters* 10 (2015).

Fearnside, Philip. "Brazil's Balbina Dam: Environment and the Legacy of the Pharaohs in Amazônia." *Environmental Management* 13, no. 4 (1989): 401–423.

A Hidrelétrica de Balbina: O faraonismo irreversível versus o meio ambiente na Amazônia. São Paulo: Instituto de Antropologia e Meio Ambiente, 1990.

"Hydroelectric Dams in the Brazilian Amazon as Sources of 'Greenhouse Gases,' *Environmental Conservation* 22, no. 1 (Spring 1995): 7–19.

"Greenhouse-Gas Emissions from Amazonian Hydroelectric Reservoirs: The Example of Brazil's Tucuruí Dam as Compared to Fossil Fuel Alternatives." *Environmental Conservation* 24, no. 1 (March 1997): 64–75.

"Social Impacts of Brazil's Tucuruí Dam." *Environmental Management* 24, no. 4 (1999): 483–495.

"Environmental Impacts of Brazil's Tucuruí Dam: Unlearned Lessons for Hydroelectric Development in Amazonia." *Environmental Management* 27, no. 3 (2001): 377–396.

"Greenhouse Gas Emissions from a Hydroelectric Reservoir (Brazil's Tucuruí Dam) and the Energy Policy Implications." *Water, Air, and Soil Pollution* 133 (2002): 69–96.

"Greenhouse Gas Emissions from Hydroelectric Dams: Controversies Provide a Springboard for Rethinking a Supposedly 'Clean' Energy Source." *Climatic Changes* 66 (2004): 1–8.

"Greenhouse Gas Emissions from Hydroelectric Dams: Reply to Rosa Et al." *Climatic Changes* 75 (2006): 103–109.

"Dams in the Amazon: Belo Monte and Brazil's Hydroelectric Development of the Xingu River Basin." *Environmental Management* 38, no. 1 (2006): 16–27.

"Greenhouse Gas Emissions from Hydroelectric Dams in Tropical Forests." In *Alternative Energy and Shale Gas Encyclopedia*, edited by Jay Lehr and Jack Keeley, 428–438. New York: Wiley and Sons, 2016.

Feitosa, Gleicyane Souza, Paulo Maurício de Alencastro Graça, and Fearnside, "Estimativa da zona de deplecionamento da hidrelétrica de Balbina por técnica de sensoriamento remoto," *Anais XIII Simpósio Brasileiro de Sensoriamento Remoto*, Florianópolis, Brasil, 21–26 abril 2007.

Ferreira, Maria Eugênia Moreira Costa. "'Doenças tropicais': O clima e a saúde coletiva. Alterações climáticas e a ocorrência de malária na área de influência do reservatório de Itaipu, PR." *Terra Live* 19 (2003): 179–191.

Ferreira, Simone do Rocio and Ennio Luz. "Malária no Estado de Paraná—Aspetos históricos e prognose." *Acta Biológica Paranaense* 32 (2003): 129–156.

Feshbach, Murray and Alfred Friendly Jr. *Ecocide in the USSR*. New York: Basic Books, 1992.

Fico, Carlos. *O grande irmão da Operação Brother Sam aos anos de chumbo: O governo dos Estados Unidos e a ditadura militar brasileira*. Rio de Janeiro: Civilização Brasileira, 2008.

Field Jr., Thomas C. "Ideology as Strategy: Military-Led Modernization and the Origins of the Alliance for Progress in Bolivia." *Diplomatic History* 36, no. 1 (2012): 147–183.

Fontes, Paulo. *Migration and the Making of Industrial São Paulo*. Durham Duke University Press, 2016.

Forcelini, Melody and Jó Klanovicz, "A Usina Hidrelétrica de Itá, Brasil e a tecnologia de força bruta na história recente do Rio Uruguai," *Revista Inclusiones* 5 (Enero-Marzo 2018): 175–191.

Franco, José Luiz de Andrade and José Augusto Drummond. *Proteção à natureza e identidade nacional no Brasil, anos 1920–1940*. Rio de Janeiro: Editora Fiocruz, 2009.

Freitas, Frederico. *Nationalizing Nature: Iguazu Falls and National Parks at the Brazil-Argentina Border*. New York: Cambridge University Press, 2021.

Frenette, M., JP Tournier, and TJ Nzakimuena. "Cas Historique de Sédimentation du Barrage Péligre, Haïti." *Canadian Journal of Civil Engineering* 9 (1982): 206–223.

Gendron, Robin S., Mats Ingulstad, and Espen Storli, eds. *Aluminum Ore: The Political Economy of the Global Bauxite Industry*. Vancouver: University of British Columbia Press, 2013.

Germani, Guiomar Inez. *Os expropriados de terra e água: O conflito de Itaipu 2ª edição*. Canoas: Editora da ULBRA, 2003.

Getirana, Augusto, Renata Libonati, and Marcio Cataldi. "Brazil Is in Water Crisis – It Needs a Drought Plan." *Nature* 600 (December 2021): 218–220.

Gilio-Whitaker, Dina. *As Long as Grass Grows: The Indigenous Fight for Environmental Justice, from Colonization to Standing Rock*. Boston: Beacon Press, 2019.

Gilley, Bruce. "Authoritarian Environmentalism and China's Response to Climate Change." *Environmental Politics* 21, no. 2 (March 2012): 287–307.

Godinho, Alexandre L., Boyd Kynard, and Hugo P. Godinho. "Migration and Spawning of Female Surubim (*Pseudoplatystoma corruscans*, Pimelodiade) in the São Francisco River, Brazil." *Environmental Biology of Fishes* 80, no. 4 (December 2007): 421–433.

Gribel, Rogério. "Os mamíferos silvestres e as grandes barragens na Amazônia." In *Bases científicas para estratégias de preservação e desenvolvimento da*

Amazônia: Fatos e perspectivas, vol. 2, edited by Efrem Jorge Gondim Ferreira, Geraldo Mendes dos Santos, Elizabeth Lima Mendes Leão, and Luíz Antônio de Oliveira, 125–140. Manaus: INPA, 1993.

Griffing, Sean Michael, Pedro Luiz Tauil, Venkatachalam Udhayakumar, and Luciana Silva- Flannery. "A Historical Perspective on Malaria Control in Brazil." *Memória Instituto Oswaldo Cruz* 110, no. 6 (2015): 701–718.

Guimarães, Anthony Érico, Rubens Pinto de Mello, Catarina Macedo Lopes, Jeronimo Alencar, Carla Gentile. "Prevalência de anofelinos (Díptera: Culicidae) no crepúsculo vespertino em área da Usina Hidrelétrica de Itaipu, no Município de Guaíra, Estado do Paraná, Brasil." *Memória Instituto Oswaldo Cruz* 92, no. 6 (1997): 745–754.

Guimarães, Roberto. *The Ecopolitics of Development in the Third World: Politics and Environment in Brazil.* Boulder: Lynne Rienner, 1991.

Hall, Anthony. *Drought and Irrigation in North-East Brazil.* New York: Cambridge University Press, 1978.

Han, Heejin. "Authoritarian Environmentalism under Democracy: Korea's River Restoration Project." *Environmental Politics* 24, no. 5 (2015): 1–20.

Harres, Marluza Marques and Fabiano Quadros Rückert. *A Natureza, o tempo, e as marcas da ação humana: Políticas públicas e ambiente em perspectiva histórica, São Leopoldo, RS.* São Leopoldo: Editora Oikos, 2011.

Harrison, Mark. "'Hot Beds of Disease': Malaria and Civilization in Nineteenth-Century British India." *Parassitologia* 40 (1998): 11–18.

Harrison-Levine, Amy, Herbert H. Covert, Marilyn A. Norconk, Ricardo Rodrigues dos Santos, Adrian A. Barnett, and Philip Fearnside. "Dams: Implications of Widespread Anthropic Flooding for Primate Populations," in *Primates in Flooded Habitats: Ecology and Conservation*, edited by Katarzyna Nowak, Barnett, and Ikki Matsuda, 285–292. New York: Cambridge University Press, 2019.

Harvey, Danny. "The Exchange between Fearnside and Rosa Concerning the Greenhouse Gas Emissions from Hydro-Electric Power Dams." *Climatic Changes* 75 (2006): 87–90.

Hecht, Susanna and Alexander Cockburn. *The Fate of the Forest: Developers, Destroyers and Defenders of the Amazon.* New York: HarperCollins, 1990.

Hill, Christopher. *South Asia: An Environmental History.* Santa Barbara: ABC-CLIO, 2008.

Hilton, Stanley E. "Vargas and Brazilian Economic Development, 1930–1945: A Reappraisal of his Attitude Toward Industrialization and Planning." *The Journal of Economic History* 35, no. 4 (1975): 754–778.

Hochstetler, Kathryn and Margaret Keck. *Greening Brazil: Environmental Activism in State and Society.* Durham: Duke University Press, 2007.

Hughes, Donald. "The Dams at Aswan: Does Environmental History Inform Decisions?" *Capitalism Nature Socialism* 11, no. 4 (2000): 73–81, republished as "Case Study C: The Aswan Dams and their Environmental Results," 213–232 in Hughes, *The Mediterranean: An Environmental History.* Santa Barbara: ABC-CLIO, 2005.

Instituto Socioambiental. *Dossiê Belo Monte: Não há condições para a Licença de Operação.* São Paulo: Instituto Socioambiental, 2015.

Intergovernmental Panel on Climate Change (IPCC). *Climate Change 2021: The Physical Science Basis. Contribution of Working Group I to the Sixth Assessment Report of the Intergovernmental Panel on Climate Change.* New York: Cambridge University Press, 2021.

International Hydropower Association (IHA), *2018 Hydropower Status Report.* London: IHA, 2018.

Joffily, Mariana. "O aparato repressivo: Da arquitetura ao desmantelamento." In *A ditadura que mudou o Brasil: 50 anos do golpe de 1964*, edited by Daniel Aarão Reis Filho, Marcelo Ridenti, and Rodrigo Patto Sá Motta, 158–171. Rio de Janeiro: Zahar, 2014.

Johnson, Adriana Michele Campos. *Sentencing Canudos: Subalternity in the Backlands of Brazil.* Pittsburgh: Pittsburgh University Press, 2010.

Johnson, Matthew P. "Swampy Sugar Lands: Irrigation Dams and the Rise and Fall of Malaria in Puerto Rico, 1898–1962," *Journal of Latin American Studies* 51, no. 2 (May 2019): 243–271.

"'Thirsty Sugar Lands': Environmental Impacts of Dams and Empire in Puerto Rico since 1898," *Environment & History* 27, no. 3 (August 2021): 337–365.

Jones, Christopher. *Routes of Power: Energy and Modern America.* Cambridge: Harvard University Press, 2014.

"Petromyopia: Oil and the energy humanities," *Humanities* 5, no. 36 (2016).

Josephson, Paul. *Resources under Regimes: Technology, Environment, and the State.* Cambridge: Harvard University Press, 2004.

Josephson, Paul, Nicolai Dronin, Ruben Mnatsakanian, Aleh Cherp, Dmitry Efremenko, and Vladislav Larin. *An Environmental History of Russia.* New York: Cambridge University Press, 2013.

Kemenes, Alexandre, Bruce R. Forsberg, and John M. Melack. "Methane Release below a Tropical Hydroelectric Dam." *Geophysical Research Letters* 34, no. 12 (June 2007).

"CO_2 Emissions from a Tropical Hydroelectric Reservoir (Balbina, Brazil)." *Journal of Geophysical Research* 116, no. G3 (September 2011).

Khagram, Sanjeev. *Dams and Development: Transnational Struggles for Water and Power.* Ithaca: Cornell University Press, 2004.

Kingston, W. Robin. "Establishment of Primates Rescued from the Tucuruí Dam Flooding in the Brazilian National Primate Centre." *Primate Eye* 28 (1986): 18–20.

Klein, Ira. "Development and Death: Reinterpreting Malaria, Economics, and Ecology in British India." *The Indian Economic and Social History Review* 38, no. 2 (2001): 147–179.

Kleinpenning, Jan. "Strong Reservations about 'New Insights into the Demographics on the Paraguayan War.'" *Latin American Research Review* 37, no. 3 (2002): 137–42.

Kohlhepp, Gerd. *Itaipú Basic Geopolitical and Energy Situation: Socio-Economic and Ecological Consequences of the Itaipú Dam on the Rio Paraná.* Braunschweig: F. Vieweg, 1987.

Kumar, Deepak, Vinita Damodaran, and Rohan D'Souza. *The British Empire and the Natural World: Environmental Encounters in South Asia.* New York: Oxford University Press, 2011.

Lagus, Charles. *Operation Noah*. New York: Coward-McCann, 1960.

Lambert, Peter. "The Myth of the Good Neighbour: Paraguay's Uneasy Relationship with Brazil." *Bulletin of Latin American Research* 35, no. 1 (2016): 34–48.

Levine, Robert M. *Vale of Tears: Revisiting the Canudos Massacre in Northeastern Brazil, 1893–1897*. Berkeley: University of California Press, 1992.

Li, Yifei and Judith Shapiro. *China Goes Green: Coercive Environmentalism for a Troubled Planet*. Cambridge: Polity Press, 2020.

Lima, Ivone Teresinha Carletto. *Itaipu: As faces de um megaprojeto de desenvolvimento*. Marechal Cândido Rondon: Editora Germânica, 2006.

Limberger, Leila and Sandra Elisa Contri Pitton. "Percepção climática de moradores lindeiros ao reservatório da UH de Itaipu." In *Estudos Regionais: Estrutura, Agentes e Processos*, edited by Edson Belo Clemente de Souza, 401–430. Cascavel: Editora Unioeste, 2012.

Luna, Francisco Vidal and Herbert S. Klein. "Transformações econômicas no período militar (1964–1985)." In *A ditadura que mudou o Brasil: 50 anos do golpe de 1964*, edited by Daniel Aarão Reis Filho, Marcelo Ridenti, and Rodrigo Patto Sá Motta, 92–111. Rio de Janeiro: Zahar, 2014.

Macfarlane, Daniel. *Fixing Niagara Falls: Environment, Energy, and Engineers at the World's Most Famous Waterfall*. Vancouver: University of British Colombia Press, 2020.

Mainwaring, Scott. *The Catholic Church and Politics in Brazil, 1916–1985*. Stanford: Stanford University Press, 1986.

Malloy, James, ed. *Authoritarianism and Corporatism in Latin America*. Pittsburgh: University of Pittsburgh Press, 1977.

Marmulla, Gerd, ed. *Dams, Fish, and Fisheries: Opportunities, Challenges, and Conflict Resolution*. FAO Fisheries Technical Paper No. 419. Rome: FAO, 2001.

Martins Costa, Ana Luiza. *Uma retirada insólita*. Rio de Janeiro: IPPUR/UFRJ, 2013.

Mascarenhas, Bento Melo and Giuseppe Puorto. "Nonvolant Mammals Rescued at the Tucuruí Dam in the Brazilian Amazon." *Primate Conservation* 9 (1988): 91–93.

McCann, Bryan. *The Throes of Democracy: Brazil since 1989*. New York: Zed Books, 2008.

McCully, Patrick. *Silenced Rivers: The Ecology and Politics of Large Dams: Enlarged and Updated Edition*. London: Zed Books, 2001.

McGregor, JoAnn. *Crossing the Zambezi: The Politics of Landscape on a Central African Frontier*. Martlesham: James Curry, 2009.

Melià, Bartomeu. *Itaipú: Deforestación y desarraigo de la nación Guaraní*. Asunción: El Lector, 2018.

Memória da Eletricidade. *Panorama do setor de energia elétrica no Brasil*. Rio de Janeiro: Memória da Eletricidade, 1988.

Octávio Marcondes Ferraz: Um pioneiro da engenharia nacional. Rio de Janeiro: Memória da Eletricidade, 1993.

Notas sobre racionamento de energia elétrica no Brasil (1940–1980). Rio de Janeiro: Memória da Eletricidade, 1996.

Notas sobre os antecedentes da criação de Itaipu Binacional. Rio de Janeiro: Memória da Eletricidade, 1999.

O meio ambiente e o setor de energia elétrica brasileiro. Rio de Janeiro: Memória da Eletricidade, 2009.

Mendelsohn, Robert and Michael J. Balick. "The Value of Undiscovered Pharmaceuticals in Tropical Forests." *Economic Botany* 49, no. 2 (1995): 223–228.

Mendes, Chico and Tony Gross. *Fight for the Forest: Chico Mendes in his Own Words*. London: Latin American Bureau, 1989.

Mesquita, Otoni Moreira de. *La belle vitrine: Manaus entre dois tempos, 1890–1900*. Manaus: Universidade Federal do Amazonas, 2009.

Miescher, Stephan *A Dam for Africa: Akosombo Stories from Ghana*. Bloomington: Indiana University Press, 2022.

Miller, Shawn William. *An Environmental History of Latin America*. Cambridge: Cambridge University Press, 2007.

Monteiro, Nilson. *Itaipu, A Luz*. Curitiba: Itaipu Binacional, Assessoria de Comunicação Social, 1999.

Mintz, Sidney. *Sweetness and Power: The Place of Sugar in Modern History*. New York: Penguin Books, 1986.

Moutinho dos Santos, Edmilson and Drielli Peyerl, "The Incredible Transforming History of a Former Oil Refiner into a Major Deepwater Offshore Operator: Blending Audacity, Technology, Policy, and Luck from the 1970s Oil Crisis up to the 2000s Pre-Salt Discoveries." In *History, Exploration & Exploitation of Oil and Gas*, edited by Silvia Fernanda Figueirôa, Gregory A. Good, and Drielli Peyerl, 87–100. New York: Springer, 2019.

Nascimento, Luiz Fernando Motta. *Paulo Afonso: Luz e força movendo o Nordeste*. Salvador: Associação Recreativa dos Servidores da CHESF, 1998.

Needham, Andrew. *Power Lines: Phoenix and the Making of the Modern Southwest*. Princeton: Princeton University Press, 2014.

Niland, Richard. "Death by Water: The Rise and Fall of Los Saltos del Guairá." *Environmental History* 23, no. 1 (2018): 56–81.

La muerte del agua: El ascenso y el descenso de Los Saltos del Guairá. Asunción: El Lector, 2021.

Nixon, Rob. *Slow Violence and the Environmentalism of the Poor*. Cambridge: Harvard University Press, 2011.

O'Donnell, Guillermo. *Modernization and Bureaucratic-Authoritarianism: Studies in South American Politics*. Berkeley: University of California Institute of International Studies, 1973.

El Estado burocrático: Triunfos, derrotas y crisis. Buenos Aires: Editorial Belgrano, 1982. Translated in English as *Bureaucratic Authoritarianism: Argentina 1966–1973 in Comparative Perspective*. Berkeley: University of California Press, 1988.

Oliveira-Ferreira, Joseli, Marcus Lacerda, Patrícia Brasil, José Ladislau, Pedro Tauil, and Cláudio Tadeu Daniel-Ribeiro. "Malaria in Brazil: An Overview." *Malaria Journal* 9 (2010).

Otoya, Natascha. "Petroleum and Science: The National Petroleum Council and the Development of Oil and Geology in Brazil." In *History, Exploration & Exploitation of Oil and Gas*, edited by Silvia Fernanda Figueirôa, Gregory A. Good, and Drielli Peyerl, 37–49. New York: Springer, 2019.

Pádua, José Augusto. *Um sopro de destruição: Pensamento político e crítica ambiental no Brasil escravista (1786–1888)*. Rio de Janeiro: Zahar, 2002.

"Environmentalism in Brazil: An Historical Perspective." In *A Companion to Global Environmental History*, edited by J.R. McNeill and Erin Mauldin, 455–473. Oxford: Wiley-Blackwell, 2012.

"The Dilemma of the 'Splendid Cradle': Nature and Territory in the Construction of Brazil." In *A Living Past: Environmental Histories of Modern Latin America*, edited by John Soluri, Claudia Leal, and José Augusto Pádua, 91–114. New York: Berghahn, 2018.

Pereira, Anthony. "The US Role in the 1964 Coup: A Reassessment." *Bulletin of Latin American Research* 37, no. 1 (January 2018): 5–17.

Pereira, Elenita Malta. *Roessler: O homem que amava a natureza*. São Leopoldo: Oikos, 2013.

Pereira, Elenita Malta, Sara Fritz, and Denis Fiuza. *José Lutzenberger: Um ambientalista global*. São José: Editora Sobre o Tempo, 2022.

Petr, T. "On Some Factors Associated with the Initial High Fish Catches in New African Man- Made Lakes." *Archives of Hydrobiology* 75, no. 1 (1975): 32–49.

Pinto, Lúcio Flávio. "De Tucuruí a Belo Monte: A história avança mesmo? *Boletim Museu Paraense Emílio Goeldi* 7, no. 3, (2012): 777–782.

Pollan, Michael. *The Omnivore's Dilemma: A Natural History of Four Meals*. New York: Penguin Books, 2006.

Prado, Daniel Porciuncula. *A figueira e o machado: Uma história das raízes do ambientalismo no Sul do Brasil e a crítica ambiental de Henrique Roessler*. Carreiros: FURG, 2011.

Priest, Tyler. "Petrobras in the History of Offshore Oil." In *New Order and Progress: Development and Democracy in Brazil*, edited by Ben Ross Schneider, 53–77. New York: Oxford University Press, 2016.

Quintero, Lisardo Osorio, Bedsy Dutary Thatcher, and Wanderli Pedro Tadei. "Ocorrência de espécies de anopheles e outros culicídeos na área de influência da hidrelétrica de Balbina – cinco anos após o enchimento do reservatório." *Acta Amazônica* 26, no. 4 (1996): 281– 296.

Rabe, Stephen. *The Killing Zone: The United States Wages Cold War in Latin America*. New York: Oxford University Press, 2012.

Reber, Vera. "The Demographics of Paraguay: A Reinterpretation of the Great War, 1865–1870." *Hispanic American Historical Review* 68, no. 2 (1988): 289–319.

Reisner, Marc. *Cadillac Desert: The American West and its Disappearing Water: Revised and Updated Edition*. New York: Penguin, 1993.

Revkin, Andrew. *The Burning Season: The Murder of Chico Mendes and the Fight for the Amazon Rainforest*. Boston: Houghton Mifflin, 1990.

Reynolds, Nancy. "Building the Past: Rockscapes and the Aswan High Dam in Egypt." In *Water on Sand: Environmental Histories of the Middle East and*

North Africa, edited by Alan Mikhail, 181–205. New York: Oxford University Press, 2012.

Rich, Bruce. *Mortgaging the Earth: The World Bank, Environmental Impoverishment, and the Crisis of Development*. Boston: Beacon Press, 1994.

Ridenti, Marcelo. "Censura e ditadura no Brasil, do golpe à transição democrática, 1964–1988." *Revista Concinnitas* 2, no. 33, (2018): 86–100.

"The Debate over Military (or Civilian-Military?) Dictatorship in Brazil in Historiographical Context." *Bulletin of Latin American Research* 37, no. 1 (January 2018): 33–42.

Roberson, Emily. *Nature's Pharmacy, Our Treasure Chest: Why We Must Conserve Our Natural Heritage*. Tucson: Center for Biological Diversity, 2008.

Rocha, Elaine Pereira. *Canal de desvio: Um estudo da experiência de agricultores e índios no confronto com a Itaipu Binacional*. São Paulo: E-Manuscrito, 2021.

Rogers, Thomas D. *Agriculture's Energy: The Trouble with Ethanol in Brazil's Green Revolution*. Chapel Hill: University of North Carolina Press, 2022.

Rosa, Luiz Pinguelli and Robert Schaeffer. "Greenhouse Gas Emissions from Hydroelectric Reservoirs." *Ambio* 23, no. 2 (March 1994): 164–165.

Rosa, Luiz Pinguelli, Roberto Schaeffer, and Marco Aurelio dos Santos. "Are Hydroelectric Dams in the Brazilian Amazon Significant Sources of 'Greenhouse' Gases?" *Environmental Conservation* 23, no. 1 (March 1996): 2–6.

Rosa, Luiz Pinguelli, Marco Aurelio dos Santos, Bohdan Matvienko, Ednaldo Oliveira dos Santos, and Elizabeth Sikar. "Greenhouse Gas Emissions from Hydroelectric Reservoirs in Tropical Regions." *Climatic Changes* 66 (2004): 9–21.

Rosa, Luiz Pinguelli, Aurelio dos Santos, Bohdan Matvienko, Oliveira dos Santos, and Elizabeth Sikar. "Scientific Errors in the Fearnside Comments on Greenhouse Gas Emissions (GHG) from Hydroelectric Dams and Response to His Political Claiming." *Climatic Changes* 75 (2006): 91–102.

Rothman, Franklin Daniel, and Pamela E. Oliver. "From Local to Global: The Anti-Dam Movement in Southern Brazil, 1979–1992." *Mobilization* 4, no. 1 (1999): 41–57.

Roy, Arundhati. *The Greater Common Good*. Bombay: India Book Distributor, 1999.

Rudd, John W.M., Reed Harris, C.A. Kelly, and R.E. Hecky. "Are Hydroelectric Reservoirs Significant Sources of Greenhouse Gases?" *Ambio* 22, no. 4 (June 1993): 246–248.

Sahar, Dan Coby. "Rejecting Eco-Authoritarianism, Again." *Environmental Values* 24 (2015): 345–366.

Saracura, VF. "History of Studies on Belo Monte Hydroelectric Plant." *Brazilian Journal of Biology* 75, no. 3 (2015): S5–S9.

Santos, Geraldo Mendes dos and Arnaldo Braga de Oliveira Jr. "A pesca no reservatório de Hidrelétrica de Balbina (Amazonas, Brasil)." *Acta Amazônica* 29, no. 1 (1999): 145–163.

Scott, Parry. *Negociações e resistências persistentes: Agricultores e a barragem de Itaparica num contexto de descaso planejado*. Recife: Editora da Universidade Federal de Pernambuco, 2009.

Scudder, Thayer. "India's Sardar Sarovar Project (SSP)." Unpublished manuscript available on his personal website, 2003.

"The Kariba Case Study." Social Science Working Paper 1227. California Institute of Technology Division of the Humanities and Social Sciences, 2005.

Sedrez, Lise. "Rubber, Trees and Communities: Rubber Tappers in the Brazilian Amazon in the Twentieth Century." In *A History of Environmentalism: Local Struggles, Global Histories*, edited by Marco Armiero and Lise Sedrez, 147–166. London: Bloomsbury Academic, 2014.

Sevá Filho, Arsenio Oswaldo, ed. *Tenotã-Mõ: Alertas sobre as consequências dos projetos do Rio Xingu*. São Paulo: International Rivers Network, 2005.

Shang Wei. "A lamentation for the Yellow River: The Three Gate Gorge Dam (Sanmenxia)." In *The River Dragon Has Come! The Three Gorges Dam and the Fate of China's Yangtze River and Its People*, edited by Dai Qing, 143–159. Armonk: M.E. Sharpe, 1998.

Shapiro, Judith. *Mao's War Against Nature: Politics and the Environment in Revolutionary China*. New York: Cambridge University Press, 2001.

Shearman, David and Joseph Smith. *The Climate Change Challenge and the Failure of Democracy*. Westport: Praeger, 2007.

Sheller, Mimi. *Aluminum Dreams: The Making of Light Modernity*. Cambridge: MIT Press, 2014.

Sigaud, Lygia. *Efeitos sociais de grandes projetos hidrelétricos: As barragens de Sobradinho e Machadinho*. Rio de Janeiro: Museu Nacional-UFRJ, 1986.

Silva, Filipe Oliveira da. "O Conselho Florestal Federal: Um parecer de sua configuração institucional (1934–1967)." *Historia Ambiental Latinoamericana y Caribeña (HALAC)* 7, no. 2 (2017): 101–129.

Silva, Margarete Pereira da. "O bispo de Juazeiro e a ditadura militar." In *Ditadura militar na Bahia: Novos olhares, novos objetivos, novos horizontes*, edited by Grimaldo Carneiro Zachariadhes, 241–258. Salvador: EDUFBA, 2009.

Silva, Orlando Sampaio. *Tuxá: Índios do Nordeste*. São Paulo: Annablume, 1997.

Silveira, Jamile Silva. "'Braços cruzados, máquinas paradas': trabalhadores, experiências, e conflitos na capital de energia (1960–1980)." *Rios Eletrônica* 16 (2018): 9–22.

Simpson, Bradley R. *Economists with Guns: Authoritarian Development and U.S.-Indonesian Relations, 1960–1968*. Stanford: Stanford University Press, 2008.

Singh, N., R.K. Mehra, and V.P. Sharma. "Malaria and the Narmada-River development in India: A Case Study of the Bargi Dam." *Annals of Tropical Medicine and Parasitology* 93, no. 5 (1999): 477–488.

Skidmore, Thomas. *The Politics of Military Rule in Brazil, 1964–85*. New York: Oxford University Press, 1988.

Brazil: Five Centuries of Change, 2nd Edition. New York: Oxford University Press, 2010.

Sneddon, Christopher. *Concrete Revolution: Large Dams, Cold War Geopolitics, and the US Bureau of Reclamation*. Chicago: University of Chicago Press, 2015.

Sneddon, Christopher, and Coleen Fox. "Struggles Over Dams as Struggles for Justice: The World Commission on Dams (WCD) and Anti-Dam Campaigns in Thailand and Mozambique." *Society and Natural Resources* 21 (2008): 625–640.

Sowers, Jeannie. "Nature Reserves and Authoritarian Rule in Egypt: Embedded Autonomy Revisited." *Journal of Environment and Development* 16, no. 4 (2007): 375–397.

Switkes, Glenn. *Foiling the Aluminum Industry: A Toolkit for Communities, Activists, Consumers, and Workers*. Berkeley: International Rivers Network, 2005.

Tadei, Wanderlei Pedro, Vera Margarete Scarpassa, and Iléa Brandão Rodrigues. "Evolução das populações de Anopheles e de mansonia, na área de influência da usina hidrelétrica de Tucuruí (Pará)." *Ciência e cultura* 30 (January 1991): 639.

Teixeira, Maria Gracinda Carvalho. *Energy Policy in Latin America: Social and Environmental Dimensions of Hydropower in Amazônia*. Aldershot: Ashgate, 1996.

Teodoro, Ueslei, Ana Lúcia Falavigna Guilherme, Ana Leuch Lozovei, Vicente la Salvia Filho, Yoshiaki Fukushigue, Roberto Palma Spinosa, Maria Eugênia Moreira Costa Ferreira, Orlando Carlos Barbosa, and Edson Maurício de Lima, "Culicídeos do Lago de Itaipu, no Rio Paraná, Sul do Brasil," *Revista Saúde Pública* 29, no. 1 (1995): 6–14.

Thomé, José Lauro. *Um grande projeto na Amazônia: Hidrelétrica de Balbina, um fato consumado*. Manaus: Editora da Universidade de Amazonas, 1999.

Tischler, Julia. *Light and Power for a Multiracial Nation: The Kariba Dam Scheme in the Central African Federation*. London: Palgrave Macmillan, 2013.

Tucker, Richard P. "Containing Communism by Impounding Rivers: American Strategic Interests and the Global Spread of High Dams in the Early Cold War." In *Environmental Histories of the Cold War*, edited by J.R. McNeill and Corinna R. Unger, 139–164. New York: Cambridge University Press, 2010.

Tundisi, José Galizia, ed. "Dossiê Crise Hídrica." *Revista Universidade de São Paulo* 106 (julho 2015): 8–94.

Turner, James Morton. *Charged: A History of Batteries and Lessons for a Clean Energy Future*. Seattle: University of Washington Press, 2022.

Turner, Terence. "The Role of Indigenous Peoples in the Environmental Crisis: The Example of the Kayapo of the Brazilian Amazon." *Perspectives in Biology and Medicine* 36, no. 3 (1993): 526–545.

"The Kayapo Resistance." In *Conformity and Conflict: Readings in Cultural Anthropology*, edited by James Spradley and David McCurdy, 391–409. Boston: Pearson, 2008.

Uekötter, Frank. *The Green and the Brown: A History of Conservation in Nazi Germany*. New York: Cambridge University Press, 2006.

United States Library of Congress Federal Research Division, *Brazil: A Country Study*. Washington DC: Library of Congress, 1998.

Urban, Teresa. *Missão (quase) impossível: Aventuras e desventuras do movimento ambientalista no Brasil*. São Paulo: Peirópolis, 2001.

Vainer, Carlos. "Águas para a vida, não para a morte: Notas para uma história do movimento de atingidos por barragens no Brasil." In *Justiça ambiental e*

cidadania, edited by Henri Acselrad, Selene Herculano, and José Augusto Pádua, 185–216. Rio de Janeiro: Relume Dumará, 2004.
Vargas Llosa, Mario. *La guerra del fin de mundo*. Barcelona: Seix Barral, 1981. Translated in English as *The War of the End of the World*. New York: Farrar, Straus and Giroux, 1984.
Vicentini, Yara. *Cidade e história na Amazônia*. Curitiba: Universidade Federal do Paraná, 2004.
Villa, Marco Antônio. *Canudos: O povo da terra*. São Paulo: Editora Ática, 1995.
Wachowicz, Ruy Christovam. *História do Paraná, 10ª edição*. Ponta Grossa: Editora UEPG, 2010.
Weinstein, Barbara. *The Amazonian Rubber Boom, 1850–1920*. Stanford: Stanford University Press, 1983.
Whigham, Thomas L. and Barbara Potthast. "The Paraguayan Rosetta Stone: New Insights into the Demographics of the Paraguayan War, 1864–1970." *Latin American Research Review* 34, no. 1 (1999): 174–86.
Whitcombe, Elizabeth. "The Environmental Costs of Irrigation in British India: Waterlogging, Salinity, and Malaria." In *Nature, Culture, Imperialism: Essays on the Environmental History of South Asia*, edited by David Arnold and Ramachandra Guha, 237–259. New York: Oxford University Press, 1995.
White, Richard. *The Organic Machine: The Remaking of the Columbia River*. New York: Hill and Wang, 1995.
World Bank and World Fisheries Trust. *Migratory Fishes of South America: Biology, Fisheries, and Conservation Status*. Washington DC: World Bank, 2003.
World Commission on Dams (WCD). *Dams and Development: A New Framework for Decision-Making*. Sterling, VA: Earthscan, 2000.
WCD Case Study Tucuruí Hydropower Complex Brazil, Final Report. Cape Town: WCD, 2000.
WCD Case Study Kariba Dam Zambia and Zimbabwe, Final Report. Cape Town: WCD, 2000.
Worster, Donald. *Rivers of Empire: Water, Aridity, and the Growth of the American West*. New York: Oxford University Press, 1985.
Zhong, Yiguang and Geoff Power. "Environmental Impacts of Hydroelectric Projects on Fish Resources in China." *Regulated Rivers: Research and Management* 12 (1996): 81–98.
Zhouri, Andréa and Raquel Oliveira. "Desenvolvimento, conflitos sociais, e violência no Brasil rural: O caso das usinas hidrelétricas." *Ambiente e Sociedade* 10, no. 2 (2007): 119–135.
Zwieten, Paul A.M. van, C. Béné, J. Kolding, R. Brummett, and J. Valbo-Jørgensen, eds. *Review of Tropical Reservoirs and Their Fisheries: The Cases of Lake Nasser, Lake Volta, and Indo-Gangetic Basin Reservoirs*. FAO Fisheries and Aquaculture Technical Paper No. 557. Rome: FAO, 2011.

Index

Other Books in the Series

Abraham H. Gibson *Feral Animals in the American South: An Evolutionary History*
Andy Bruno *The Nature of Soviet Power: An Arctic Environmental History*
David A. Bello *Across Forest, Steppe, and Mountain: Environment, Identity, and Empire in Qing China's Borderlands*
Erik Loomis *Empire of Timber: Labor Unions and the Pacific Northwest Forests*
Peter Thorsheim *Waste into Weapons: Recycling in Britain during the Second World War*
Kieko Matteson *Forests in Revolutionary France: Conservation, Community, and Conflict, 1669–1848*
Micah S. Muscolino *The Ecology of War in China: Henan Province, the Yellow River, and Beyond, 1938–1950*
George Colpitts *Pemmican Empire: Food, Trade, and the Last Bison Hunts in the North American Plains, 1780–1882*
John L. Brooke *Climate Change and the Course of Global History: A Rough Journey*
Paul Josephson et al. *An Environmental History of Russia*
Emmanuel Kreike *Environmental Infrastructure in African History: Examining the Myth of Natural Resource Management*
Gregory T. Cushman *Guano and the Opening of the Pacific World: A Global Ecological History*
Sam White *The Climate of Rebellion in the Early Modern Ottoman Empire*
Edmund Russell *Evolutionary History: Uniting History and Biology to Understand Life on Earth*
Alan Mikhail *Nature and Empire in Ottoman Egypt: An Environmental History*
Richard W. Judd *The Untilled Garden: Natural History and the Spirit of Conservation in America, 1740–1840*
James L. A. Webb Jr. *Humanity's Burden: A Global History of Malaria*
Myrna I. Santiago *The Ecology of Oil: Environment, Labor, and the Mexican Revolution, 1900–1938*
Frank Uekotter *The Green and the Brown: A History of Conservation in Nazi Germany*
Matthew D. Evenden *Fish versus Power: An Environmental History of the Fraser River*
Alfred W. Crosby *Ecological Imperialism: The Biological Expansion of Europe, 900–1900, second edition*
Nancy J. Jacobs *Environment, Power, and Injustice: A South African History*
Edmund Russell *War and Nature: Fighting Humans and Insects with Chemicals from World War I to Silent Spring*
Adam Rome *The Bulldozer in the Countryside: Suburban Sprawl and the Rise of American Environmentalism*
Judith Shapiro *Mao's War against Nature: Politics and the Environment in Revolutionary China*
Andrew Isenberg *The Destruction of the Bison: An Environmental History*
Thomas Dunlap *Nature and the English Diaspora*
Robert B. Marks *Tigers, Rice, Silk, and Silt: Environment and Economy in Late Imperial South China*
Mark Elvin and Tsui'jung Liu *Sediments of Time: Environment and Society in Chinese History*

Richard H. Grove *Green Imperialism: Colonial Expansion, Tropical Island Edens and the Origins of Environmentalism, 1600–1860*
Thorkild Kjærgaard *The Danish Revolution, 1500–1800: An Ecohistorical Interpretation*
Donald Worster *Nature's Economy: A History of Ecological Ideas, second edition*
Elinor G. K. Melville *A Plague of Sheep: Environmental Consequences of the Conquest of Mexico*
J. R. McNeill *The Mountains of the Mediterranean World: An Environmental History*
Theodore Steinberg *Nature Incorporated: Industrialization and the Waters of New England*
Timothy Silver *A New Face on the Countryside: Indians, Colonists, and Slaves in the South Atlantic Forests, 1500–1800*
Michael Williams *Americans and Their Forests: A Historical Geography*
Donald Worster *The Ends of the Earth: Perspectives on Modern Environmental History*
Robert Harms *Games against Nature: An Eco-Cultural History of the Nunu of Equatorial Africa*
Warren Dean *Brazil and the Struggle for Rubber: A Study in Environmental History*
Samuel P. Hays *Beauty, Health, and Permanence: Environmental Politics in the United States, 1955–1985*
Arthur F. McEvoy *The Fisherman's Problem: Ecology and Law in the California Fisheries, 1850–1980*
Kenneth F. Kiple *The Caribbean Slave: A Biological History*

For EU product safety concerns, contact us at Calle de José Abascal, 56–1°,
28003 Madrid, Spain or eugpsr@cambridge.org.

www.ingramcontent.com/pod-product-compliance
Lightning Source LLC
Chambersburg PA
CBHW032359110725
29505CB00002B/5
* 9 7 8 1 0 0 9 4 2 8 6 9 9 *